Reliability-Based Design in Soil and Rock Engineering

Reliability-Based Design in Soil and Rock Engineering

Enhancing Partial Factor Design Approaches

Bak Kong Low

CRC Press
Taylor & Francis Group
Boca Raton London New York

CRC Press is an imprint of the
Taylor & Francis Group, an **informa** business

First edition published 2022
by CRC Press
6000 Broken Sound Parkway NW, Suite 300, Boca Raton, FL 33487-2742

and by CRC Press
2 Park Square, Milton Park, Abingdon, Oxon, OX14 4RN

© 2022 Bak Kong Low

CRC Press is an imprint of Taylor & Francis Group, LLC

Library of Congress Cataloging-in-Publication Data
Names: Low, Bak Kong, author.
Title: Reliability-based design in soil and rock engineering / Bak Kong Low.
Description: First edition. | Boca Raton, FL : CRC Press, [2022] |
    Includes bibliographical references and index.
Identifiers: LCCN 2021022817 (print) | LCCN 2021022818 (ebook) |
    ISBN 9780367631390 (hbk) | ISBN 9780367631406 (pbk) |
    ISBN 9781003112297 (ebk)
Subjects: LCSH: Geotechnical engineering. | Soil mechanics. | Rock mechanics. |
    Earthwork--Reliability.
Classification: LCC TA705 .L69 2022 (print) |
    LCC TA705 (ebook) | DDC 624.1/51--dc23
LC record available at https://lccn.loc.gov/2021022817
LC ebook record available at https://lccn.loc.gov/2021022818

ISBN: 978-0-367-63139-0 (hbk)
ISBN: 978-0-367-63140-6 (pbk)
ISBN: 978-1-003-11229-7 (ebk)

DOI: 10.1201/9781003112297

Typeset in Sabon
by SPi Technologies India Pvt Ltd (Straive)

Access the Support Materials: www.routledge.com/9780367631390

Contents

Abbreviations and symbols   xiii
Author   xv

1   Introduction   1

    1.1   Intended audience 1
    1.2   Overview of chapters 2

PART I
Background chapters on probabilistic
methods, LRFD and EC7   7

2   Hasofer-lind index, FORM, reliability-based design and
    SORM   9

    2.1   Suggested books and Internet resources for
           understanding basic statistical terms 9
    2.2   A simple hands-on Hasofer–Lind method in spreadsheet 10
    2.3   Intuitive grasp of the Hasofer–Lind index via
           expanding dispersion ellipsoid perspective 14
    2.4   FORM for correlated non-Gaussian variables 17
    2.5   A simple example of RBD-via-FORM
           and correlated sensitivities 24
    2.6   How mean-value point determines the sign of reliability
           index 29
    2.7   Probabilistic consolidation settlement analysis 30
           2.7.1   Method 1: Efficient spreadsheet method 32
           2.7.2   Method 2: Traditional iterative algorithm 34
    2.8   Probabilistic analysis of earthquake-induced cyclic
           shear stress in saturated sand 35
    2.9   Second-order reliability method (SORM) in spreadsheet 38
    2.10 Correlation matrix must be positive definite 40

2.11 System FORM, response surface method
and importance sampling 41

3 Civil and environmental applications of reliability
analysis and design 43

3.1 Introduction 43
3.2 Example 1: Probability of traffic congestion 45
   3.2.1 Solution based on mean and standard deviation
         of safety margin 45
   3.2.2 Alternative solution procedure based
         on Hasofer–Lind index 46
3.3 Example 2: Design of traffic capacity for
    a target reliability index 48
3.4 Example 3: Probabilities of arriving late and arriving
    in time 50
   3.4.1 Alternative solution using spreadsheet function
         for the single random variable case of T_3 52
3.5 Example 4: Required travel time from RBD-via-FORM 53
3.6 Example 5: Supply and demand of irrigation water 57
3.7 Example 6: Thermal pollution in a river 61
3.8 Example 7: Reliability of a storm sewer system 63
3.9 Example 8: Reliability of a breakwater at La
    Spezia harbor in northern Italy 66
3.10 Example 9: Spillway capacity 69
3.11 Example 10: Two ways of increasing spillway
     capacity, and RBD with cost considerations 71
3.12 Example 11: Column subjected to biaxial bending
     moments and axial force 73
3.13 Example 12: Moment capacity of a reinforced concrete beam 76
3.14 Example 13: Reliability analysis of an asymmetrically
     loaded beam on Winkler medium 76
3.15 Example 14: A strut with complex supports
     and implicit performance function 79

4 Eurocode 7, LRFD and links with the first-order
reliability method 83

4.1 Introduction 83
4.2 Load and resistance factor design approach, and links
    with FORM design point 83
4.3 Eurocode 7 approach, and links with the design point
    of RBD-via-FORM 87
4.4 Example RBD of foundation settlement with
    back-calculated partial factors 90

4.5   Probabilistic settlement analysis of Hong Kong land
      reclamation test fill 93
      4.5.1   Sensitivity considerations 96
      4.5.2   Limit state surfaces and performance
              functions pertaining to magnitude and
              rate of soft clay settlement 99
      4.5.3   Distinguishing positive and negative
              reliability indices 99
      4.5.4   Reliability analysis for different
              limiting state surfaces 101
      4.5.5   Obtaining probability of failure (P_f)
              and CDF from β indices 102
      4.5.6   Obtaining PDF curves from β indices 105
4.6   Conduct Monte Carlo simulations wisely 108
      4.6.1   Distortions caused by negative values
              of random numbers 108
      4.6.2   Distortions caused by physically incompatible
              random numbers in Monte Carlo simulations 110

PART II
Reliability-based design applied to soil engineering   113

5   Spread foundations   115
    5.1   Introduction 115
    5.2   Hasofer–Lind index applied to the RBD of a spread
          foundation with two random variables 116
    5.3   Hasofer–Lind index reinterpreted via
          expanding ellipsoid perspective 119
    5.4   Effect of parametric correlations on failure probability 123
    5.5   First-order reliability method (FORM) 124
    5.6   An EC7 design of strip footing width based on
          characteristic values and partial factors 128
    5.7   Form analysis of EC7 design to estimate its
          reliability and probability of failure 130
    5.8   RBD-via-FORM for target reliability index, to compare
          with EC7 footing width design 133
    5.9   RBD-via-FORM resolves load-resistance duality of a
          retaining wall foundation 135
          5.9.1   Information and insights at the design point
                  of RBD-via-FORM 137
          5.9.2   Comparison of RBD-via-FORM with Monte
                  Carlo simulation and MVFOSM 140
    5.10 RBD-via-FORM insights for LRFD 141

5.10.1   *Back-calculations of LF and RF from the design point of RBD-via-FORM 141*

5.10.2   *Load-resistance duality revealed in load factors back-calculated from RBD-via-FORM 144*

5.11   *Reliability analysis of serviceability limit state involving the Burland and Burbidge method 145*

5.11.1   *Deliberations by Tomlinson (2001) on the ULS and SLS of a spread foundation 145*

5.11.2   *Statistical properties of compressibility m_v in Burland and Burbidge data 149*

5.11.3   *FORM analysis for probability of settlement exceedance 151*

5.11.4   *From normally distributed log(m_v) to lognormally distributed m_v for FORM analysis 152*

5.12   *Distinguishing positive from negative reliability indices 155*

6   Pile foundations   157

6.1   *Introduction 157*

6.2   *A driven pile in stiff clay below a jetty 157*

6.2.1   *Deterministic design of the length of a jetty pile 157*

6.2.2   *Statistical inputs, and RBD-via-FORM 160*

6.2.2.1   *Quantifying the scatter/uncertainty in c_u 160*

6.2.2.2   *Quantifying the scatter/ uncertainty in adhesion factor α 160*

6.2.2.3   *Reliability-based design of pile length in soil 162*

6.2.2.4   *Discretized random field 164*

6.3   *A driven pile in sand 166*

6.3.1   *Maximum allowable design load based on EC7-DA1b 166*

6.3.2   *RBD of design load for the driven pile in sand 168*

6.3.2.1   *Refined formulation prior to RBD 168*

6.3.2.2   *RBD involving discretized random field 169*

6.4   *Randolph and Wroth method of estimating pile settlement, extended probabilistically 172*

6.4.1   *Deterministic estimation of pile settlement using the Randolph and Wroth method 172*

6.4.2   *Extending the Randolph and Wroth method to reliability analysis of pile settlement 175*

6.5   *A laterally loaded pile in soil with strain-softening and depth-dependent p-y curves 177*

6.5.1   *Deterministic numerical procedure for a laterally loaded pile involving strain-softening p-y curves 179*

6.5.2 From deterministic numerical procedure to probabilistic analysis of laterally loaded piles 183

6.5.3 Pile with 26 m cantilever length 184

6.5.4 Fully embedded pile with no cantilever length 189

6.6 Laterally loaded pile: Verification of spreadsheet numerical procedure with Hetenyi solution for linear p-y model 191

7 Earth retaining structures 193

7.1 Introduction 193

7.2 Rotational ultimate limit state of a semi-gravity wall 194

7.3 Reliability-based design of a semi-gravity retaining wall 197

7.3.1 Limit state functions with respect to rotation and sliding 198

7.3.2 Hasofer-Lind reliability index for correlated normal variates 199

7.4 Comparison with EC7 DA1b design of base width b for rotation ULS 206

7.5 Vector components of active earth thrust are allied, not adversarial 209

7.6 RBD-via-FORM for an anchored sheet pile wall 214

7.6.1 Possible multiple outcomes in EC7 (2004) DA1b design of anchored sheet pile involving stabilizing–destabilizing unit weight 218

7.7 Positive reliability index only if the mean-value point is in the safe domain 220

7.8 RBD-via-FORM for an example modified from Terzaghi et al. (1996) 222

7.8.1 Deterministic analysis with search for the critical quadrilateral wedge 222

7.8.2 Design resistance from the perspective of LRFD and EC7-DA2 224

7.8.3 Extending the modified Terzaghi et al. (1996) deterministic example to RBD-via-FORM 225

7.8.4 RBD-via-FORM of tie rod force to resist lateral thrust 226

7.8.5 Another failure mode involving surface rolling along the slope of the stockpiled iron ore 226

7.9 Retaining walls with steady-state seepage towards a vertical drainage layer behind the wall 229

7.9.1 Deterministic example from Clayton et al. (2013), extended into probability-based design 229

7.9.2 Extending the Clayton et al. example to probability-based design 229

7.9.3   *Deterministic example from Lambe and Whitman (1979), extended probabilistically 232*

8   Soil slope stability   233

8.1   *Introduction 233*
8.2   *A deterministic example from Terzaghi et al. (1996), extended probabilistically 233*
   8.2.1   *Analytical $\phi_u = 0$ procedure 233*
   8.2.2   *From deterministic analysis to reliability-based design of the Terzaghi et al. problem 238*
   8.2.3   *Perspectives, insights and information provided by RBD-via-FORM 240*
8.3   *An excavated slope in London Clay that failed despite a high computed factor of safety 243*
   8.3.1   *Deterministic $\phi_u = 0$ slip circle analysis, with summation of moments from horizontal strips 244*
   8.3.2   *Probabilistic analysis incorporating discretized random field for London Clay 246*
8.4   *Probabilistic analyses of a slope failure in San Francisco Bay Mud 249*
   8.4.1   *Reliability analysis with correlated lognormals and statistical inputs based on in situ and lab tests 249*
   8.4.2   *Reliability-based design of slope angle and excavation depth 253*
8.5   *Reliability analysis of a Norwegian slope accounting for spatial autocorrelation 253*
8.6   *System reliability analysis for multiple failure modes 256*
8.7   *Lambe and Whitman sloping core dam example, refined deterministically and extended probabilistically 258*
   8.7.1   *Deterministic analysis for a specific slip surface of $\theta_1 = 10°$ 260*
   8.7.2   *Search for the deterministic critical slip surface by varying the inclination angle θ_1 260*
   8.7.3   *From deterministic critical slip surface to reliability analysis 262*
8.8   *Spencer method reformulated for spreadsheet, for non-circular slip surface 264*
   8.8.1   *Deterministic reformulations and example analyses 264*
   8.8.2   *Same template for Spencer, Bishop Simplified and force equilibrium methods but different constraints 270*
   8.8.3   *Reformulated Spencer method extended probabilistically 271*

8.8.4   *Effect of closer discretization of random field 273*
8.9   *Finite element reliability analysis via*
    response surface methodology 274

PART III
Reliability-based design applied to rock engineering   275

9   Plane sliding in rock slopes   277

9.1   *Introduction 277*
9.2   *Plane sliding stability analysis, an example from*
    Goodman (1989) 277
    9.2.1   *Deterministic analysis using an alternative*
        procedure 278
    9.2.2   *RBD-via-FORM of potential plane sliding 281*
9.3   *An example of two-block stability analysis in rock slopes 288*
    9.3.1   *Goodman (1989) closed form equation for*
        limiting equilibrium of a two-block mechanism 288
    9.3.2   *An alternative deterministic procedure for*
        two-block stability analysis 290
    9.3.3   *Reliability-based design of support force for*
        a two-block mechanism in rock slope 294
        9.3.3.1   *Comparison with Monte*
            Carlo simulation 294
        9.3.3.2   *Information and insights at the*
            FORM design point, and implications
            for Eurocode 7 and LRFD 296
9.4   *Deterministic formulations and reliability-based design*
    of Sau Mau Ping slope in Hong Kong 297
    9.4.1   *LRFD considerations: Non-uniqueness of*
        inferred LF and RF from RBD-via-FORM 302
    9.4.2   *Combined methods of drainage, slope reprofiling,*
        and reinforcement to improve slope reliability 303
9.5   *Reliability analysis of a failed slope in a limestone quarry 303*

10   Rock slopes with three-dimensional tetrahedral wedges   309

10.1   *Introduction 309*
10.2   *Obtaining β_1, δ_1, β_2 and δ_2 angles from dip directions*
    and dips 311
10.3   *Kinematic requirement for formation of wedge*
    mechanism in rock slopes 311
10.4   *Limit equilibrium equations for the factor of safety*
    of wedge in rock slope 312

      10.4.1   *Sliding along both planes, that is, along the line of intersection 312*

      10.4.2   *Sliding along plane 1 only 313*

      10.4.3   *Sliding along plane 2 only 316*

      10.4.4   *Contact is lost on both planes 316*

      10.4.5   *Water pressure coefficients G_{w1} and G_{w2}, and relationship with water pressures u_1 and u_2 317*

  10.5 *Comparison with stereographic method 318*

  10.6 *Verification using the vectorial method described in Hoek and Bray 320*

      10.6.1   *Verification of an example with BiPlane failure mode 320*

      10.6.2   *Verification of an example with sliding failure along a single plane 321*

      10.6.3   *Verification for all four failure modes of wedge mechanism in rock slope 321*

  10.7 *Reliability analysis and RBD of a tetrahedral rock wedge with a dominant failure mode 324*

  10.8 *Reliability-based design of drainage to enhance stability of wedge in rock slope 326*

  10.9 *Reliability analysis reveals a more critical* Plane1 *mode behind the mean-value* BiPlane *mode 328*

  10.10 *Sensitivity computation in the ellipsoid approach 333*

  10.11 *Chapter summary 334*

11 Underground excavations in rock   **335**

  11.1 *Introduction 335*

  11.2 *Deterministic ground-support interaction analysis of a shaft excavated in sandstone 335*

  11.3 *Reliability analysis involving ground-support interaction of a shaft excavated in sandstone 339*

  11.4 *Roof wedge in tunnel, a tale of two factors of safety 342*

  11.5 *Reliability analysis of tunnel roof wedge reveals context-dependent sensitivities 346*

  11.6 *Deterministic verification using Bobet and Einstein (2011) formulation of reinforced tunnel 349*

  11.7 *Reliability-based design of the length and spacing of rockbolts for a target β value 351*

Appendix: An efficient spreadsheet algorithm for FORM   357

References   363

Further reading   371

Index   377

Abbreviations and symbols

C	covariance matrix
CDF	cumulative distribution function
c.o.v.	coefficient of variation, i.e., standard deviation/mean, or σ/μ
CV	coefficient of variation, same as c.o.v. above
FORM	first-order reliability method
EC7	Eurocode 7
$g(\mathbf{x})$	performance function
LF	load factor
LF_m	load factor based on mean value, i.e., Q^*/μ_Q
LRFD	load and resistance factor design
LSS	limit state surface, defined by $g(\mathbf{x}) = 0$
MCS	Monte Carlo simulation
MPP	most probable point (of failure)
n_i	dimensionless sensitivity indicator of variable $n_i = \left(x_i^* - \mu_{xi}\right)/\sigma_{xi}$
n	vector of dimensionless sensitivity indicators
Nonnormal	Non-Gaussian
Normal variable	A variable with normal (Gaussian) probability distribution.
PDF	probability density function
PerFunc	performance function
P_f	probability of failure
Q	load
Q_h	horizontal load
Q_v	vertical load
Q^*	design value of load at the MPP of failure of FORM
R	resistance
R^*	design value of resistance at the MPP of failure of FORM
R	matrix of correlation coefficients
R^{-1}	inverse of correlation matrix
RBD	reliability-based design

RF	resistance factor
RSM	response surface method
SD	standard deviation
SLS	serviceability limit state
SORM	second-order reliability method
ULS	ultimate limit state
VBA	Visual Basic for Applications, programming language in Microsoft Excel
β	reliability index, based on FORM
μ	mean value
μ^N	equivalent normal mean of a nonnormal distribution based on Rackwitz–Fiessler transformation
ρ	correlation coefficient between two random variables
σ	standard deviation
σ^N	equivalent normal standard deviation of a nonnormal distribution based on Rackwitz–Fiessler transformation
Φ	standard normal cumulative distribution

Author

Bak Kong Low, PhD, earned a BS and an MS at the Massachusetts Institute of Technology and a PhD at the University of California, Berkeley.

He is a Fellow of the American Society of Civil Engineers, a registered PE of Malaysia, and for 35 years taught at Nanyang Technological University, Singapore. He has conducted research while on sabbatical visits at the Hong Kong University of Science and Technology, the University of Texas at Austin, and the Norwegian Geotechnical Institute. In 2019, he received the Thomas A. Middlebrooks Award from ASCE.

Chapter 1

Introduction

1.1 INTENDED AUDIENCE

The probabilistic analyses and reliability-based designs (RBDs) in this book contain insights for enhancing the Eurocode 7 (EC7) and the load and resistance factor design (LRFD) methods. The soil and rock engineering RBD examples are suitable for geotechnical graduate students, researchers, professors and practitioners. In addition, the succinct solution procedures presented for the multidisciplinary examples in Chapter 3, on traffic, environmental, ecological and structural engineering, are very suitable for undergraduate engineering students, to introduce them to probabilistic applications, spreadsheet VBA programming and the Excel Solver tool for constrained optimization, which are all very useful in diverse engineering and non-engineering disciplines regardless of the major of the students.

The deterministic analyses in the soil and rock engineering portion of the book (Chapters 4–11), prior to the probabilistic contents there, can be used with benefits by civil engineering undergraduates even if they are not ready for the deeper probabilistic contents there. This is because there are often new perspectives and innovative procedures in this book's deterministic calculations which precede the first-order reliability method (FORM) and RBD contents. These can enhance students' conceptual understanding of geotechnical calculations and stimulate them to thoughtful and creative solution methodologies. The inclusion of many renowned deterministic geotechnical cases from highly regarded soil and rock engineering books (and extending them probabilistically) will also motivate graduate and undergraduate students alike. Deterministic and probabilistic solutions of the book's examples are available at the Publisher's website, ideal for hands-on participation for a deeper appreciation.

In short, this book can be used with benefits by many.

DOI: 10.1201/9781003112297-1

1.2 OVERVIEW OF CHAPTERS

Chapter 2, on *Hasofer–Lind index, FORM, reliability-based design and SORM*, explains the efficient spreadsheet *Solver* procedure for the Hasofer–Lind index and the first-order reliability method (FORM), as background for understanding the multidisciplinary reliability examples of Chapter 3, and the other examples in the chapters on soil and rock engineering. Simple examples are used to facilitate hands-on experience, for a deeper appreciation of the intuitive dispersion ellipsoid perspective (or equivalent ellipsoid for nonnormal variates) and the meaning of the most probable failure point. The phenomenon of correlated sensitivities is demonstrated and explained. The spreadsheet FORM procedure obtains the same solution as the classical FORM procedure, in a more succinct and lucid way. Further illustrations are given in an example of soft clay consolidation settlement, and another example of earthquake-induced cyclic shear stress in sand, both from Ang and Tang (1984). Also discussed are the extension of FORM into second-order reliability method (SORM), system FORM for multiple failure modes and the use of response surface methods to couple spreadsheet-based FORM and stand-alone numerical packages.

Chapter 3, on *Civil and environmental applications of reliability analysis and design*, is intentionally *multidisciplinary*, using 14 probability-based design examples in civil and environmental engineering from established publications on applied probability in engineering. Alternative lucid procedures for obtaining the same solutions efficiently are presented and explained, to facilitate understanding of reliability analysis and reliability-based design by all who are interested. The probability-based design examples, progressing from simple to more complex, concern traffic congestion, traffic capacity, design of travel time, supply and demand of irrigation water, thermal pollution in a river, reliability of a storm sewer system, sliding of a harbour breakwater, spillway capacity with cost considerations, axially loaded column with biaxial moments, moment capacity of a reinforced concrete beam, asymmetrically loaded beam on Winkler medium, and a strut with complex supports. The chapter suits deterministic engineers and researchers who are considering probabilistic approach, undergraduate and graduate students in civil and environmental engineering and all who want to know how probabilistic approach can be applied in many daily-life situations.

Chapter 4, on *Eurocode 7, LRFD and links with first-order reliability method*, provides an overview of the LRFD used in North America, and similarities and differences between LRFD and RBD via the first-order reliability method (RBD-via-FORM). This is followed by an overview of Eurocode 7 (EC7), and similarities and differences between EC7 and RBD-via-FORM. An example of probability-based design of foundation settlement based on partial factors derived from FORM is studied, with original classical FORM procedure neatly explained by Ang and Tang (1984), for comparison with an alternative FORM procedure using spreadsheet. Also presented are deterministic and probabilistic settlement analyses of the Chek Lap Kok land reclamation test fill in Hong Kong, involving the use of pre-fabricated vertical drains to accelerate soft clay consolidation. The chapter concludes with a simple example illustrating the need to apply Monte Carlo simulations with understanding in order to avoid potential pitfalls.

Reliability-based analysis and design of shallow foundations are presented in Chapter 5, *Spread foundations*. Focus is on information and insights which can enhance partial factor design methods like EC7 and LRFD. Ultimate limit states are considered first, including reliability analysis for estimating the failure probability of a strip footing, reliability-based design of the footing width for a target reliability level, and comparison with EC7 design. This is followed by a deterministic example of retaining wall foundation design from Tomlinson (2001) involving eccentric and inclined loads, extended into RBD to demonstrate the ability of RBD to deal with context-dependent sensitivities and the stabilizing–destabilizing duality of the vertical load when horizontal load and overturning moment are also acting. Another reliability analysis example extends probabilistically the Burland and Burbidge settlement estimation method applied to the foundation of a water storage tank, including statistical analysis of the SPT test data extracted from Terzaghi et al. (1996).

Probability-based designs of axially loaded piles and laterally loaded piles are illustrated in Chapter 6, *Pile foundations*, with modelling of spatially autocorrelated shear strength and model inaccuracy. The first case involves statistical analysis and reliability-based design of the length of a driven pile in stiff clay below a jetty, and comparison with the deterministic design in Tomlinson (2001). The second case deals with the probabilistic design load on a tubular pile driven into sand, for comparison with the design load based on EC7-DA1b as presented in Knappett and Craig (2019). The third case deals with the probabilistic estimation of the long-term pile settlement based on the Randolph and Wroth Method, for a bored pile in clay. A fourth case deals with probabilistic analysis of the lateral displacement and bending limit state of a laterally loaded pile in a breasting dolphin, based on the strain-softening and depth-dependent Matlock p-y curves data from Tomlinson and Woodward (2015). The context-dependent sensitivities of the parameters affecting pile bending capacity and displacement are demonstrated. Connections are made with Eurocode 7 and the LRFD method.

Chapter 7, *Earth retaining structures*, explains reliability-based designs for semi-gravity retaining walls, anchored sheet pile walls and walls with seepage in the retained fill towards a vertical drainage layer. The probabilities of failure inferred from reliability indices are compared with Monte Carlo simulations. System reliability involving multiple failure modes is illustrated. Comparisons are made with LRFD and EC7 design methods. The cases include a problem after Terzaghi et al. (1996) that requires searching for the critical quadrilateral soil wedge in both deterministic analysis and probability-based designs, and deterministic examples from Clayton et al. (2013) and Lambe and Whitman (1979) which are extended probabilistically. Insights are presented for LRFD and EC7-DA2 to resolve a paradox that arises due to segregating the active earth thrust into a destabilizing horizontal component and a stabilizing vertical component. Possible multiple outcomes in partial factor design of an anchored sheet pile wall involving stabilizing–destabilizing duality of soil unit weight are explained.

Deterministic solutions are presented for several well-known cases of actual soil slopes in Chapter 8, *Soil slope stability*, before extending the analysis probabilistically to account for uncertainty in input values. Focus is on insights and information from the design point of reliability analysis, for enhancing partial factor design approach. There are novelties in this book's deterministic solution procedures (which underlies the performance function in reliability analysis), which could be of broader appeal beyond the probabilistic focus of this book. The cases include an excavated slope failure in stiff fissured London clay, an underwater slope failure in San Francisco Bay Mud, a clay slope in Norway with anisotropic and spatially autocorrelated undrained shear strength, a sloping core dam example from Lambe and Whitman (1979) which is refined deterministically and extended probabilistically, and reliability analysis of a hypothetical embankment on soft ground with discretized 1D random field and with search for critical noncircular slip surface based on a reformulated Spencer method in spreadsheet.

Chapter 9, *Plane sliding in rock slopes*, illustrates RBD of two-dimensional slopes in rock mass containing one or multiple joint planes along which rock blocks may slide. Deterministic cases, and some Monte Carlo simulation cases, from established publications, are presented and solved, before extending them into RBD accounting for uncertainty in input variables. The first-order reliability method (FORM) and RBD-via-FORM are conducted using the Low and Tang (2007) spreadsheet procedure, with lucid outcomes. The discussions on insights and information attainable from the design point of RBD-via-FORM, and connections with partial factor design methods, constitute the gist which the author hopes to share with readers. It does not matter even if readers are used to conducting FORM and RBD-via-FORM on other platforms or using stand-alone packages. Comparisons are made with second-order reliability method (SORM) and Monte Carlo simulations. The cases include a two-block slope from Goodman (1989), a Hong Kong slope from Hoek (2007) and reliability analysis for comparison with a Monte Carlo simulation study by Wyllie (2018) of a failed slope in a limestone quarry.

Rock slopes with three-dimensional tetrahedral wedges is explained in Chapter 10, deterministically and probabilistically. An analytical solution is given for assessing the stability of rock slopes with 3D wedges. The closed form solution embodies completed vector operations, so that no vector products or stereographic projections are required. The intrinsic geometrical angles used in the closed form solution are obtained easily from dip directions and dips of the slope face and the two discontinuity planes. The factor of safety values calculated for four failure modes are identical to those from the Hoek and Bray sequential vectorial procedure. Reliability analysis and probability-based design are then illustrated for a case with uncertain discontinuity orientations, shear strength parameters and water pressures on the two discontinuity planes. It is demonstrated that the same uncertain entity, for example, friction angle, or water pressure, exhibits different sensitivity levels when the two discontinuity planes have different steepness. Another case of reliability analysis detects a much more critical sliding-on-single-plane mode even though the mean values of parameters give a safer sliding-on-both-planes mode. Monte Carlo simulations on the closed form solutions obtain failure probabilities virtually identical to those from Monte Carlo simulations done using the Hoek–Bray vectorial procedure.

Chapter 11 deals with *Underground excavations in rock*. A procedure involving spreadsheet-automated iterations for ground–support interaction analysis is first verified deterministically on an example of a shaft excavated in sandstone from Hoek et al. (2000), followed by probabilistic analysis using the first-order reliability method (FORM) to reveal the *in situ* stress and Young's modulus as the two parameters of influence on the excavation behaviour. This is followed by reliability analysis of a symmetric roof wedge in a circular tunnel, where the ambiguity caused by two different definitions of factor of safety is resolved, and the valuable context-dependent sensitivity information offered by reliability analysis is discussed. A final example applies probability-based design to a circular tunnel supported with elastic rockbolts in a homogeneous and isotropic elasto-plastic ground obeying the Coulomb failure criterion, with deterministic formulations from Bobet and Einstein (2011). The performance function is implicit. Connections with partial factors of limit state design are discussed. Comparisons are made with second-order reliability method (SORM), and Monte Carlo simulations with importance sampling.

In summary, this book aims to explain probabilistic analysis and RBD-via-FORM lucidly, with efficient implementation and applications in soil and rock engineering in Chapters 4–11, and multi-disciplinarily in Chapter 3. The primary emphasis is on how the valuable information and insights at the design point of FORM can enhance partial factor design approach like LRFD in North America, and Eurocode 7. Comparisons are often made with other probabilistic methods in the examples.

Part I

Background chapters on probabilistic methods, LRFD and EC7

Chapter 2

Hasofer–Lind index, FORM, reliability-based design and SORM

2.1 SUGGESTED BOOKS AND INTERNET RESOURCES FOR UNDERSTANDING BASIC STATISTICAL TERMS

Readers of this book are assumed to be familiar with basic statistical terms and concepts such as mean (μ), median, mode, standard deviation (σ), coefficient of variation (CV or c.o.v.), correlation coefficient (ρ), probability distribution, probability density function (PDF), cumulative distribution (CDF), and normal (or Gaussian) distribution. These are explained, for example, in Benjamin and Cornell (1970), Ang and Tang (1984, 2006) and Baecher and Christian (2003), which deal with statistics and probability concepts, with emphasis on applications to civil and geotechnical engineering, and includes Bayesian decision theory, Monte Carlo simulations, site characterization and spatial variability and other higher level topics.

A convenient Internet resource on definitions of statistical terms is Wikipedia. According to two complimentary articles in the Economist.com issue of January 9, 2021:

> Wikipedia is not free from honest mistakes. .... The nature of crowd-sourcing means its quality varies. The most popular articles receive the most scrutiny, and tend to be the best.

And:

> The site's open nature and its popularity help ensure that errors in well-read articles are usually spotted and fixed quickly. .... it is, on the whole, fairly accurate. An investigation by Nature in 2005 compared the site with "Britannica", and found little difference in the number of errors that experts could find in a typical article. Other studies, conducted since, have mostly endorsed that conclusion.

Apart from the above three books and an Internet resource, readers may already be using other references like the venerable https://www.britannica.

DOI: 10.1201/9781003112297-3

com/, the *Cambridge Dictionary of Statistics* by Everitt and Skrondal (2010), the *Oxford Dictionary of Statistics* by Upton and Cook (2014), the *Pocket Dictionary of Statistics* by Sahai and Khurshid (2001), for example.

The non-normal (i.e., non-Gaussian) distributions, and other more advanced topics, will be explained in context as we encounter them in the reliability-based design (RBD) examples of this book.

We are now ready to explore the meanings, alternative perspectives and efficient computational procedures, of the Hasofer–Lind index and the first-order reliability method (FORM), as background for understanding the multidisciplinary RBD examples of the next chapter, and the RBD examples in the chapters on soil and rock engineering.

It is suggested that in probabilistic analysis, whether using FORM (which includes the Hasofer–Lind method as a special case), Monte Carlo simulations, or other methods, estimating the probability of failure should not be the only aim, because it is illuminating to draw other valuable information, for example, from the design point of RBD-via-FORM. The focus of this book is on information and insights attainable from the most probable point (MPP) of failure in RBD-via-FORM, and how RBD-via-FORM can play a valuable complementary role to partial factor design methods like the Eurocode 7 (EC7) and the load and resistance factor design (LRFD) method.

Although the Low and Tang (2004, 2007) efficient FORM procedures in spreadsheet, described below, are used in this book, the same outcome can be obtained on other platforms or stand-alone FORM software. The information and insights discussed, which constitute the gist of this book, can be appreciated by readers regardless of how and on what platform (e.g., Matlab, Mathcad, R, python) they implement FORM.

2.2 A SIMPLE HANDS-ON HASOFER–LIND METHOD IN SPREADSHEET

The spreadsheet reliability evaluation approach will be illustrated first for a case with two normally distributed random variables. Readers who want a better understanding of the procedure and deeper appreciation of the ellipsoidal perspective are encouraged to go through the procedure from scratch on a blank Excel worksheet.

The example concerns the bearing capacity of a continuous footing 1.2 m wide, Figure 2.1, supported on an overconsolidated clay for which the mean values of the shear strength parameters c' and ϕ' are $\mu_{c'} = 20$ kPa and $\mu_{\phi'} = 15°$, with standard deviations of 5 kPa and 2°, respectively. The unit weight of the soil is $\gamma = 20$ kN/m³. The ground water table is 4 m below ground surface. The depth of foundation is 0.9 m, hence the vertical pressure at foundation level is $p_0 = 0.9$ m $\times$ 20 kN/m³ $= 18$ kPa. The footing sustains a non-eccentric vertical load Q_v of 200 kN/m, treated as a deterministic value in this case, to facilitate understanding of Hasofer–Lind index in the

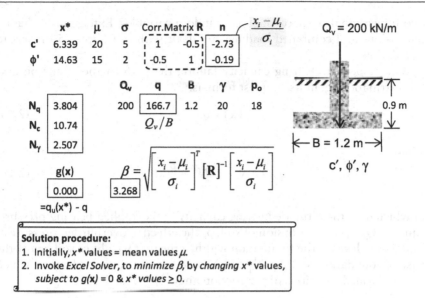

(a)

In the Solver Parameters dialog box:

Set Objective:	the β cell
To:	Minimum
By Changing Variable Cells:	x* values (select the column)
Subject to the Constraints:	g(x) = 0, and x* values >= 0

Note: leave the following box *unchecked*:

☐ Make Unconstrained Variables Non-Negative

The default "GRG Nonlinear" solving method is okay.

Click the "Options" button,

Set the "Constraint Precision" to ⟨ 0.000001 ⟩

and tick the box of ☑ "Use Automatic Scaling"

(b)

Figure 2.1 (a) Method 1 for Hasofer–Lind index in spreadsheet, by changing the x* values, (b) setting the Excel Solver tool to change the x* values.

two-dimensional space of Gaussian random variables. Extensions to higher dimensions and combined loadings are dealt with in Chapter 5 on spread foundations.

With respect to bearing capacity failure, the performance function $g(\mathbf{x})$ for a strip footing, in its simplest form, is:

$$g(\mathbf{x}) = q_u - q \tag{2.1a}$$

where

$$q_u = c'N_c + p_oN_q + \frac{B}{2}\gamma N_\gamma \tag{2.1b}$$

in which q_u is the ultimate bearing capacity, q the applied bearing pressure, equal to Q_v/B, c' the cohesion of soil, p_o the effective overburden pressure at foundation level, B the foundation width, γ the unit weight of soil below the base of foundation and N_c, N_q and $N\gamma$ are bearing capacity factors, which are established functions of the friction angle (ϕ') of soil:

$$N_q = e^{\pi \tan\phi'} \tan^2\left(45 + \frac{\phi'}{2}\right) \tag{2.2a}$$

$$N_c = (N_q - 1)\cot(\phi') \tag{2.2b}$$

$$N_\gamma = 2(N_q + 1)\tan\phi' \tag{2.2c}$$

Several expressions for N_γ exist. The above N_γ is attributed to Vesic in Bowles (1996).

The statistical parameters and correlation matrix of c' and ϕ' are shown in Figure 2.1, with a negative correlation of -0.5 between c' and ϕ', which means high values of c' tend to go with low values of ϕ', and vice versa. The parameters c' and ϕ' in Eqs. 2.1b and 2.2 read their values from the column labelled x^*, which were initially set equal to the mean values. These x^* values, and the functions dependent on them, change during the Excel Solver optimization search for the most probable point (MPP) of failure.

The Hasofer–Lind (1974) index β is:

$$\beta = \min_{\mathbf{x}\in F}\sqrt{(\mathbf{x} - \mathbf{\mu})^T \mathbf{C}^{-1}(\mathbf{x} - \mathbf{\mu})} \tag{2.3a}$$

or, equivalently:

$$\beta = \min_{\mathbf{x}\in F}\sqrt{\left[\frac{x_i - \mu_i}{\sigma_i}\right]^T \mathbf{R}^{-1}\left[\frac{x_i - \mu_i}{\sigma_i}\right]} \tag{2.3b}$$

where $\mathbf{x}$ is a vector of random variables x_i, $\boldsymbol{\mu}$ the vector of mean values μ_i, $\mathbf{C}$ the covariance matrix, $\mathbf{R}$ the correlation matrix, σ_i the standard deviation and F the failure domain. Equation 2.3b containing the correlation matrix $\mathbf{R}$ is easier to set up and conveys the correlation structure more explicitly than the covariance matrix $\mathbf{C}$. The meaning of Eq. 2.3b will be explained soon.

For now, the hands-on continues as follows:

1. The Excel formula of the cell labelled β in Figure 2.1 is Eq. 2.3b:

$$`= \mathrm{sqrt}\Big(\mathrm{mmult}\big(\mathrm{transpose}(\mathbf{n}),\mathrm{mmult}\big(\mathrm{minverse}(\mathbf{R}),\mathbf{n}\big)\big)\Big)`.$$

The arguments $\mathbf{n}$ and $\mathbf{R}$ are entered by selecting the corresponding numerical cells of the column vector $(x_i - \mu_i)/\sigma_i$ and the correlation matrix $\mathbf{R}$, respectively. This array formula is then entered by pressing 'Enter' while holding down the 'Ctrl' and 'Shift' keys. Microsoft Excel's built-in matrix functions *mmult*, *transpose* and *minverse* have been used in this step. Each of these functions contains Excel built-in programme codes for matrix operations. Matrix of size 50 by 50 can be dealt with by these spreadsheet functions.

2. The formula of the performance function is Eq. 2.1a, $g(\mathbf{x}) = q_u - q$, where the equation for q_u is Eq. 2.1b and reads its input values of c' and ϕ' from the x^* values.

3. Microsoft Excel's built-in constrained optimization programme Solver is invoked (via the menu tab Data\Solver), to *Minimize* β, *By Changing* the x^* values, *Subject To* the cell $g(\mathbf{x}) \leq 0$, and x^* values ≥ 0. (If used for the first time, Solver needs to be activated once via File\Options\Add-ins\Excel Add-ins\Solver Add-in.)

The β value obtained is 3.268. The spreadsheet approach is simple and intuitive because it works in the original space of the variables. It does not involve the orthogonal transformation of the correlation matrix, and iterative numerical partial derivatives are done automatically on spreadsheet objects which may be implicit or contain codes.

Instead of invoking the Solver tool to change the x^* values in Figure 2.1, one can invoke Solver to change the $\mathbf{n}$ values, as shown in Figure 2.2. The results are the same. The approach in Figure 2.2 allows easy randomization of the initial values under the $\mathbf{n}$ column if desired.

(Note: the free Solver tool which resides in Microsoft Excel can deal with constraints on a maximum of 100 cells and can change up to 200 decision variables or changing cells. Excel matrix functions like MMult (for multiplying matrices), MInverse (for inversing matrix) and MDeterm (for computing the determinant of a matrix) can deal with a matrix of maximum size about 50 × 50.

2.3 INTUITIVE GRASP OF THE HASOFER–LIND INDEX VIA EXPANDING DISPERSION ELLIPSOID PERSPECTIVE

The 'x*' values, obtained by either Method 1 or Method 2, in Figures 2.1 and 2.2, respectively, represent the MPP of failure on the limit state surface

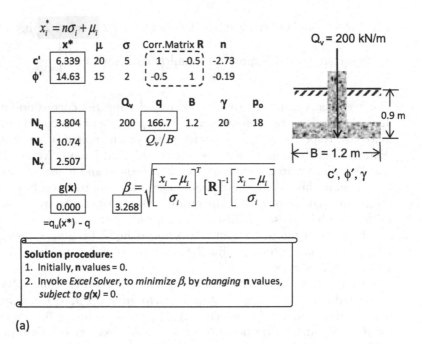

$$x_i^* = n\sigma_i + \mu_i$$

	x*	μ	σ	Corr.Matrix R		n
c'	6.339	20	5	1	-0.5	-2.73
φ'	14.63	15	2	-0.5	1	-0.19

		Qᵥ	q	B	γ	Pₒ
N_q	3.804	200	166.7	1.2	20	18
N_c	10.74		Q_v/B			
N_γ	2.507					

g(x)	
0.000	

=qᵤ(x*) - q

$$\beta = \sqrt{\left[\frac{x_i - \mu_i}{\sigma_i}\right]^T [\mathbf{R}]^{-1} \left[\frac{x_i - \mu_i}{\sigma_i}\right]}$$

β
3.268

Q_v = 200 kN/m

0.9 m

$\leftarrow$ B = 1.2 m $\rightarrow$

c', φ', γ

Solution procedure:
1. Initially, n values = 0.
2. Invoke *Excel Solver*, to *minimize β*, by *changing n* values, subject to g(**x**) = 0.

(a)

In the Solver Parameters dialog box:

Set Objective:	the β cell
To:	Minimum
By Changing Variable Cells:	n values (select the column)
Subject to the Constraints:	g(**x**) = 0

<u>Note</u>: leave the following box *unchecked*:

☐ Make Unconstrained Variables Non-Negative

The default "GRG Nonlinear" solving method is okay.

(b)

Figure 2.2 (a) Method 2 for Hasofer–Lind index in spreadsheet, by changing the n values, (b) setting the Excel Solver tool to change the n values.

(LSS). It is the point of tangency (Figure 2.3) of the expanding dispersion ellipsoid with the bearing capacity LSS. The following may be noted:

(a) The x^* values shown in Figures 2.1 and 2.2 render the performance function $g(x)$, Eq. 2.1a, equal to zero. Hence, the point represented by these x^* values lies on the bearing capacity LSS, which separates the safe domain from the unsafe domain of combinations of c' and ϕ'. The one-standard-deviation (1σ) ellipse and the β-ellipse in Figure 2.3 are rotated because the correlation coefficient between c' and ϕ' is -0.5, as shown in the correlation matrix $\mathbf{R}$ in Figures 2.1 and 2.2. The MPP of failure, also called the *design point* if width B was designed for a target β (e.g., 2.5 or 3.0), is where the expanding dispersion ellipse touches the LSS, at the point represented by the x^* values of Figures 2.1 and 2.2, and plotted to scale in Figure 2.3.

(b) As a multivariate normal dispersion ellipsoid expands, its expanding surfaces are contours of decreasing probability values, according to the established probability density function of the multivariate normal distribution:

$$f(\mathbf{x}) = \frac{1}{(2\pi)^{\frac{n}{2}}|\mathbf{C}|^{0.5}} \exp\left[-\frac{1}{2}(\mathbf{x}-\boldsymbol{\mu})^T \mathbf{C}^{-1}(\mathbf{x}-\boldsymbol{\mu})\right] \qquad (2.4a)$$

$$= \frac{1}{(2\pi)^{\frac{n}{2}}|\mathbf{C}|^{0.5}} \exp\left[-\frac{1}{2}\beta^2\right] \qquad (2.4b)$$

where β is defined by Eq. 2.3a or 2.3b, without the 'min.' Hence, to minimize β (or β^2 in the above multivariate normal distribution) is to maximize the value of the multivariate normal probability density function, and to find the smallest ellipsoid tangent to the limit state surface is equivalent to finding the MPP of failure (the *design point*). This intuitive and visual understanding of the *design point* is consistent with the more mathematical approach in Shinozuka (1983, Equations 4, 41, and associated figure), in which all variables were transformed into their standardized forms and the limit state equation had also to be written in terms of the standardized variables.

(c) Therefore, the design point, being the first point of contact between the expanding ellipsoid and the LSS in Figure 2.3, is the MPP of failure with respect to the safe mean-value point at the centre of the expanding ellipsoid, where $F_s \approx 2.0$ against bearing capacity failure. The reliability index β is the axis ratio (R/r) of the ellipse that touches the limit state surface and the one-standard-deviation dispersion ellipse. By geometrical properties of ellipses, this co-directional axis ratio is the same along any 'radial' direction.

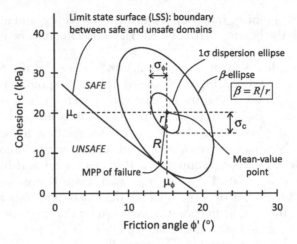

Figure 2.3 The most probable point (MPP) of failure, the mean-value point and expanding ellipsoidal perspective of the reliability index in the original space of the Gaussian random variables.

(d) For each parameter, the ratio of the mean value μ to the x^* value is similar in nature to the partial factors in limit state design (e.g., Eurocode 7). However, in a reliability-based design one does not specify the partial factors. The design point values (x^*) are determined automatically (i.e., obtained as a corollary of reliability index β) and reflect parameter sensitivities, standard deviations, correlation structure and probability distributions in a way that prescribed partial factors cannot reflect.

(e) In Figures 2.1 and 2.2, the mean value point, at (20 kPa and 15°), is safe against bearing capacity failure; but bearing capacity failure occurs when the c' and ϕ' values are decreased to the values shown: (6.339, 14.63). The distance from the safe mean-value point to this MPP of failure, in units of directional standard deviations, is the reliability index β, equal to 3.268 in this case. These normalized distances are the values under the column labelled **n**, which shows $n_{c'} = -2.73$ and $n_{\phi'} = -0.19$. One may regard these **n** values as sensitivity indicators of the variables. In this case, the MPP of failure value of c' is at $2.73\sigma_{c'}$ *below* the mean value of c', while the MPP of failure value of ϕ' is at $0.19\sigma_{\phi'}$ *below* the mean value of ϕ', suggesting that the RBD is much more sensitive to c' than to ϕ', for the case in hand. This sensitivity inference should not be generalized, because a different combination of the statistics for c' and ϕ' (e.g., lower mean and/or standard deviation of cohesion c', and higher mean and/or standard deviation for ϕ') can reverse the sensitivities of c' and ϕ', even for the same bearing capacity problem, as demonstrated in Low (2017). The ability of

RBD-via-FORM to reflect context-dependent sensitivities of parameters from case to case is one of its merits.

(f) The probability of failure (P_f) can be estimated from the reliability index β. Microsoft Excel's built-in function NormSDist(.) can be used to compute the cumulative distribution $\Phi(.)$ and hence probability of failure P_f:

$$P_f = 1 - \Phi(\beta) = \Phi(-\beta) = \text{NormSDist}(-\beta) \qquad (2.5)$$

Equation 2.5 is accurate if the LSS in planar and the variables are normally distributed, and approximate otherwise. Thus, for the bearing capacity problem of Figure 2.1, P_f = NormSDist(−3.268) = 0.054%. This value compares remarkably well with the range of values 0.051–0.060% obtained from several Monte Carlo simulations each with 800,000 trials using the commercial simulation software @RISK (http://www.palisade.com). The correlation matrix was accounted for in the simulation. The excellent agreement between 0.054% from reliability index and the range 0.051–0.060% from Monte Carlo simulation is hardly surprising given the almost linear limit state surface and normal variates shown in Figure 2.3. For other examples in this book involving more random variables and nonnormal distributions, the equivalent hyper-ellipsoids and limit state hypersurfaces can only be visualized in the mind's eye. Nevertheless, the probabilities of failure inferred from reliability indices are generally in good agreement with Monte Carlo simulations. Computing the reliability index and $P_f = \Phi(-\beta)$ by FORM takes only a few seconds. It is also a simple matter to investigate sensitivities by re-computing the reliability index β (and P_f) for different mean values and standard deviations in numerous what-if scenarios. If performance function g(**x**) requires numerical methods, response surface method can be used in conjunction with FORM.

2.4 FORM FOR CORRELATED NON-GAUSSIAN VARIABLES

The Hasofer–Lind (1974) index for cases with correlated Gaussian random variables and the first-order reliability method (FORM) for cases with correlated non-Gaussian variables are well explained in Ditlevsen (1981), Shinozuka (1983), Ang and Tang (1984), Der Kiureghian and Liu (1986), Madsen et al. (1986), Haldar and Mahadevan (2000), Baecher and Christian (2003), Melchers and Beck (2018), for example. The potential inadequacies of the FORM in some cases have been recognized, and more refined alternatives proposed in the literature. On the other hand, the usefulness and accuracy of the FORM in most applications are well recognized, for instance by Rackwitz (2001).

FORM extends the Hasofer–Lind index to deal with correlated nonnormal random variables and hence includes the earlier Hasofer–Lind index as a special case. In FORM, Eq. 2.3b can be rewritten as follows:

$$\beta = \min_{x \in F} \sqrt{\left[\frac{x_i - \mu_i^N}{\sigma_i^N} \right]^T \mathbf{R}^{-1} \left[\frac{x_i - \mu_i^N}{\sigma_i^N} \right]} \qquad (2.6)$$

where μ_i^N and σ_i^N are *equivalent normal* mean and *equivalent normal* standard deviation values, which can be calculated by the Rackwitz and Fiessler (1978) transformation:

$$\text{Equivalent normal standard deviation:} \quad \sigma^N = \frac{\phi\left\{ \Phi^{-1}\left[F(x) \right] \right\}}{f(x)} \qquad (2.7)$$

$$\text{Equivalent normal means:} \quad \mu^N = x - \sigma^N \times \Phi^{-1}\left[F(x) \right] \qquad (2.8)$$

where x is the original nonnormal variate, $\Phi^{-1}[.]$ is the inverse of the cumulative probability (CDF) of a standard normal distribution, $F(x)$ is the original nonnormal CDF evaluated at x, $\phi\{.\}$ is the PDF of the standard normal distribution, and $f(x)$ is the original nonnormal probability density ordinates at x. Figure 2.4 shows the spreadsheet procedure for FORM using the Microsoft Excel Solver. The equivalent normal transformation (Eqs. 2.7 and 2.8) requires only a few lines of Excel VBA codes for each probability type, as shown in Function EqvN(DistributionName, ...) in the figure. The complete code for Function EqvN(...) of Figure 2.4 is in the online Excel solution files at the publisher's website.

One can regard the computation of the FORM β by Eq. 2.6 as that of finding the smallest *equivalent* hyperellipsoid (centred at the *equivalent* normal mean-value point μ^N and with *equivalent* normal standard deviations σ^N) that is tangent to the LSS, Figure 2.5. Hence, for correlated nonnormals, the ellipsoidal perspective still applies in the original coordinate system, except that the nonnormal distributions are replaced by an equivalent normal ellipsoid.

Equations 2.6–2.8 were used in the Low and Tang (2004) spreadsheet-automated FORM procedure, as summarized in Figure 2.4.

An alternative was given in Low and Tang (2007), summarized in Figure 2.6, which uses the following equation:

$$\beta = \min_{x \in F} \sqrt{\mathbf{n}^T \mathbf{R}^{-1} \mathbf{n}} \quad \left(\text{Obviating the calculations of } \mu_i^N \text{ and } \sigma_i^N \right) \qquad (2.9)$$

where $\mathbf{n}$ is the dimensionless vector defined by the bracketed term in Eq. 2.6. For each value of $\mathbf{n}$ tried by Solver, a short and simple Excel VBA function

Low and Tang 2004 FORM procedure:
minimize β by directly varying **x**

$$\beta = \min_{\mathbf{x} \in F} \sqrt{\left[\frac{x_i - \mu_i^N}{\sigma_i^N}\right]^T \mathbf{R}^{-1} \left[\frac{x_i - \mu_i^N}{\sigma_i^N}\right]}$$

Rackwitz-Fiessler equivalent normal transformation

$$\sigma^N = \frac{\phi\{\Phi^{-1}[F(x)]\}}{f(x)}$$

$$\mu^N = x - \sigma^N \times \Phi^{-1}[F(x)]$$

Use Excel's *Solver* to change the **x** vector.

Subject to g(**x**) = 0

```
Function EqvN(DistributionName, para, x, code) As Double
  del = 0.0001
  a1 = para(1):  a2 = para(2):  a3 = para(3):  a4 = para(4)
  Select Case UCase(Trim(DistributionName))
    Case "NORMAL":     If code = 1 Then EqvN = a1
      If code = 2 Then EqvN = a2
    Case "LOGNORMAL":  If x < del Then x = del
      lamda = Log(a1) - 0.5 * Log(1 + (a2 / a1) ^ 2)
      If code = 1 Then EqvN = x * (1 - Log(x) + lamda)
      If code = 2 Then EqvN = x * Sqr(Log(1 + (a2 / a1) ^ 2))
    Case "EXTVALUE1": ...
    Case "EXPONENTIAL": ...
    Case "UNIFORM":   ...
    Case "TRIANGULAR": ...
    Case "WEIBULL":   ...
    Case "GAMMA": ...
    Case "PERTDIST":  ...
    Case "BETADIST": ...
  End Select
End Function
```

Figure 2.4 Excel Solver procedure for FORM, Low and Tang (2004).

code x_i(DistributionName, para, ni) automates the computation of x_i from n_i, for use in the constraint g(**x**) = 0, via the following equation:

$$x_i = F^{-1}\left[\Phi(n_i)\right] \qquad (2.10)$$

For uncorrelated variables, the correlation matrix **R** is an identity matrix, and Eq. 2.9 reduces to:

$$\beta = \min_{\mathbf{x} \in F} \sqrt{\sum n_i^2} \quad \text{(Independent or uncorrelated variables } x_i) \quad (2.11)$$

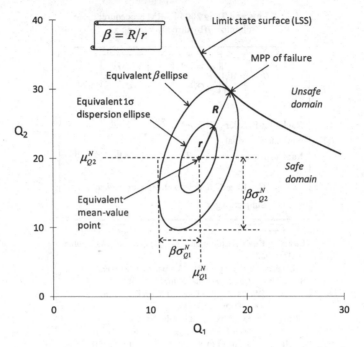

Figure 2.5 Equivalent ellipsoid perspective for non-Gaussian variables, with centre at the equivalent Gaussian mean values, and of size defined by equivalent Gaussian standard deviations. Q_1 and Q_2 are positively correlated in this case.

For cases involving a single random variable only (illustrated by Example 3.3d), there is no optimization of direction involved, and the reliability index is simply $\beta = n$, in which n is the value producing an x value (by Eq. 2.10) causing failure.

The objective of Eq. 2.10 is to find the value x_i such that the non-Gaussian cumulative probability distribution $F(x_i)$ at x_i is equal to the standard normal cumulative distribution $\Phi(n_i)$. The requirement that x_i be on the failure surface (limit state surface), $\mathbf{x} \in F$, is imposed as a constraint $g(\mathbf{x}) = 0$ in Excel Solver (Figure 2.6). When this is satisfied and β by Eq. 2.9 is at its minimum, the $\mathbf{x}$ values become the design point $\mathbf{x}^*$ values. The complete code for automatically determining x_i by Eq. 2.10 is in the Appendix, and also in the online Excel files of this book's FORM and RBD examples at the publisher's website.

Figure 2.7 provides an illustration of the Low and Tang (2004, 2007) FORM procedures involving three correlated non-Gaussian variables, for which the performance function is $g(\mathbf{x}) = VW - Z$. The Weibull distribution (22, 41) of variable V has a range 0 to ∞, a mean of 40 and a standard deviation of 2.26. The Triangular (30, 50, 70) distribution of variable W has a mean of 50 and a standard deviation of 8.17. The four-parameter beta

Low and Tang 2007 FORM procedure:
minimize β by varying **n**, on which **x** depends

$$\beta = \min_{\mathbf{x}\in F} \sqrt{\mathbf{n}^T \mathbf{R}^{-1}\mathbf{n}}$$

Use Excel's Solver to *minimize β* by changing the **n** vector, *subject to the constraint* g(**x**) = 0.

For each value of n_i tried by Solver, a short and simple Excel VBA code automates the computation of x_i from n_i, via
$$x_i = F^{-1}[\Phi(n_i)]$$
for use in the constraint g(**x**) = 0.

```
Function x_i(DistributionName, para, ni) As Double
  a1 = para(1):  a2 = para(2):  a3 = para(3):  a4 = para(4)
  Select Case UCase(Trim(DistributionName))
    Case "NORMAL":    x_i = a1 + ni * a2
    Case "LOGNORMAL"
      lamda = Log(a1) - 0.5 * Log(1 + (a2 / a1) ^ 2)
      zeta = Sqr(Log(1 + (a2 / a1) ^ 2))
      x_i = Exp(lamda + zeta * ni)
    Case "EXTVALUE1": ...
    Case "EXPONENTIAL": ...
    Case "UNIFORM":   ...
    Case "TRIANGULAR": ...
    Case "WEIBULL":   ...
    Case "GAMMA": ...
    Case "PERTDIST":  ...
    Case "BETADIST":  ...
  End Select
End Function
```

Figure 2.6 Excel Solver procedure for FORM, Low and Tang (2007).

distribution of variable Z, BetaDist (2, 4, 600, 1800), is bounded within the range 600–1800, has a mean of 1000 and has a standard deviation of 213.8. The PDF curves of these three distributions are shown in Figure 2.7c. The beta distribution can display diverse shapes depending on its first two parameters which are shape parameters.

The 2007 method at the top of Figure 2.7, based on changing the n values by the Excel Solver tool, obtains virtually identical results in β, **x*** values and **n** values, as the 2004 method below it based on changing the **x*** values by Solver. The 2007 method does not need equivalent normal mean μ^N and equivalent normal standard deviation σ^N of Eqs. 2.7 and 2.8.

For the case in hand, where the performance function is g(**x**) = VW – Z, it will be appreciated that V and W contribute to resistance, and Z represents

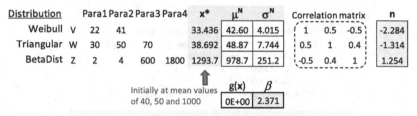

Cell G3: "=@x_i(A3,C3:F3,L3)", autofill to cell G5.
where x_i(...) is the user-defined function in Low and Tang (2007).

(a) Low and Tang (2007) *Excel Solver* method; variable cells are the **n** values.

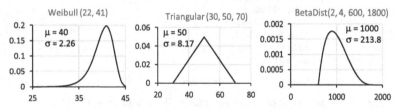

Equivalent normal Rackwitz-Fiessler transformation (μ^N and σ^N)
done in simple Excel VBA codes for various distributions.

(b) Low and Tang (2004) *Excel Solver* method; variable cells are the **x*** values.

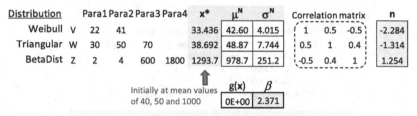

(c) Probability density function plots of Weibull, Triangular and Beta distributions

Figure 2.7 Illustrative example of FORM, using two efficient spreadsheet-based constrained optimization methods of Low and Tang (2004, 2007).

an unfavourable load. The **n** values of the three sensitivity indicators indicate that the most probable failure values of V and W (33.44 and 38.69) are at 2.28 and 1.31 times their respective *equivalent normal standard deviations* σ^N values *below* their respective μ^N values. In contrast, the unfavourable load Z has its most probable failure value (1293.7) at 1.25 times its *equivalent normal standard deviation* σ^N value *above* its μ^N value. The values under the **x*** column is the MPP of failure, with a reliability index $\beta = 2.371$, or probability of failure about 0.89%, obtained from $P_f = \Phi(-\beta)$. This compares with a failure probability of 0.71% from Monte Carlo simulation with 200,000 realizations.

Although the mean-value point (40, 50, 1000), Figure 2.7, is safe (because mean resistance is $40 \times 50 = 2000$, which is greater the mean load 1000), failure is reached (at a small probability of 0.89%) when V and W decrease from their mean values to their most probable failure values of 33.44 and 38.69, and the load Z increases from its mean value of 1000 to its most probable failure value of 1293.7. The MPP of failure (33.44, 38.69, 1293.7) is the point where the expanding 3D equivalent dispersion ellipsoid first touches the LSS. This MPP is a failure state because $33.44 \times 38.69 = 1293.7$. The higher absolute value of the sensitivity indicator n_V, -2.284 in Figure 2.7a, indicates that random variable V is the most sensitive parameter. The other two variables W and Z, respectively, with n_W and n_Z values of -1.31 and $+1.25$, are about equal in their sensitivity measures.

For correlated Gaussian variables or correlated non-Gaussian variables, the *unrotated* n-space FORM computational approach of Eq. 2.9 is related to the *uncorrelated* u *vector* in the *rotated space* of the classical u-space FORM computational approach. The vectors n and u can be obtained, one from the other, using the equations below (e.g., Low et al., 2011):

$$\mathbf{n} = \mathbf{Lu} \qquad (2.12a)$$

and

$$\mathbf{u} = \mathbf{L}^{-1}\mathbf{n} \qquad (2.12b)$$

in which L is the lower triangular matrix of Cholesky decomposition of R. The vector u is equal to the vector n only when the random variables are uncorrelated, because then $\mathbf{L}^{-1} = \mathbf{L} = \mathbf{I}$ (the identity matrix). In general, u is not equal to n. Melchers (1999) noted that, because the classical u-space FORM computational approach transforms correlated random variables to *uncorrelated* random variables in a *rotated* u space, the physical meanings of the original parameters are lost. We may note here that the *correlated* but *unrotated* n-space computational approach of Eq. 2.9 preserves the physical meanings of the random variables, with ellipsoidal perspective and meaningful sensitive indicators n.

The example RBD cases in this book use the Low and Tang 2007 spreadsheet FORM procedure of Eq. 2.9, which are summarized in Figures 2.6 and 2.7. The procedure allows normal distribution, lognormal distribution, beta distribution, exponential distribution, and also Weibull, gamma, extreme value type 1, triangular, uniform and PERT distributions.

RBD-via-FORM is next illustrated, which, unlike the FORM analyses of Figures 2.1, 2.2 and 2.7, aims for a design (e.g., width of spread foundation) that achieves a target reliability index value, for example, $\beta = 2.5$ or 3.0. An intuitive ellipsoidal perspective in the *dimensionless space of correlated sensitivity indicators* n is also presented, as an alternative to the perspective in *dimensional* space of Figures 2.3 and 2.5.

2.5 A SIMPLE EXAMPLE OF RBD-VIA-FORM AND CORRELATED SENSITIVITIES

This section implements RBD-via-FORM for the overturning ULS involving only two random variables, to convey an intuitive understanding of the MPP of failure (design point) and to investigate the effect of parametric correlation on design point and reliability index. How the sensitivity of a parameter is affected by its correlation to another parameter is demonstrated.

For the tall lightweight structure shown in Figure 2.8a, acted by a horizontal load Q_h, a superstructure vertical load G_v, and the weight of the square foundation pad G_{pad}, the performance function g(x) against the rotational ULS is:

Rotational ULS: $g(\mathbf{x}) = (G_v + G_{pad}) \times B/2 - Q_h \times 12\,\mathrm{m} = M_R - M_o$ (2.13)

where

$$G_{pad} = B^2 \times 2\,\mathrm{m} \times 24.5\,\mathrm{kN/m^3}$$

The horizontal load Q_h and superstructure vertical load G_v are the random variables, with mean values 250 kN and 545 kN, and coefficients of variation (σ/μ) equal to 0.2 and 0.1, respectively.

Physical considerations may warrant correlation between Q_h and G_v. For example, larger G_v may imply more surface area exposed to wind. In Figure 2.8a, Q_h and G_v are assumed positively correlated with a correlation coefficient ρ of 0.5 as shown in the correlation matrix **R**. The **x*** column invokes short VBA codes from Low and Tang (2007) for calculating Q_h and G_v from the **n** values. Initially the **n** column values were zeros. Then:

a) A trial value (e.g., 7 m) was input for B such that the cell labelled g(**x**) displays a positive value.
b) Microsoft Excel built-in routine *Solver* (hereafter referred to as the *Solver tool*) was invoked, to *minimize* the cell β, by automatically changing the two **n** column cells, subject to the constraint that the performance function cell 'g(**x**)' be equal to 0.0.
c) Repeat the above steps until a width B achieves the target reliability index (3.0 in this case).

A width B of 5.32 m achieves the target reliability index β of 3.0. The **x*** values of '441.79' and '606.21' in Figure 2.8a represent the design point where the expanding *equivalent* dispersion ellipsoid first touches the LSS. The following are noteworthy:

(1) The mean value point (Q_h = 250 and G_v = 545) is in the safe domain if cell g(**x**) displays positive value (for a trial width B) when the **n** column values were initially zeros in step (a) above. The closest unsafe

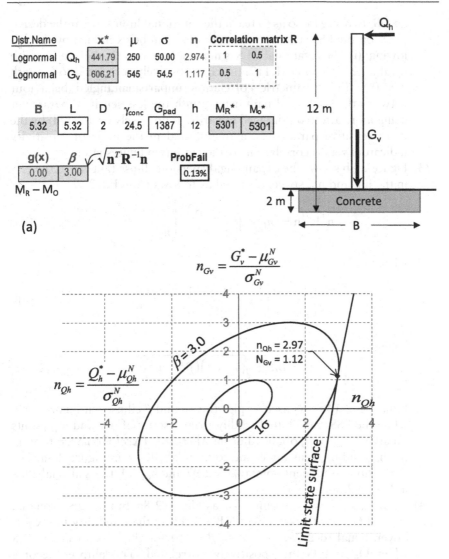

Figure 2.8 (a) RBD-via-FORM for the rotational ULS of a tall lightweight structure, with correlated Q_h and G_v, (b) equivalent dispersion ellipses,
limit state surface and design point, in the unrotated space of correlated dimensionless sensitivity indicators n_{Qh} and n_{Gv}.

point on the LSS (for target $\beta = 3.0$) is represented by $Q_h = 441.79$ and
$G_v = 606.21$.

(2) The design value of G_v, at 606.21, is 1.117 times its equivalent normal
standard deviation *above* its equivalent normal mean value, where
1.117 is the second value under the column labelled n in Figure 2.8a.
This challenges our intuition, which expects Q_h to increase (it does)

and G_v to *decrease* so as to reach the rotational limit state at the design point on the LSS. This apparent paradox can be explained by the following: (i) The rotational ULS is more sensitive to Q_h than to G_v, with n values 2.974 versus 1.117; (ii) G_v is positively correlated to Q_h with $\rho = 0.5$; (iii) parametric correlation imparts entangled behaviour between the correlated parameters, with the less sensitive parameter being more affected ('dragged towards,' if positively correlated) by the more sensitive parameters, as manifested in the plots of sensitivity indicators versus correlation coefficient ρ two figures later.

(3) Figure 2.8b shows the expanding dispersion ellipse tangent to the LSS in the n_{Gv} and n_{Qh} space, obtained as follows from Eq. 2.9:

$$\mathbf{n}^T \mathbf{R}^{-1} \mathbf{n} = \begin{bmatrix} n_1 & n_2 \end{bmatrix} \begin{bmatrix} 1 & \rho \\ \rho & 1 \end{bmatrix}^{-1} \begin{bmatrix} n_1 \\ n_2 \end{bmatrix} = \beta^2 \qquad (2.14a)$$

where

$$\begin{bmatrix} 1 & \rho \\ \rho & 1 \end{bmatrix}^{-1} = \frac{1}{1-\rho^2} \begin{bmatrix} 1 & -\rho \\ -\rho & 1 \end{bmatrix} \qquad (2.14b)$$

Hence

$$n_2 = \rho n_1 \pm \sqrt{(1-\rho^2)(\beta^2 - n_1^2)} \qquad (2.14c)$$

Equation 2.14c was used to plot the *1σ* and β ellipses in Figure 2.8b, while the LSS was plotted readily from a series of n_{Qh} and n_{Gv} points satisfying $g(\mathbf{x}) = 0$, obtained using *Microsoft Excel GoalSeek* tool on the n_{Qh} and n_{Gv} cells under the column labelled **n** in Figure 2.8a. The dimensionless tangent point ($n_{Qh} = 2.97$ and $n_{Gv} = 1.12$) is also labelled in Figure 2.8b.

(4) Figure 2.9a shows the same case as Figure 2.8a, but for uncorrelated Q_h and G_v. For the same width B = 5.32 m, the reliability index β is lower, equal to 2.78, instead of $\beta = 3.0$ for the case in Figure 2.8 when Q_h and G_v are positively correlated. The ellipses become canonical and reduce to circles (or spheres in higher dimensions), and **n** = **u** by Eq. 2.12a when **L** reduces to the identity matrix for uncorrelated random variables. The tangent point in Figure 2.9b shows that the MPP value of G_v on the LSS is at 0.38 times its equivalent normal standard deviation σ_{Gv}^N *below* its equivalent normal mean value μ_{Gv}^N, testifying to the unambiguous stabilizing nature of G_v (in the overturning ULS) when it is not positively correlated with the destabilizing effect of Q_h.

(5) Figure 2.10a shows β increases with increasing correlation ρ between G_v and Q_h for the case in hand, which is a case of favourable (stabilizing) G_v and unfavourable (destabilizing) Q_h when the overturning

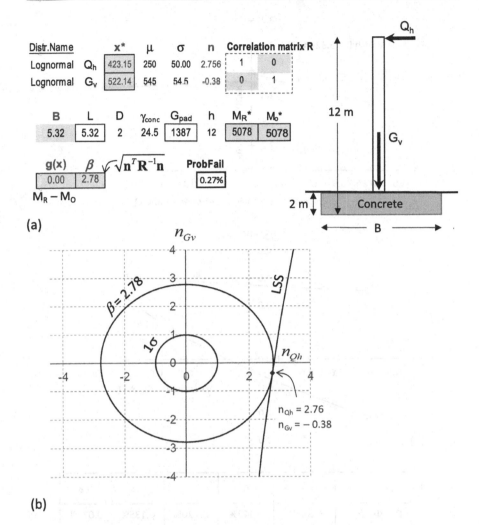

(a)

(b)

Figure 2.9 (a) For uncorrelated Q_h and G_v, FORM analysis yields $\beta = 2.78$, for the design width B = 5.32 m from Figure 2.8a, (b) equivalent dispersion ellipses (reduces to circles), limit state surface and design point.

limit state is concerned. Other scenarios of overestimation and under-estimation of reliability index β when parametric correlation is justi-fied but ignored are shown in Table 2.1. Figure 2.10b shows the entangled sensitivity of G_v when it is positively correlated with the more sensitive Q_h.

(6) Figure 2.10c provides verification from Monte Carlo simulations (using @RISK) that, even when entangled sensitivities are involved, RBD-via-FORM correctly evaluates the probabilities of failure P_f. Note that Monte Carlo simulation is valuable for checking P_f, but not

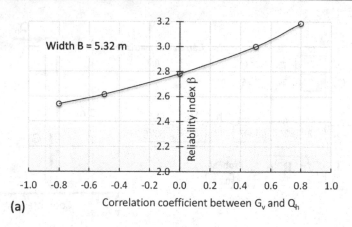

(a)

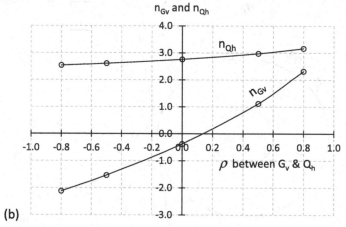

(b)

	$\rho = -0.8$	-0.5	0	0.5	0.8
$P_f = \Phi(-\beta)$	0.554%	0.443%	0.270%	0.135%	0.072%
Monte Carlo P_f	0.560%	0.432%	0.269%	0.136%	0.071%

(c)

Figure 2.10 Overturning limit state (a) reliability index varies with correlation between unfavourable Q_h and favourable G_v, (b) effect of correlation coefficient ρ on sensitivity indicator n_{Gv}, (c) probabilities of failure from β nearly identical with Monte Carlo simulations each 300,000 trials.

likely to provide information and insights like those attainable from the FORM design point.

(7) The sensitivities conveyed in the dimensionless indicators n_i are with respect not only to the LSS (defined in this case by $g(\mathbf{x}) = M_R - M_o = 0$,

Table 2.1 Overestimation and underestimation of reliability index if parametric correlations are warranted but not modelled.

	Resistance parameter (or favourable load)	Unfavourable load parameter
Resistance parameter (or Favourable load)	• Overestimate β if ignore positive correlation. • Underestimate β if ignore negative correlation.	• Underestimate β if ignore positive correlation (e.g., case in Figure 2.9). • Overestimate β if ignore negative correlation.
Unfavourable load parameter	(same as upper right)	• Overestimate β if ignore positive correlation. • Underestimate β if ignore negative correlation.

Eq. 2.13), but also to the size and orientation of the equivalent dispersion ellipse/ellipsoid, as defined by the statistics of the random variables in the quadratic form in the equation of β (Eq. 2.9). This can be readily appreciated from the fact the MPP of failure (design point) depends on the interplay of the expanding dispersion ellipsoid and the LSS. This imparts greater context-dependent sensitivities to RBD-via-FORM than the sensitivities with respect to LSS only.

(8) Monte Carlo simulation software can also provide probability of failure and sensitivity analysis. In the @RISK software (www.Palisade.com) that the author is familiar with, there are four different kinds of sensitivity analysis: (1) change in output statistics method, (2) regression method, (3) correlation method and (4) contribution to variance method, each of which performs sensitivity analysis in a different way and with a different emphasis.

The simple overturning limit state is used above to facilitate understanding of FORM, design point, parametric correlations and sensitivity indicators n_{Qh} and n_{Gv}. In practice, other ULS also need to be checked using FORM, as shown in Low (2017, 2020a), for example, and in Chapter 5.

2.6 HOW MEAN-VALUE POINT DETERMINES THE SIGN OF RELIABILITY INDEX

In RBD-via-FORM (e.g., Figure 2.8) and FORM analysis (e.g., Figure 2.9), one must distinguish negative from positive reliability indices. The computed β index can be regarded as positive only if the mean-value point is in the safe domain. Although the discussions in the next paragraph assume

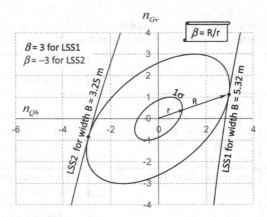

Figure 2.11 Distinguishing negative from positive reliability index: reliability index β is 3 for LSS1, and negative 3 for LSS2.

normally distributed random variables, they are equally valid for the equivalent normals of nonnormal random variables in FORM.

For the design footing width B = 5.32 m shown in Figure 2.8a, the mean-value point is in the safe domain. This can be confirmed by noting that the bottom-left cell of the performance function, $g(x) = M_R - M_o$, displays a positive value when the two cells under the column labelled n are at their initial values (i.e., zeros) prior to invoking the Solver tool to obtain the solution β = 3.0 by Eq. 2.9. However, if the footing width B is 3.25 m (instead of 5.32 m), the Solver-computed β is also 3.0 by Eq. 2.9, which must be given a negative sign (to become β = –3) because the performance function (*PerFn*) cell displays a negative value when the cells under the column labelled n are at their initial values of 0.0 prior to invoking the Solver tool. The β = 3 ellipse and the limit state surface (LSS) for width B = 5.32 m, denoted as LSS1, and that for width B = 3.25 m, denoted as LSS2, are shown in Figure 2.11. The mean-value point (the centre of the ellipse, or ellipsoid in higher dimensions) is in the safe domain (hence positive β) with respect to LSS1 for width B = 5.32 m, but in the unsafe domain with respect to LSS2 for width B = 3.25 m.

2.7 PROBABILISTIC CONSOLIDATION SETTLEMENT ANALYSIS

An example problem from Ang and Tang (1984, pp. 372–374) involves the settlement of a structure on a layer of normally consolidated clay; the secondary compression of the clay layer is assumed to be negligible. The primary consolidation settlement S is:

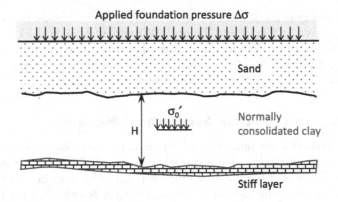

$$g(\mathbf{x}) = s_L - N \frac{C_c}{1+e_o} H \log \frac{\sigma_0' + \Delta\sigma}{\sigma_0'}, \qquad N = \text{model error factor}$$

Statistics of Variables affecting primary consolidation of NC clay

	Mean	c.o.v.	Standard Deviation
N	1.0	0.10	0.10
C_c	0.396	0.25	0.099
e_o	1.19	0.15	0.1785
H	168 in. (4.27 m)	0.05	8.4
σ_0'	3.72 ksf (178 kPa)	0.05	0.186
$\Delta\sigma$	0.50 ksf (24 kPa)	0.20	0.10

Figure 2.12 An example from Ang and Tang (1984).

$$S = \frac{C_c}{1+e_o} H \log \frac{\sigma_0' + \Delta\sigma}{\sigma_0'} \qquad (2.15)$$

where C_c is the compression index, e_o is the initial void ratio, H is the thickness of the clay layer, σ_0' is the original effective pressure at mid-depth of the clay layer and $\Delta\sigma$ is the increase in pressure at mid-depth. The variables have the statistics shown in Figure 2.12; they are assumed to be uncorrelated and normally distributed.

A correction factor N has been introduced to account for the idealizations underlying the analytical model. It is supposed that satisfactory performance requires that the settlement be less than 2.5 inches (6.4 cm). The performance function is therefore given by:

$$g(\mathbf{x}) = 2.5 - N \frac{C_c}{1+e_o} H \log \frac{\sigma_0' + \Delta\sigma}{\sigma_0'} \qquad (2.16)$$

The reliability index β will be evaluated using two methods: (1) the ellipsoid perspective and Low and Tang (2007) spreadsheet FORM approach (which reduces to the Hasofer–Lind method in this case of normal variates); (2) the traditional FORM procedure involving explicit analytical partial derivatives.

2.7.1 Method 1: Efficient Spreadsheet Method

Since the variables are uncorrelated, the equation for the six-dimensional ellipsoid is an extension of the canonical form of the equation for two-dimensional ellipse. The reliability index Eq. 2.9 reduces to Eq. 2.11. There is no need to set up the 6-by-6 correlation matrix $\mathbf{R}$ (which is the identity matrix for independent variables). The problem is therefore formulated as follows:

$$\text{Minimize:}\quad \beta = \sqrt{\sum_{i=1}^{i=6} n_i^2} \tag{2.17a}$$

$$\text{Subject to:}\quad g(\mathbf{x}) = 2.5 - N\frac{C_c}{1+e_o}H\log\frac{\sigma_0' + \Delta\sigma}{\sigma_0'} = 0 \tag{2.17b}$$

where

$$x_i = \mu_i + n_i\sigma_i \tag{2.17c}$$

in which $x_1, x_2, ..., x_6$ are N, C_c, e_o, H, σ_0' and $\Delta\sigma$ respectively, and μ_i and σ_i are the mean and standard deviation of the respective x_i's.

One first sets up the spreadsheet as shown in Figure 2.13a, in which σ_i = c. o. v. × μ_i, the β cell is entered with Eq. 2.17a, the g(x) with Eq. 2.17b and referring to values from the cells under the $\mathbf{x}^*$ column, which are based on Eq. 2.17c. With initial $\mathbf{n}$ values of zeros, the g(x) cell displays a positive value, indicating that the mean-value point is in the safe domain; hence, the β value obtained shortly will be treated as a positive reliability index.

The spreadsheet *Solver* tool is then invoked, *To Set* the β cell *to minimum*, *By Changing* the $\mathbf{n}$ column (which is selected at one go), *Subject to the Constraints* that the performance function cell g(x) be equal to 0.0. The default generalized reduced gradient (GRG) Nonlinear method is efficient. The solution shown in Figure 2.13b is obtained immediately, with β = 1.270. Besides β, it is illuminating to ponder over the information and insights provided by the most probable failure point (the $\mathbf{x}^*$ column) and the sensitivity indicators (the $\mathbf{n}$ column). If the variables are correlated (not the case here), correlated sensitivities may occur, as explained in the previous example in connection with Figures 2.8, 2.9 and 2.10 and Table 2.1.

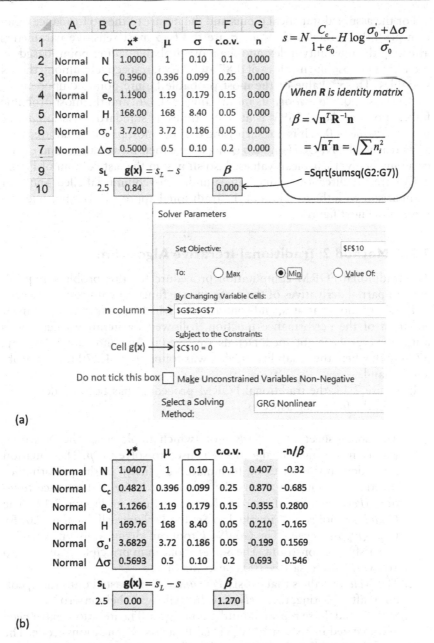

	A	B	C	D	E	F	G
1			x*	μ	σ	c.o.v.	n
2	Normal	N	1.0000	1	0.10	0.1	0.000
3	Normal	C_c	0.3960	0.396	0.099	0.25	0.000
4	Normal	e_0	1.1900	1.19	0.179	0.15	0.000
5	Normal	H	168.00	168	8.40	0.05	0.000
6	Normal	σ_0'	3.7200	3.72	0.186	0.05	0.000
7	Normal	$\Delta\sigma$	0.5000	0.5	0.10	0.2	0.000
9		s_L	$g(x) = s_L - s$			β	
10		2.5	0.84			0.000	

$$s = N\frac{C_c}{1+e_0}H\log\frac{\sigma_0'+\Delta\sigma}{\sigma_0'}$$

When R is identity matrix

$$\beta = \sqrt{\mathbf{n}^T\mathbf{R}^{-1}\mathbf{n}}$$

$$= \sqrt{\mathbf{n}^T\mathbf{n}} = \sqrt{\sum n_i^2}$$

=Sqrt(sumsq(G2:G7))

Solver Parameters

Set Objective: $F$10

To: ○ Max   ● Min   ○ Value Of:

By Changing Variable Cells:

n column ──→ $G$2:$G$7

Subject to the Constraints:

Cell g(x) ──→ $C$10 = 0

Do not tick this box ☐ Make Unconstrained Variables Non-Negative

Select a Solving   GRG Nonlinear
Method:

(a)

		x*	μ	σ	c.o.v.	n	-n/β
Normal	N	1.0407	1	0.10	0.1	0.407	-0.32
Normal	C_c	0.4821	0.396	0.099	0.25	0.870	-0.685
Normal	e_0	1.1266	1.19	0.179	0.15	-0.355	0.2800
Normal	H	169.76	168	8.40	0.05	0.210	-0.165
Normal	σ_0'	3.6829	3.72	0.186	0.05	-0.199	0.1569
Normal	$\Delta\sigma$	0.5693	0.5	0.10	0.2	0.693	-0.546
	s_L	$g(x) = s_L - s$			β		
	2.5	0.00			1.270		

(b)

Figure 2.13 Excel spreadsheet method for reliability index β: (a) before running Solver, (b) solution obtained by Excel Solver.

For the assumed statistical inputs and the nature of the performance function $g(\mathbf{x})$, the variables C_c and $\Delta\sigma$ are the two most sensitive parameters affecting the reliability index and the most probable failure point, based on their highest absolute n values of 0.87 and 0.693, respectively, among the six variables. The $\mathbf{x}^*$ value of the most probable failure point component C_c (0.4821) is at $0.87\sigma_{C_c}$ *above* its mean value of 0.396, and the most probable failure point of $\Delta\sigma$ (0.5693) is at 0.693 times its standard deviation *above* its mean value of 0.5. It is not surprising that the most probable failure point $\mathbf{x}^*$ is reached by N, C_c, H and $\Delta\sigma$ going above their mean values, and e_0 and σ'_0 going below their mean values. Also shown in the last column of Figure 2.13b are the direction cosine values under the column labelled $-n/\beta$, for comparison with the α_i values of the traditional approach (last but one column in the next figure).

2.7.2 Method 2: Traditional Iterative Algorithm

The traditional FORM computation procedure for this problem requires explicit partial derivatives of the performance function with respect to each of the six random variates, and solution of β which appear in the nonlinear equation of the performance function, followed by iterative calculations until convergence is obtained. Details are given in Ang and Tang (1984, pp. 372–374), where the reliability index β was found to be 1.270 for the problem in hand.

In Figure 2.14, the traditional FORM procedure has been rendered less tedious in the following aspects:

1. The spreadsheet's *GoalSeek* tool (which implements the Newton–Raphson root-finding algorithm) is used to solve for β. The equation for β does not need to be formulated explicitly in the performance function. One merely enters the equation for the performance function (*PerFunc*) based on the *New_x_i* values in Figure 2.14. The *GoalSeek* tool is then invoked to *Set* the *PerFunc* cell equal to 0.0, *By Changing* the β cell. (The *GoalSeek* tool can change only one variable to satisfy one constraint. The *Solver* tool can in principle change up to 100 variables to satisfy 100 constraints.)
2. The *New_x_i* values that satisfy *PerFunc* = 0 are pasted onto the x_i column; after pasting, the *PerFunc* value will again be nonzero.
3. Steps 2 and 3 are repeated until β converges. The iterations have been automatized in a short VBA (Visual Basic for Applications) code. The module (*AutoGoalSeek*) is also shown in Figure 2.14.

Nevertheless, because it requires explicit analytical partial derivatives from the user, this traditional method lacks the simplicity and convenience of the spreadsheet FORM approach of Figure 2.13, in which iterations and numerical derivatives were done automatically by the Solver tool.

Traditional FORM procedure requiring explicit partial derivatives

	x	μ	c.o.v.	σ	$\partial g/\partial x_i'$	α_i	$\dfrac{\mu_i - \beta\alpha_i\sigma_i}{\text{New x}}$
N	1.0407	1	0.1	0.10	-0.2402	-0.3205	1.0407
C_c	0.4821	0.396	0.25	0.099	-0.5134	-0.6850	0.4821
e_0	1.1266	1.19	0.15	0.1785	0.2098	0.2800	1.1266
H	169.76	168	0.05	8.40	-0.1237	-0.1650	169.76
σ_0'	3.68	3.72	0.05	0.186	0.1176	0.1569	3.6829
$\Delta\sigma$	0.5693	0.5	0.2	0.10	-0.4090	-0.5458	0.5693
					0.7495		0.0000

Initially, x = μ

$$\beta \qquad \sqrt{\sum(\partial g/\partial x_i)^2}$$

PerFunc

Auto GoalSeek
(click to run)

1.270

Initially β = 1.0

```
Sub AutoGoalSeek()
'Excel VBA Macro for iterating GoalSeek, followed by updating x values
    For i = 1 To 8
        Range("PerFunc").GoalSeek Goal:=0, ChangingCell:=Range("beta")
        Range("xvector").Value = Range("New_x").Value
    Next i
End Sub
```

Figure 2.14 Traditional FORM procedure requires explicit partial derivatives.

The traditional FORM procedure will be even more complicated mathematically when dealing with correlated nonnormal random variables, due to rotation of coordinate system using eigenvalue and eigenvector techniques.

The reliability analysis of consolidation settlement of Figure 2.13 deals with uncorrelated normal random variables. The next example investigates the risk that earthquake-induced cyclic shear stress in saturated sand will exceed its shear resistance, involving eight nonnormal variables, some of which are correlated.

2.8 PROBABILISTIC ANALYSIS OF EARTHQUAKE-INDUCED CYCLIC SHEAR STRESS IN SATURATED SAND

This example is based on problem 6.14 of Ang and Tang (1984), solved here using the efficient Low and Tang (2007) spreadsheet FORM procedure. The case involves eight nonnormal variables, some of which are correlated due to physical considerations. The traditional u space FORM procedure (as explained in Ang and Tang 1984, for example) requires eight analytical partial derivatives, rotation of coordinate system and iterations. In the spreadsheet FORM procedure, numerical partial derivatives and iterations are required but are performed automatically by the Solver tool, such that the results are presented succinctly and the MPP of failure can be visualized in the mind's eye as the contact point between an eight-dimensional expanding equivalent dispersion ellipsoid and the LSS in the original space of the random variables.

Problem 6.14 of Ang and Tang (1984) considered an element of saturated sand at depth. The earthquake-induced cyclic shear stress τ_A at a depth of h in a saturated sand deposit was modelled by:

$$\tau_A = S_L r_d \gamma h \frac{a_{\max}}{g} \tag{2.18}$$

where S_L is the amplitude (as fraction of peak stress) of equivalent uniform cycles, r_d is the stress reduction due to flexibility of the sand column, γ is the unit weight of the sand, h the depth of the sand element, a_{max} is the maximum ground acceleration and g is the gravitational acceleration.

The shear resistance for the sand element is modelled in Ang and Tang (1984) by:

$$\tau_R = N_f N_S C_r R \sigma'_v D_r \tag{2.19}$$

where C_r is the discrepancy between *in situ* strength and laboratory-measured strength, R is the normalized laboratory strength parameter for a given type of test and failure strain criterion, σ_v' is the effective vertical stress acting on the soil element *in situ*, D_r is the relative density of the soil *in situ*, N_s is the correction accounting for secondary factors such as frequency of cyclic loading, grain shape, etc., and N_f is the correction for additional error associated with the simplified model.

The performance function is:

$$g(\mathbf{x}) = \tau_R - \tau_A = N_f N_S C_r R \sigma'_v D_r - S_L r_d \gamma h \frac{a_{\max}}{g} \tag{2.20}$$

Figure 2.15a adopts the above Ang and Tang performance function g(x), and statistics for a critical location at a site. The gravitational acceleration g and the three variables (A, h and S_L), with zero c.o.v., are shown with their respective deterministic values in the second row of Figure 2.15. Seven of the eight random variables are assumed to follow the lognormal distribution, as suggested in Ang and Tang (1984). However, for the relative density D_r, which by definition has a theoretical range of 0–1.0, the beta distribution, with shape parameters 8.022 and 4.263, and lower and upper bounds of 0.0 and 1.0, is used instead of lognormal distribution because the latter extends beyond the theoretically permissible upper bound of $D_r = 1.0$, as shown in Figure 2.15b.

Probability of earthquake-induced cyclic shear stress exceeding its shear resistance

g	a_{max}	h	S_L
9.81	0.981	7.62	0.75
(m/s²)	(m/s²)	(m)	

		x^*	Para1	Para2	Para3	Para4	c.o.v.	n		Correlation matrix $\mathbf{R}$							
Lognormal	γ	18.86	18.9	0.246			0.013	-0.1619	γ	1	0	0	0	0	0.5	0.5	0.5
Lognormal	r_d	0.948	0.948	0.017			0.018	0.0128	r_d	0	1	0	0	0	0	0	0
Lognormal	N_f	0.997	1	0.05			0.05	-0.0358	N_f	0	0	1	0	0	0	0	0
Lognormal	N_S	0.988	1	0.1			0.1	-0.0714	N_S	0	0	0	1	0	0	0	0
Lognormal	C_r	0.577	0.58	0.035			0.06	-0.0429	C_r	0	0	0	0	1	0	0	0
Lognormal	R	0.369	0.4	0.09			0.225	-0.2462	R	0.5	0	0	0	0	1	0.5	0.5
Lognormal	σ_v'	77.55	78	2.34			0.03	-0.1773	σ_v'	0.5	0	0	0	0	0.5	1	0.5
BetaDist	D_r	0.627	8.022	4.263	0	1	0.2	-0.2474	D_r	0.5	0	0	0	0	0.5	0.5	1
										γ	r_d	N_f	N_S	C_r	R	σ_v'	D_r

τ_R	τ_A	g(x)	β	P_f
10.2	10.2	0.00	0.300	38.2%

(a)

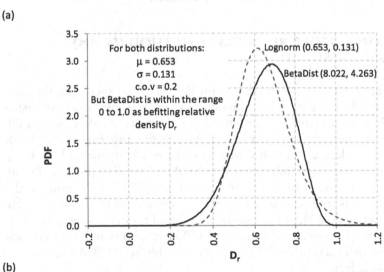

(b)

Figure 2.15 (a) Spreadsheet FORM results involving earthquake-induced cyclic shear stress in saturated sand, (b) bounded beta distribution is preferred to lognormal distribution for relative density D_r of sand.

Also, based on physical considerations, it is logical to assume that γ, σ_v', R and D_r are positively correlated. A correlation coefficient of 0.5 between each pair combination of the four is assumed, as shown in the correlation matrix $\mathbf{R}$.

A reliability index of $\beta = 0.300$ is obtained in Figure 2.15a using the spreadsheet procedure for FORM, corresponding to a P_f of 38.2%. The x* values are referred to as the MPP of failure instead of the design point, since, unlike Figure 2.8 which is RBD with a target β, this case is FORM analysis for the probability of failure, with no target design β value.

It is not surprising that D_r and R, with the two largest absolute n values, are the most sensitive parameters affecting β and the most probable failure point. This is attributable in part to their bigger c.o.v. values (0.2 and 0.225) than the other six random variables. Note however that parameter sensitivities are also affected by the correlation structure, the mean values and the role played by each parameter in the performance function. It is an important merit of FORM that it automatically reveals context-dependent sensitivities in the n values of the design point without the need for parametric studies.

FORM and RBD-via-FORM can play a valuable complementary role to EC7 and LRFD because it is difficult for the partial factor design approach to account for miscellaneous variables (as in this case, and other cases in this book), and to account for parameter correlations.

2.9 SECOND-ORDER RELIABILITY METHOD (SORM) IN SPREADSHEET

The failure probability estimate based on $P_f = \Phi(-\beta)$ is exact when the limit state surface (LSS) is a plane and the random variables follow normal distributions. Inaccuracies in P_f estimation may arise when the LSS is curved and/or nonnormal distributions are involved. More refined alternatives are available, for example, the established second-order reliability method (SORM). SORM analysis requires the β value of FORM and the most probable failure point as inputs and therefore is an extension of FORM. In general, the SORM attempts to assess the curvatures of the LSS at the most probable failure point of FORM, in the dimensionless and rotated u-space. The failure probability is calculated from the FORM reliability index β and estimated principal curvatures of the LSS using established SORM equations of the following form:

$$P_f(SORM) = f(\beta_{FORM}, \text{Curvatures at the MPP of failure}) \quad (2.21)$$

These SORM formulas have been used by Chan and Low (2012a), who presented a practical and efficient spreadsheet-automated approach of implementing SORM using an approximating paraboloid (Der Kiureghian et al., 1987) fitted to the LSS in the neighborhood of the most probable failure point located by FORM. Complex mathematical operations associated with Cholesky factorization, Gram–Schmidt orthogonalization and inverse

Seven curvature components
for 8 variables, computed using
spreadsheet SORM method of
Chan and Low (2011).

	Curvatures	
	α	K_i
γ	-0.5400	-0.017
r_d	0.0427	-0.001
N_f	-0.1193	-0.004
N_s	-0.2381	-0.014
C_r	-0.1431	-0.004
R	-0.6363	-0.016
$\sigma_v{}'$	-0.1686	-0.001
D_r	-0.4265	

FORM $\beta = 0.300$
In previous figure
$\downarrow$

$$P_{f,SORM} = f(\text{FORM } \beta, \text{ Curvatures})$$

SORM formulas	$P_{f,SORM}$
Tvedt	39.29%
Breitung	38.54%
Hohenbichler_Rackwitz	39.31%
Koyluoglu_Nielsen	39.27%
Cai_Elishakoff	39.29%
Hong(P_3)	39.29%
Hong(P_4)	39.30%
AVERAGE	39.18%

(a)

	FORM $\beta = 0.300$	SORM (after FORM) 7 components of curvature	Monte Carlo simulations sample size 60,000
Failure probability	38.2%	39.2%	39.3%%

(b)

Figure 2.16 (a) Second-order reliability method (SORM) estimates the curvatures of the LSS at the most probable failure point of FORM, and (b)
comparison of failure probabilities based on FORM, SORM and
Monte Carlo simulations.

transformation are relegated to relatively simple short function codes in the
Microsoft Excel spreadsheet platform.

The $P_{f,SORM}$ value of 39.2% shown in Figure 2.16a was obtained using the
Chan and Low (2012a) spreadsheet SORM approach. Fourteen points near
the most probable failure point of FORM in the eight-dimensional random
variable u-space are used for estimating the seven components of curvature.
The FORM β and the seven components of curvature at the MPP of failure
are then used in several SORM P_f formulas (Figure 2.16a, right) in the
literature.

The Breitung formula is known to be accurate at higher β values (e.g., $\beta \geq$
2.0) which are required in RBD. In this case of low β value of 0.300, the
Breitung formula estimated SORM P_f of 38.5%, slightly lower than the other
six formulas for SORM P_f which all predicted SORM P_f of about 39.3%.

The accuracy of this FORM-then-SORM P_f estimation is confirmed by Monte Carlo simulation with 60,000 Latin Hypercube sampling, yielding a probability of failure of 39.3%, practically the same as the average $P_{f,SORM}$ value of 39.2%. The number of sampling points (14) for estimating curvatures at the FORM design point is three orders of magnitude smaller than the 60,000 in a Monte Carlo simulation.

Figure 2.16b shows that in this case the P_f of 38.2% based on FORM β is only slightly different from the P_f of 39.2% based on SORM, or 39.3% based on Monte Carlo simulations. Some RBD cases in the other chapters of this book show that SORM P_f can sometimes differ more significantly from FORM P_f, when the nonnormal distributions are asymmetric (skewed) and the values of the coefficient of variation are higher. For example, SORM P_f of 0.20% versus FORM P_f of 0.13% (for target β = 3.0), or reliability of 99.8% based on SORM versus reliability of 99.87% based on FORM.

One can readily extend the FORM analysis into the SORM analysis on the spreadsheet platform, for cases with skewed nonnormal distributions and nonlinear performance function. Should the curvatures at the most probable failure point of the LSS turn out to be negligible (as in this case), all the SORM formulas will approach $P_f = \Phi(-\beta)$, with the result that the computed SORM probability of failure will be similar to FORM probability of failure.

Readers who are not ready to extend FORM into SORM on the spreadsheet platform, when skewed nonnormal distributions and nonlinear performance functions are involved, may take the following pragmatic standpoint. The target reliability index for ultimate limit states is typically 2.5 or 3.0, or higher depending on the importance of the structure and the consequence of failure. When a design achieves the target β of 3.0, the estimated P_f is $\Phi(-\beta) = 0.13\%$. Even if the 'exact' P_f is 0.20% (based on SORM or Monte Carlo simulations), the reliability against failure is 99.87% (=1−0.13%) based on FORM β, compared with the 'exact' 99.80% (=1−0.20%). This pragmatic FORM standpoint is in line with the spirit of Eurocode 7 and LRFD, both aiming at achieving a sufficiently safe design, not at a specific 'exact' value of low P_f.

2.10 CORRELATION MATRIX MUST BE POSITIVE DEFINITE

Although the correlation coefficient between two random variables has a range $-1 \le \rho_{ij} \le 1$, one is not totally free in assigning any values within this range for the correlation matrix. This was explained in Ditlevsen (1981, p. 85), where a 3×3 correlation matrix with $\rho_{12} = \rho_{13} = \rho_{23} = -1$ implies a contradiction and is hence inconsistent. A full-rank (i.e., no redundant variables) correlation matrix has to be positive definite. More detailed discussions are in Low and Tang (2004).

The Monte Carlo simulation programme @RISK (http://www.palisade.com) will also display a warning of 'Invalid correlation matrix' and offer to correct the matrix if simulation is run involving an inconsistent correlation matrix.

2.11 SYSTEM FORM, RESPONSE SURFACE METHOD, AND IMPORTANCE SAMPLING

The previous section illustrates the extension of FORM into SORM. Other possible extensions of FORM are summarized in Figure 2.17, including system FORM for multiple failure modes, response surface method as a bridge between FORM and stand-alone numerical packages, and importance sampling near the most probable failure point of FORM. Some of these extensions are illustrated in the chapters of Parts II and III.

Programmes can be written in spreadsheet to handle implicit limit state functions (e.g., Low et al. 1998, 2007; Low 2003a, 2003b; Low and Tang 2004, p. 87; Chan and Low 2012b). However, there are situations where serviceability limit states can only be evaluated using stand-alone finite

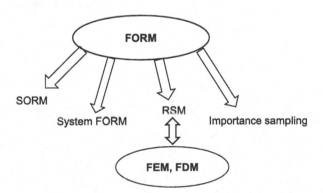

Figure 2.17 FORM can be extended into SORM, system FORM, importance sampling and coupled with numerical methods via the response surface method.

element or finite difference programmes. In these circumstances, reliability analysis and reliability-based design by the present approach can still be performed, provided one first obtains a response surface function (via the established response surface methodology) which closely approximates the outcome of the stand-alone finite element or finite difference programmes. Once the closed-form response functions have been obtained, performing reliability-based design for a target reliability index is straightforward and fast. Performing Monte Carlo simulation on the closed form approximate response surface function also takes little time but lacks the insights and information at the FORM design point.

The response surface method (and other surrogate methods) was illustrated in Li (2000) for consolidation analysis of soft clay, Tandjiria et al. (2000) and Chan and Low (2012b) for laterally loaded single piles, Shahin et al. (2002) for predicting settlement of shallow foundations, Xu and Low (2006) for finite element analysis of embankments on soft ground, and Lü and Low (2011) on underground rock excavations, among others.

Chapter 3

Civil and environmental applications of reliability analysis and design

3.1 INTRODUCTION

Parts II and III of this book focus on probability-based design and analysis in soil and rock engineering, which can be appreciated by practitioners and researchers trained in geotechnical engineering.

This chapter aims to illustrate probabilistic applications in diverse areas of civil and environmental engineering. The examples can be appreciated by most engineers regardless of their background. The beginning examples of capacity-demand scenarios are often encountered in human life, such as probability of arriving on time, reliability-based design of travel time, environmental pollution, irrigation, flood control and mitigation, probabilistic cost–benefit analysis and so on, which even laypeople can appreciate.

Geotechnical engineers and researchers whose interests are in probabilistic soil and rock engineering are nevertheless strongly encouraged to go through the non-geotechnical examples of this chapter, because the RBD-via-FORM principles and probability concepts can be grasped more readily in this chapter of familiar issues, as preparation for the RBD-via-FORM in soil and rock engineering of Parts II and III.

The original solution procedures of the examples in this chapter were explained in the cited sources, with commendable details, based mostly on the traditional mathematical FORM procedure. Only the questions are excerpted from the cited sources; the lengthy original solutions are not reproduced here. Instead, the example cases are solved in this chapter using the efficient Low and Tang (2007) spreadsheet-automated FORM procedure, with succinct outputs, thereby providing interesting comparisons with, and facilitating understanding of, the traditional computational procedures in the cited sources.

The spreadsheet FORM procedure is explained in the previous chapter and illustrated in Figure 3.1. The reliability index equation is:

$$\beta = \min_{\mathbf{x} \in F} \sqrt{\mathbf{n}^T \mathbf{R}^{-1} \mathbf{n}} \quad \left(\text{Obviating the calculations of } \mu_i^N \text{ and } \sigma_i^N\right) \quad (3.1)$$

DOI: 10.1201/9781003112297-4

Low and Tang (2007) spreadsheet FORM procedure

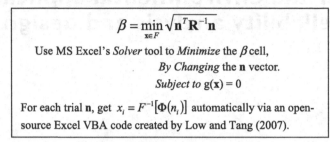

(a)

$$\beta = \min_{x \in F} \sqrt{\mathbf{n}^T \mathbf{R}^{-1} \mathbf{n}}$$

Use MS Excel's *Solver* tool to *Minimize* the β cell,

By Changing the **n** vector.

Subject to g(**x**) = 0

For each trial **n**, get $x_i = F^{-1}[\Phi(n_i)]$ automatically via an open-source Excel VBA code created by Low and Tang (2007).

(b)

	A	B	C	D	E	F	G	H	I	J	K	L
2	Distribution		Para1	Para2	Para3	Para4	x*		Correlation matrix R			n
3	Weibull	V	22	41			33.44		1	0.5	-0.5	-2.284
4	Triangular	W	30	50	70		38.69		0.5	1	0.4	-1.314
5	BetaDist	Z	2	4	600	1800	1293.7		-0.5	0.4	1	1.2536
6												
7			g(x) = VW - Z		g(x)	β	P_f		VBA function			Initially
8			based on x* values		0.00	2.371	0.89%					zeros
9									$\sqrt{\mathbf{n}^T \mathbf{R}^{-1} \mathbf{n}}$			

(c)

Solver Parameters

Set Objective: $F$8

To: ○ Max   ● Min   ○ Value Of:

By Changing Variable Cells:

$L$3:$L$5

Subject to the Constraints:

$E$8 <= 0

Leave this box unchecked

☐ Make Unconstrained Variables Non-Negative

Select a Solving Method: GRG Nonlinear

Figure 3.1 The efficient and succinct spreadsheet FORM solutions for the example cases of this book are based on the Low and Tang (2007) open-source VBA code and the *MS Excel Solver* tool.

which can be entered easily as an Excel array formula using matrix functions *mmult*, *transpose* and *minverse*, and, for each value of n_i tried by the Excel Solver tool, a short and simple open-source Excel VBA function code (created by Low and Tang 2007) automates the computation of x_i from n_i via the following equation:

$$x_i = F^{-1}\left[\Phi(n_i)\right] \tag{3.2}$$

based on $F(x_i) = \Phi(n_i)$, for use in the constraint $g(\mathbf{x}) = 0$, in which $F(x)$ is the original nonnormal CDF, and $\Phi(n)$ is the standard normal CDF.

For uncorrelated variables, $\mathbf{R}$ is an identity matrix, and Eq. 3.1 reduces to:

$$\beta = \min_{\mathbf{x}\in F}\sqrt{\sum n_i^2}\ \text{(for independent or uncorrelated variables } x_i) \tag{3.3}$$

which can be entered conveniently as '=sqrt(sumsq(n))' in spreadsheet, where $\mathbf{n}$ is entered at one go by selecting the values under the $\mathbf{n}$ column. There is no need to set up the identity matrix $\mathbf{R}$ when the variables are independent, for example, in some of the later examples below.

3.2 EXAMPLE I: PROBABILITY OF TRAFFIC CONGESTION

This problem is from Harr (1987, Example 3.2.3) but solved here using FORM.

Studies indicate that the expected traffic at a given intersection at a given time of day is 1000 vehicles per hour with a coefficient of variation of 10%. It is estimated that 1200 vehicles per hour with a coefficient of variation of 20% can be accommodated. What is the probability of traffic congestion if the correlation coefficient between capacity (C) and demand (D) is $\rho = 0$?

Harr's solution based on mean and standard deviation of normally distributed safety margin and its relationship to the probability of failure is summarized first, to acquaint the readers with some basic probability concepts. An alternative procedure producing the same solution based on Hasofer–Lind reliability index calculation in spreadsheet is then presented and extended to spreadsheet-automated FORM for correlated nonnormal capacity and demand of traffic. The FORM procedure in spreadsheet will be efficient and valuable for later examples involving correlated nonnormal variables.

3.2.1 Solution based on mean and standard deviation of safety margin

Assuming normal distributions for the traffic capacity C and demand D, the standard deviations are computed from:

$$\sigma_C = \text{c.o.v.} \times \mu_C = 0.2 \times 1200 = 240$$

$$\sigma_D = \text{c.o.v.} \times \mu_D = 0.1 \times 1000 = 100$$

Congestion occurs if the demand exceeds the capacity, $D \geq C$.

The safety margin (S) is $S = C - D$, which is normally distributed because it is a linear function of two normal variables. The mean of S is $\mu_S = \mu_C - \mu_D$ = 1200–1000 = 200, and the standard deviation of S is:

$$\sigma_S = \sqrt{\sigma_C^2 + \sigma_D^2} = \sqrt{100^2 + 240^2} = 260$$

The probability of congestion is the probability that the safety margin $S \leq 0$, which can be obtained from a table of standard normal probability cumulative distribution function (CDF), $\Phi(n_x)$. For this case:

$$Probability[\,S \leq 0\,] = \Phi(n_x) = \Phi\left(\frac{0 - \mu_S}{\sigma_S}\right) = \Phi\left(\frac{0 - 200}{260}\right)$$

$$= \Phi(-0.77) = 1 - \Phi(0.77) = 1 - 0.779 = 0.221$$

where $\Phi(0.77) = 0.779$ was read from Table A.1. of Ang and Tang (1976, 2006), for example.

More conveniently, one can use the Excel function Norm.Dist(...)as follows:

$$Probability\ of\ congestion = Norm.Dist(x, \mu_S, \sigma_S, true)$$

$$= Norm.Dist(0, 200, 260, True) = 22.09\%.$$

where 'true' indicates cumulative distribution function (instead of probability density).

The solution means that 22.09% of the area under the probability density function curve of safety margin (S, with mean 200 and standard deviation 260) is in the negative domain (unsafe) where $S \leq 0$.

The Excel function Norm.S.Dist for standard normal distribution can also be used (instead of looking up in Tables and interpolating):

$$Norm.S.Dist\big((0 - 200)/260, true\big) = 22.09\%$$

3.2.2 Alternative solution procedure based on Hasofer–Lind index

The probability of traffic congestion at the intersection can be obtained efficiently via the Low and Tang 2007 spreadsheet procedure for FORM, which includes the Hasofer–Lind index as a special case. Figure 3.2a was created as described in the previous chapter. The performance function is $g(x) = C - D$. The β cell for reliability index contains the array formula of ellipse (ellipsoid in higher dimensions). The two x^* cells invoke a simple

Example 1: Probability of traffic congestion:
Spreadsheet-automated FORM solution

		x*	μ	σ	n	Matrix R	
Normal	C	1029.59	1200	240	-0.7101	1	0
Normal	D	1029.59	1000	100	0.2959	0	1

	g(x)	β	P_f
	0.000	0.769	22.09%

(a) Traffic capacity and demand are independent and normally distributed.

		x*	μ	σ	n	Matrix R	
Lognormal	C	1029.38	1200	240	-0.6754	1	0
Lognormal	D	1029.38	1000	100	0.3402	0	1

	g(x)	β	P_f
	0.000	0.756	22.48%

(b) Traffic capacity and demand are independent lognormal variables

		x*	μ	σ	n	Matrix R	
Normal	C	990.826	1200	240	-0.8716	1	0.5
Normal	D	990.826	1000	100	-0.0917	0.5	1

	g(x)	β	P_f
	0.000	0.958	16.91%

(c) Traffic capacity and demand are correlated normal variables

		x*	μ	σ	n	Matrix R	
Lognormal	C	995.451	1200	240	-0.8446	1	0.5
Lognormal	D	995.45	1000	100	0.0042	0.5	1

	g(x)	β	P_f
	0.000	0.978	16.41%

(d) Traffic capacity and demand are correlated lognormal variables

Figure 3.2 Probability of traffic congestion at an intersection.

VBA code to calculate x* values based on distribution type, statistical inputs (just mean and standard deviation for this case) and n values, as explained in the previous chapter. The Excel Solver tool was then used to minimize the β cell, by changing the n values (initially zeros), subject to the constraint that $g(x) = 0$.

Solutions are also obtained for independent and lognormally distributed C and D (Figure 3.2b), normally distributed and correlated ($\rho = 0.5$) C and

D (Figure 3.2c), and lognormally distributed and correlated ($\rho = 0.5$) C and D (Figure 3.2d). The following are noted:

(a) At the mean-value point, $\mu_C = 1200$ and $\mu_D = 1000$, traffic congestion will not occur. However, given the uncertain capacity and demand, as characterized by $\sigma_C = 240$ and $\sigma_D = 100$, congestion will occur when capacity decreases to, and demand increases to, the values shown under the $\mathbf{x}^*$ column in Figure 3.2a–3.2d. The chances of this happening are 22.1%, 22.5%, 16.9% and 16.4%, respectively, as indicated by the value of the failure (i.e., congestion) probability (P_f) next to the β cell for each case. The value of P_f was obtained using an Excel function: $P_f = \mathrm{NormSDist}(-\beta)$.

(b) The probability of traffic congestion is more sensitive to whether C and D are correlated, than to whether C and D are normally distributed or lognormally distributed.

(c) The four solutions in Figure 3.2 were obtained promptly by the Solver tool after changing the distribution type and the correlation matrix $\mathbf{R}$ in Figure 3.2a. Although only normal and lognormal distributions are illustrated, other distributions can be used with the same ease of implementation.

(d) For comparison, Monte Carlo simulations were done using the @ RISK software, with 50,000 realizations each for the four cases in Figure 3.2. The probabilities of traffic congestion are 22.1%, 22.5%, 16.8%, 16.4%, which are practically the same as the four P_f values based on reliability index values.

3.3 EXAMPLE 2: DESIGN OF TRAFFIC CAPACITY FOR A TARGET RELIABILITY INDEX

Assuming the same demand and the same coefficients of variation, to what value should the mean capacity be increased so that the probability of congestion is reduced to 0.05?

The required reliability index is $\beta_{target} = \mathrm{Norm.S.Inv}(1.0 - 0.05) = 1.645$.

Figure 3.3a shows that the required mean capacity, μ_C, is 1529 vehicles, for uncorrelated and normally distributed capacity and demand. That is, 329 more vehicles would have to be accommodated during the time period of interest.

The β ellipse and the 1σ ellipse are canonical (i.e., not rotated) when capacity and demand are uncorrelated, Figure 3.3a. The $\mathbf{n}$ values (−1.563, 0.511) mean that the most probable scenario of congestion is when the uncertain capacity is 1.563 times its standard deviation *below* its mean capacity of 1529 vehicles, and the uncertain demand is 0.511 times its standard deviation *above* its mean demand of 1000 vehicles, so that $C^* = D^* = 1051$. The chance of this congestion occurring is $\Phi(-1.644) = \mathrm{Norm.S.Dist}(-1.644) = 5\%$.

Example 2: Design of traffic capacity for a target reliability index

Spreadsheet-automated FORM solution

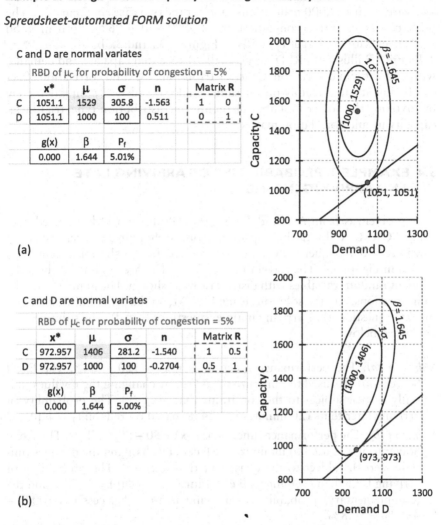

(a)

(b)

Figure 3.3 Reliability-based design of mean capacity (μ_C) to reduce probability of congestion to 5% (a) independent C and D, (b) correlated C and D.

The β ellipse and the 1σ ellipse are rotated (Figure 3.3b) when capacity and demand are correlated with ρ = 0.5. The n values (–1.540, –0.270) mean that the most probable scenario of congestion is when the uncertain capacity is 1.540 times its standard deviation *below* its mean capacity of 1406 vehicles, and the uncertain demand is 0.270 times its standard deviation *below* its mean demand of 1000 vehicles, so that C* = D* = 973. The chance of this congestion occurring is Φ(–1.644) = Norm.S.Dist(–1.644) = 5%. In this case, D* value of 973 is smaller than its mean value, due to the positive correlation ρ = 0.5 between capacity and demand.

For comparison, Monte Carlo simulations were done using the @RISK software, with 100,000 realizations each for the two cases in Figure 3.2. The probabilities of traffic congestion are 4.99% and 4.96%, both in good agreement with the target P_f of 5% in Figure 3.3a and 3.3b.

Figure 3.3 illustrates RBD for normally distributed capacity and demand, with and without correlation. Other distribution types and correlation effect can be investigated readily on the same single spreadsheet template merely by invoking the Solver tool after changing the distribution type, trial mean capacity μ_C and correlation matrix.

3.4 EXAMPLE 3: PROBABILITIES OF ARRIVING LATE AND ARRIVING IN TIME

This problem is from Ang and Tang (2006, Problem 5.4) but solved here using FORM. A shuttle bus operates from a shopping centre, travels to towns A and B sequentially, and then returns to the shopping centre, as shown in Figure 3.4. The travel times T_1, T_2 and T_3 are assumed to be independent random variables with distributions as shown. The figure shows the four questions, and the solutions using FORM, on the probabilities of arriving late (question a) or arriving in time (questions b, c, d). The solutions are explained below:

Solution (a): The performance function is $g(x) = 120 - \Sigma x_i^*$. The Excel Solver tool was used to minimize the β cell, by changing the n values (initially zeros), subject to the constraint that $g(x) = 0$. The probability of arriving later than 120 minutes is 3.24%, based on reliability $= \Phi(\beta)$.

Solution (b): The performance function is $g(x) = 60 - (T_2^* + T_3^*)$. The Excel Solver tool was used to minimize the β cell, by changing the n values (initially zeros), subject to the constraint that $g(x) = 0$. The probability of arriving later than 60 minutes (i.e., 'failure') $= 1 - \Phi(\beta) = 43.2\%$, and the (complementary) probability of arriving in time ('success') is 100% – 43.2% = 56.8%.

Solution (c): During rush hours, the mean and standard deviation of T_2 are 30 minutes and 6 minutes (instead of 20 minutes and 4 minutes under normal traffic). The performance function is $g(x) = 60 - (T_2^* + T_3^*)$. Prior to invoking Solver, the $g(x)$ cell shows a negative value when n_2 and n_3 cells were initially zeros, which indicates that the mean-value point is already in the 'unsafe' domain. Hence, the β value of 0.656 obtained by Solver must be treated as a negative reliability index of –0.656. The reliability, or probability of arriving in time, is $\Phi(-\beta) = \Phi(-(-0.656)) = 25.6\%$. For passengers from Town B, T_3 is the only variable involved. The *Excel Goalseek* tool was used to obtain the n value of 1.528 which renders $T_3^* = 60$ minutes.

Example 3: Probabilities of arriving at a destination in time

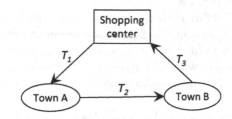

Travel Time	Distribution	Mean (minutes)	c.o.v.
T_1	Gaussian	30	0.30
T_2	Lognormal	20, normal traffic. (30 if rush hours)	0.20
T_3	Lognormal	40	0.30

The travel times T_1, T_2 and T_3 are assumed to be independent random variables. Determine the following:

(a) What is the probability that the shuttle bus will fail to complete one round trip within 2 hours under normal traffic?

(b) What is the probability that a passenger, originating in Town A, will arrive at the shopping center within 1 hour under normal traffic?

(c) The number of passengers originating from Town B is twice that of the number originating from Town A. What percentage of the passengers will arrive at the shopping center within 1 hour during rush hours?

(d) A passenger starting from Town B has an appointment at 3:00 PM at the shopping center. If the bus left Town B at 2:00 PM but has not arrived at the shopping center at 2:45 PM, what is the probability that he or she will arrive on time for the appointment. (2:00-3:00 PM is normal traffic hour.)

Solutions using spreadsheet FORM:

Distribution		μ	σ	x_i^*	n_i	Matrix R
Normal	T_1	30	9	37.25	0.81	1 0 0
Lognormal	T_2	20	4	21.12	0.37	0 1 0
Lognormal	T_3	40	12	61.63	1.62	0 0 1

	g(x)	β	P_f
(a)	0.00	1.847	3.24%

Distribution		μ	σ	x_i^*	n_i	Matrix R
Lognormal	T_2	20	4	19.82	0.054	1 0
Lognormal	T_3	40	12	40.18	0.162	0 1

	g(x)	β	P_f	$P(T_2+T_3 \leq 60)$
(b)	0.00	0.171	43.2%	56.8%

Distribution		μ	σ	x_i^*	n_i	Matrix R
Lognormal	T_2	30	6	27.58	-0.33	1 0
Lognormal	T_3	40	12	32.42	-0.57	0 1

	g(x)	β	$P(T_2+T_3 \leq 60)$
	0.00	0.656	-0.656	25.6%

Distribution		μ	σ	x_i^*	n_i	$P(T_3 \leq 60)$
Lognormal	T_3	40	12	60.00	1.528	93.67%

(c) $(2 \times 93.7\% + 1 \times 25.6\%)/3$ =71% of passengers

Distribution		μ	σ	x_i^*	n_i	$P(T_3 \leq 45)$
Lognormal	T_3	40	12	45.00	0.548	70.82%

(d) $(0.9367 - 0.7082)/(1 - 0.7082) = 0.783$ =78.3%

Figure 3.4 Example 3 on probabilities of: (a) arriving late; and (b, c, d) arriving in time, using FORM.

For cases involving single random variable only, the solution of the n value is the reliability index β.

This means the reliability of Town B passengers arriving at the shopping centre in time is $\Phi(1.528) = 93.7\%$. Since the number of passengers originating from Town B is twice that of the number originating from Town A, the percentage of passengers who will arrive at the shopping centre within 1 hour during rush hours is the weighted average of 71%, as shown.

Solution (d): Only one time variable is involved, namely T_3. Using the same *Goalseek* tool as explained above, the probability of arriving within 45 minutes (i.e., $T_3 \leq 45$) is 70.8%. The probability of $T_3 \leq 60$ is 93.7%, from (c). Hence the probability of arriving between 45 and 60 minutes is:

$$P\left(T_3 < 60 \,/\, T_3 > 45\right) = \frac{P\left(45 < T_3 < 60\right)}{P\left(T_3 > 45\right)} = \frac{0.9367 - 0.7082}{1 - 0.7082} = 78.3\% \quad (3.4)$$

For comparison, the probabilities using @Risk Monte Carlo simulation with 100,000 realizations are 3.66%, 55%, 70.3% and 78.4%, to questions a, b, c and d.

3.4.1 Alternative solution using spreadsheet function for the single random variable case of T_3

It is interesting to compare the FORM solution of T_3 in Figure 3.4c and 3.4d with an alternative approach using the Excel lognormal distribution Lognorm.Dist(...). To do this, one must take note that the mean μ and standard deviation (SD) σ of a lognormally distributed random variable x can be given in two ways, either the mean and SD of $\ln x$ (i.e., $\mu_{\ln x}$ and $\sigma_{\ln x}$), or the mean and SD of x (i.e., μ_x and σ_x). The two statistical functions LOGNORM. DIST and LOGNORM.INV in Microsoft Excel require inputs in $\mu_{\ln x}$ and $\sigma_{\ln x}$, whereas Example 3 uses μ_x and σ_x as inputs of the lognormal distribution. One can obtain μ_x and σ_x from $\mu_{\ln x}$ and $\sigma_{\ln x}$, or $\mu_{\ln x}$ and $\sigma_{\ln x}$ from μ_x and σ_x, using the following established relationships (e.g., Ang and Tang 1976):

$$\mu_{\ln x} = \ln \mu_x - 0.5\left[\ln\left(1 + \left(\sigma_x \,/\, \mu_x\right)^2\right)\right] \quad (3.5a)$$

$$\sigma_{\ln x} = \sqrt{\ln\left(1 + \left(\sigma_x \,/\, \mu_x\right)^2\right)} \quad (3.5b)$$

$$\mu_x = \exp\left(\mu_{\ln x} + 0.5\sigma_{\ln x}^2\right) \quad (3.6a)$$

$$\sigma_x = \sqrt{\left[\exp\left(\sigma_{\ln x}^2\right) - 1\right]\exp\left(2\mu_{\ln x} + \sigma_{\ln x}^2\right)} \quad (3.6b)$$

For the variable T_3 with mean = 40 minutes and SD = 12 minutes, Eqs. 3.5a and 3.5b yields:

$$\mu_{\ln T_3} = \ln \mu_{T_3} - 0.5\left[\ln\left(1+\left(\sigma_{T_3}/\mu_{T_3}\right)^2\right)\right] = \ln 40 - 0.5\left[\ln\left(1+0.3^2\right)\right] = 3.646$$

and

$$\sigma_{\ln x} = \sqrt{\ln\left(1+\left(\sigma_x/\mu_x\right)^2\right)} = \sqrt{\ln\left(1+0.3^2\right)} = 0.2936$$

Equation 3.4 can then be solved as follows, with inputs of $\mu_{\ln T3} = 3.646$ and $\sigma_{\ln T3} = 0.2936$:

$$P\left(T_3 < 60 / T_3 > 45\right) = \frac{P\left(45 < T_3 < 60\right)}{P\left(T_3 > 45\right)} = \frac{CDF_{60} - CDF_{45}}{1 - CDF_{45}} = 78.3\% \quad (3.7a)$$

where CDF_{60} and CDF_{45} are computed conveniently from the spreadsheet Lognorm.Dist function as follows:

$$CDF_{60} := \text{Lognorm.Dist}\left(60, 3.646, 0.2936, \text{true}\right) = 0.9366 \quad (3.7b)$$

and

$$CDF_{45} := \text{Lognorm.Dist}\left(45, 3.646, 0.2936, \text{true}\right) = 0.7079 \quad (3.7c)$$

which leads to practically the same solution (78.3%) as the FORM solution in Figure 3.4d and Eq. 3.4 which makes use of the two FORM β values of 1.528 and 0.548.

Equations 3.4 and 3.7a can be appreciated from the Bayesian theorem perspective. Alternatively, Figure 3.5 explains in an intuitive manner how the original (prior) probability of arriving in time (93.7%) is revised to a lower 78.3% given that the passenger has not arrived at the destination 45 minutes after the bus departed from Town B.

3.5 EXAMPLE 4: REQUIRED TRAVEL TIME FROM RBD-VIA-FORM

This question is from Ang and Tang (1984, Example 6.22). It concerns the reliability-based design of total travel time T_0, and associated design time factors, for a target reliability of 97.5% (or $\beta = 1.96$). The total travel time T_0 is the sum of three correlated time components which are assumed normally distributed in Ang and Tang (1984). The performance function is:

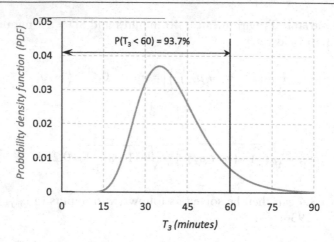

(a) Prior probability of $T_3 \leq 60$ minutes is 93.7%

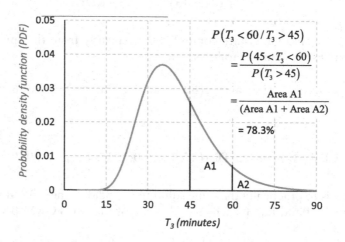

(b) Updated probability that $T_3 \leq 60$ minutes is 78.3%, given that the passenger has not shown up at destination at $T_3 = 45$ minutes.

Figure 3.5 Example 3d: Given that the passenger has not arrived at $T_3 = 45$ minutes, the updated probability of arriving in time is 78.3%, smaller than the prior probability of 93.7%.

$$g(\mathbf{x}) = T_0 - (T_1 + T_2 + T_3) \qquad (3.8)$$

where the mean and standard deviation of T_1, T_2 and T_3 and their correlations ρ_{12}, ρ_{13} and ρ_{23} are shown in Figure 3.6. The solution in Figure 3.6a uses the Low and Tang (2007) FORM procedure, for comparison with the traditional approach in Ang and Tang (1984). Assumptions different from Ang and Tang (1984) are also investigated in the RBD-via-FORM of Figure 3.6b and 3.6c.

Example 4: Reliability-based design of travel times

A transportation network between cities A and B consists of three branches as shown below:

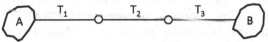

T_1, T_2 and T_3 are the travel times over the three respective branches. Because of uncertainties in the traffic and weather conditions, these travel times are random variables with mean and standard deviation of (40, 10), (15, 5), and (50, 10), respectively.

Also, since the T_i's will be subject to common environmental factors, they are expected to be partially correlated. Suppose these correlations are:

$$\rho_{12} = \rho_{13} = \rho_{23} = 0.5$$

The performance function is:

$$g(\mathbf{x}) = T_0 - (T_1 + T_2 + T_3)$$

where T_0 = The scheduled total travel time between A and B.

Determine T_0 for a target reliability of 97.5%, and back-calculate the design factors of T_1, T_2 and T_3.

Spreadsheet RBD-via-FORM solutions for three different scenarios: $x_i = F^{-1}\left[\Phi(n_i)\right]$

(a)

Distribution		μ	σ	x*	Correlation matrix R			n_i	Design factor γ
Normal	T_1	40	10	56.64	1	0.5	0.5	1.664	1.42
Normal	T_2	15	5	22.13	0.5	1	0.5	1.426	1.48
Normal	T_3	50	10	66.64	0.5	0.5	1	1.664	1.33

	T_0	g(x)	β	Reliability
	145.4	0.00	1.96	97.5% Φ(β)

(b)

Distribution		μ	σ	x*	Correlation matrix R			n_i	Design factor γ
Lognormal	T_1	40	10	58.63	1	0.5	0.5	1.676	1.47
Lognormal	T_2	15	5	22.69	0.5	1	0.5	1.438	1.51
Lognormal	T_3	50	10	67.88	0.5	0.5	1	1.643	1.36

	T_0	g(x)	β	Reliability
	149.2	0.00	1.96	97.5% Φ(β)

(c)

Distribution		μ	σ	x*	Correlation matrix R			n_i	Design factor γ
Normal	T_1	40	10	53.07	1	0	0	1.307	1.33
Normal	T_2	15	5	18.27	0	1	0	0.653	1.22
Normal	T_3	50	10	63.07	0	0	1	1.307	1.26

	T_0	g(x)	β	Reliability
	134.4	0.00	1.96	97.5% Φ(β)

Figure 3.6 RBD-via-FORM of travel time and back-calculated design factors (a) correlated normal, (b) correlated lognormals, (c) underestimation of design travel time T_0 if correlations are ignored.

The following are noted:

(1) The neat but intricate traditional solution in Ang and Tang (1984, pp. 423–426) assumed normally distributed T_1, T_2 and T_3 and involved iterative transformation of the correlated travel times to a set of uncorrelated variates via eigenvalues and eigenvectors, and analytical partial derivatives of the transformed performance function. The derived closed-form solution for target β and design factors of time are valid only for the assumed correlations among T_1, T_2 and T_3 of $\rho_{12} = \rho_{13} = \rho_{23} = 0.5$.

(2) The solution in Figure 3.6a, obtained using the Low and Tang (2007) spreadsheet procedure in the original coordinate system, is efficient and versatile. Iterations and partial derivatives are performed automatically by the Solver tool. Other scenarios, such as various different probability distributions (Figure 3.6b demonstrates lognormals), and different correlations (Figure 3.6c), can be instigated quickly using the same template of Figure 3.6a merely by changing the distribution names in the first column and changing the correlation matrix $\mathbf{R}$. More variables can also be treated readily by increasing the number of rows (T_1, T_2, T_3, ..., T_n) and the size of the correlation matrix.

(3) It is unconservative to ignore positive correlations among unfavorable variables (which are T_1, T_2 and T_3 in this case) when these correlations are justified by physical considerations. The required total travel time for the uncorrelated normal T_1, T_2 and T_3 of Figure 3.6c is smaller than that for positively correlated normal T_1, T_2 and T_3 of Figure 3.6a.

(4) With the three design travel times of T_0 in Figure 3.6a, 3.6b and 3.6c, the mean-value point (40, 15, 50) is in the safe domain because the performance function (Eq. 3.8) is positive. Nevertheless, the allocated design time T_0 is exhausted when the component times T_1, T_2 and T_3 increase to their values shown under the $\mathbf{x}^*$ columns, which constitute the most probable point (MPP) of failure, also referred to as the design point when there is a target reliability index as for this case where $\beta = 1.96$.

(5) The design factors for the three time components of T_1, T_2 and T_3 are back-calculated from:

$$\gamma_{T_1} = T_1^* / \mu_{T_1}; \quad \gamma_{T_2} = T_2^* / \mu_{T_2}; \quad \gamma_{T_3} = T_3^* / \mu_{T_3} \tag{3.9}$$

where T_1^*, T_2^* and T_3^* are the design values under the $\mathbf{x}^*$ column.

From the perspective of design factors γ_{T_i}, the MPP of failure time is brought about by:

$$T_1^* + T_2^* + T_3^* = \gamma_{T_1}\mu_{T_1} + \gamma_{T_2}\mu_{T_2} + \gamma_{T_3}\mu_{T3} = T_0 \tag{3.10}$$

The probability of this occurrence is equal to $1 - \Phi(\beta) = \Phi(-\beta)$, and the reliability is $\Phi(\beta)$.

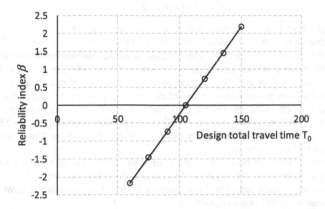

Figure 3.7 Positive and negative reliability index values for the case in Figure 3.6a. The sign of β is determined by whether the mean-value point is in the safe or unsafe domain.

(6) The design travel time T_0 must be selected such that the performance function cell shows a positive value when the n values are initially zeros prior to obtaining the β solution using the Solver tool. For the case in Figure 3.6a, the cell g(**x**) is positive (when the n values are initially zeros) for $T_0 > 105.0$, and negative for $T_0 < 105.0$. For the case in Figure 3.6a, if the allocated $T_0 = 64.6$ minutes, the computed β is also 1.96, to which a negative sign must be assigned, because the g(**x**) cell was negative prior to obtaining the solution using the Solver tool. A β value of –1.96 means a reliability of only 2.5% (or a probability of 97.5% of exceeding T_0). The mean-value-point (for $T_0 = 64.6$ minutes) is already in the unsafe domain, and 1.96 is the distance from the unsafe mean-value point to the nearest safe boundary, in units of directional standard deviations. Figure 3.7 shows the importance of distinguishing negative from positive reliability indices.

(7) For comparison, the probabilities of not exceeding the design travel time T_0, using @Risk Monte Carlo simulation with 200,000 realizations, are 97.48%, 97.18%, 97.46%, for the cases in Figure 3.6a, 3.6b and 3.6c, respectively, which agree well with the target of 97.5% in each case.

3.6 EXAMPLE 5: SUPPLY AND DEMAND OF IRRIGATION WATER

This is another capacity-demand situation which laypeople can understand, at least on the nature of the question and the need to account for input uncertainties and correlations. The three parts of the question (a, b and c) in Figure 3.8 are based on Kottegoda and Rosso (2008, Examples 9.4, 9.6

Example 5: Supply and demand of irrigation water

(Questions after Kottegoda and Rosso, 2008, Examples 9.4, 9.6 and 9.7).

During the growing season the expected demand, D, from an irrigation scheme is 10 units with a coefficient of variation (c.o.v.) of 50%, which accounts for fluctuations associated with weather variability. The mean available water supply, S, which is diverted from a river barrage, is 20 units, with a coefficient of variation of 20%, which accounts for fluctuations associated with hydrologic variability in that season. Because of the relationship between hydrology and climate, the natural water availability often tends to decrease when the demand increases, so that the correlation coefficient between S and D is negative. The estimated value of ρ_{SD} is −0.5.

(a) Estimate the reliability of the system assuming that both capacity S and demand D are normally distributed variates.

(b) In order to increase the reliability of the system to to 95%, the diversion of a neighboring stream is considered for the purpose of increasing the mean capacity S. Assuming that the coefficient of variation values and correlation coefficient ρ_{SD} remain the same, determine the required increase in mean capacity μ_S.

(c) Estimate the reliability of the system at the original mean supply of μ_S = 20 units, if it is assumed that correlation between capacity S and demand D can be neglected.

Solution using spreadsheet FORM procedure: $x_i = F^{-1}\left[\Phi(n_i)\right]$

(a)

		μ	σ	c.o.v.	n	x*	Matrix R	
Normal	S	20	4	0.2	-1.066	15.74	1	-0.5
Normal	D	10	5	0.5	1.148	15.74	-0.5	1

g(x)	β	P_f
0.00	1.280	10.0%

$\sqrt{\mathbf{n}^T\mathbf{R}^{-1}\mathbf{n}}$ 1-$\Phi(\beta)$

(b)

		μ	σ	c.o.v.	n	x*	Matrix R	
Normal	S	24	4.8	0.2	-1.419	17.19	1	-0.5
Normal	D	10	5	0.5	1.438	17.19	-0.5	1

Target β	g(x)	β	Reliability
1.645	2E-10	1.649	95.0%

$\Phi(\beta)$

(c)

		μ	σ	c.o.v.	n	x*	Matrix R	
Normal	S	20	4	0.2	-0.976	16.10	1	0
Normal	D	10	5	0.5	1.220	16.10	0	1

g(x)	β	P_f
0	1.562	5.9%

$\sqrt{\mathbf{n}^T\mathbf{R}^{-1}\mathbf{n}}$ $\Phi(-\beta)$

Figure 3.8 (a) Risk of insufficient irrigation water, (b) RBD of mean capacity for increased reliability, (c) underestimation of the risk of failure if ignore negative correlation between supply and demand.

and 9.7). The solutions are the same as the source, but obtained here using the FORM approach.

The performance function is:

$$g(\mathbf{x}) = S - D \tag{3.11}$$

where S is the supply of irrigation water, and D is the demand for the irrigation water. Both S and D are assumed to be normal (Gaussian) random variables, with mean (μ) and standard deviation (σ = c.o.v. $\times$ μ) shown in Figure 3.8. It is likely that low supply tends to occur with high demand, and vice versa. For example, dry weather may reduce the amount of water retained in a reservoir but increase the demand for irrigation water. Hence, it is logical to assume that S and D are negatively correlated; a negative correlation coefficient of ρ = -0.5 is modelled in the correlation matrix $\mathbf{R}$ in Figure 3.8a.

Equation 3.11 is entered in the spreadsheet cell labelled g($\mathbf{x}$) and read its S and D values from the cells under the $\mathbf{x}^*$ column. The cells under the $\mathbf{x}^*$ column invoke simple VBA codes to calculate $\mathbf{x}^*$ values from the n values (initially zeros). Alternatively, for normal distribution, one can simply enter the formula '= $n_i\sigma_i + \mu_i$' in the $\mathbf{x}^*$ cells. The β cell contains the array formula $\sqrt{\mathbf{n}^T\mathbf{R}^{-1}\mathbf{n}}$, which uses Excel matrix functions *mmult*, *transpose* and *minverse*, as explained in Chapter 2. The solution of β was obtained quickly by the Solver tool, with useful corollary information of most probable point (MPP) of failure under the $\mathbf{x}^*$ column, and the sensitivity indicators under the n column.

An alternative and relatively straightforward analytical solution method based on computing the reliability index as the ratio of the mean to the standard deviation of the performance function g($\mathbf{x}$), $\beta = \mu_g/\sigma_g$, is possible for this case with linear performance function and normal variates, as explained well in Kottegoda and Rosso (2008). The approach requires computing μ_g and σ_g as functions of the means and standard deviations of the underlying variables and their correlations, which for this case are μ_S, σ_S, μ_D, σ_D and ρ_{SD}. This analytical solution method would be complicated or not feasible for complicated nonlinear performance functions and nonnormal (non-Gaussian) distributions. The definition $\beta = \mu_g/\sigma_g$, which is referred to as the mean-value first-order second moment (MVFOSM) method, can lead to non-unique β values for nonlinear performance function formulated differently (but are mechanically equivalent).

In contrast, the FORM solution method illustrated in Figure 3.8 can deal with correlated non-Gaussian variables and complicated nonlinear performance functions, as illustrated in the sections below, and in other chapters of this book.

The FORM method for correlated nonnormal variates reduces to the Hasofer–Lind index for this case with correlated normal (Gaussian) variates. For linear performance function and normally distributed variables, such as this case, the probability of failure based on $P_f = 1 - \Phi(\beta) = \Phi(-\beta)$ is theoretically exact.

Solution (a) in Figure 3.8 shows that, when the mean capacity μ_S of irrigation water is 20 units, the computed β index is 1.28, corresponding to a probability of demand exceeding supply of 10%, or, put another way, the reliability is 90%. The $\beta = 1.28$ dispersion ellipse which just touches the limit state surface (LSS) at $D^* = 15.74$ and $S^* = 15.74$ (the MPP of failure) is shown by the lower ellipse in Figure 3.9, with centre at the mean-value point of $\mu_D = 10$ and $\mu_S = 20$.

Solution (b) in Figure 3.8 shows that a higher mean capacity μ_S of 24 units is required to achieve a higher reliability of 95% against the risk of demand exceeding supply. The $\beta = 1.65$ dispersion ellipse which just touches the LSS at $D^* = 17.19$ and $S^* = 17.19$ (the MPP of failure) is shown by the upper bigger ellipse in Figure 3.9, with centre at the mean-value point of $\mu_D = 10$ and $\mu_S = 24$.

Solution (c) in Figure 3.8 shows that uncorrelated capacity and demand (similar to uncorrelated resistance and destabilizing load in engineering examples) leads to lower estimated probability of failure (and correspondingly higher estimated reliability), than when the capacity and

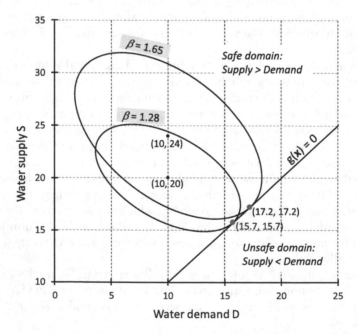

Figure 3.9 With mean irrigation water supply at 20 units and mean demand 10 units, the reliability index is 1.28. An increased mean supply of 24 units (with the same c.o.v. of 0.2) is required to increase the reliability index to 1.65.

demand are negatively correlated. The solutions of Figure 3.8a and 3.8c illustrate that an engineer who disregards the logical *negative correlation* between *capacity and demand* of irrigation water in this case can come to the misleadingly higher estimation of 94% reliability (= 1 − P_f) when the mean capacity is 20 units.

3.7 EXAMPLE 6: THERMAL POLLUTION IN A RIVER

The discharge D from the cooling system of a thermal power plant is normally distributed with mean 2 m^3/s and standard deviation of 0.4 m^3/s. It is required that the discharge D does not exceed a faction of the natural flow of the river Q, for example Q/a, where Q/a < Q. The statistical inputs here follow those in Kottegoda and Rosso (2008).

The performance function is:

$$g(\mathbf{x}) = Q/a - D \tag{3.12}$$

with values Q/a and D read from the x* column.

Part (a) of the question assumes that the discharge limit Q/a is exponentially distributed with mean equal to 8 m^3/s, which is entered as a single input in the second row of Figure 3.10a. As shown in Figure 3.10c, the standard deviation of the exponential distribution is equal to its mean value. The reliability index β is 0.76, obtained by invoking the Solver tool to *minimize* the β cell, *by changing* the n column values (initially zeros), *subject to the constraint* that the cell g(x) be equal to 0.0. This β value corresponds to a probability of Φ(−β) = Norm.S.Dist(−0.762) = 22.3% that the discharge will exceed the permissible limit, or a reliability of about 77.7% that the discharge will be below the thermal pollution limit.

Part (b) of the question assumes that the Q/a is gamma distributed with mean 8 m^3/s, and its coefficient of variation is $1/\sqrt{2}$, which means a standard deviation σ of $8/\sqrt{2} = 5.657$ m^3/s. The argument values of the gamma distribution (Figure 3.10c, right) which will produce a mean of 8 m^3/s and a standard deviation of 5.657 m^3/s are α = 2 and λ = 4, which have been entered under the Para1 and Para2 column of Figure 3.10b. The same simple template of Figure 3.10a is used, but with the probability distribution type of Q/a changed to 'Gamma,' and the column headings μ and σ changed to Para1 and Para2, because the inputs of 2 and 4 for the Gamma distribution are not mean and standard deviation values.

The Solver tool obtains a β value of 1.318, corresponding to a probability of about 9.4% that the discharge will exceed the permissible limit, or a reliability of about 90.6% that the discharge will be below the thermal pollution limit.

Detailed calculations of the traditional approach of iterative equivalent normal transformation are presented in Kottegoda and Rosso (2008,

Example 6: Thermal pollution in a river

(Question after Kottegoda and Rosso, 2008, Examples 9.9).

The discharge D from the cooling system of a thermal power plant flows into a river. To prevent thermal pollution in the river, it is desirable that D does not exceed a fraction of natural flow Q in the river, say, $C = Q/a$, where a denotes a constant that depends on the difference in temperature between the two flows. An engineer wishes to evaluate the risk that thermal pollution occurs in the river. Assume that D is normally distributed with mean 2 m³/s and coefficient of variation of 20%.

For the period in which the river receives the discharge, the natural flow Q can be approximated by an exponential distribution with mean 40 m³/s, and $a = 5$. Inflow D and streamflow Q are further assumed to be independent variates.

(a) Determine the reliability against river pollution.
(b) Determine the reliability against river pollution, if, instead of exponential distribution, the Q/a is gamma distributed with mean 8 m³/s, and its coefficient of variation is 0.707.

Solution using spreadsheet FORM procedure: $x_i = F^{-1}\left[\Phi(n_i)\right]$

		x*	μ	σ	n	Matrix R	
Normal	D	2.039	2	0.4	0.0976	1	0
Exponential	Q/a	2.039	8		-0.755	0	1

(a)

$\sqrt{\mathbf{n}^T \mathbf{R}^{-1} \mathbf{n}}$	g(x)	β	P_f	Reliability
	0.000	0.762	22.3%	77.7%

		x*	Para1	Para2	n	Matrix R	
Normal	D	2.0935	2	0.4	0.2336	1	0
Gamma	Q/a	2.093	2	4	-1.297	0	1

(b)

g(x)	β	P_f	Reliability
0.000	1.318	9.38%	90.6%

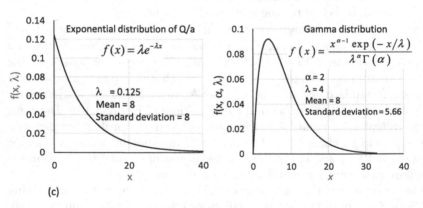

(c)

Figure 3.10 (a) Exponential distribution of permissible discharge x = Q/a, (b) gamma distribution of permissible discharge x = Q/a, (c) probability density plots of exponential and gamma distributions.

pp. 555–558), obtaining $\beta = 0.76$ for the case in Figure 3.10a, and $\beta = 1.32$ for the case in Figure 3.10b, practically the same as the succinct spreadsheet-automated solutions in Figure 3.10.

3.8 EXAMPLE 7: RELIABILITY OF A STORM SEWER SYSTEM

This example is solved efficiently using the spreadsheet FORM procedure (with discussions on the most probable failure point and parametric sensitivities), for comparison with the traditional reliability computational approach detailed in Tung et al. (2006, Examples 4.10 and 4.11).

It was mentioned in Tung et al. (2006, Example 4.1) that, in the design of a storm sewer system, the sewer flow-carrying capacity Q_C and the inflow Q_L to the sewer need to be considered. The sewer flow-carrying capacity Q_C (ft^3/s) is determined by the Manning's formula:

$$Q_C = \frac{0.463}{n} \lambda_C D^{8/3} S^{0.5} \tag{3.13}$$

where n is Manning's roughness coefficient, λ_c is the model correction factor to account for the model uncertainty of Q_c, D is the actual pipe diameter (ft) and S is the pipe slope (ft/ft).

The inflow Q_L to the sewer is the surface runoff whose peak discharge can be estimated by the formula:

$$Q_L = \lambda_L CiA \tag{3.14}$$

in which λ_L is the correction factor for model uncertainty, C is the runoff coefficient, i is the rainfall intensity and A is the runoff contributing area.

The performance function can be expressed as:

$$g(\mathbf{x}) = Q_c - Q_L = 0.463 \lambda_c \frac{D^{8/3} S^{0.5}}{n} - \lambda_L CiA \tag{3.15}$$

The statistical inputs in Figure 3.11a and 3.11b follow those in Tung et al. (2006, Examples 4.10 and 4.11). The inflow Q_L to the sewer was treated as a deterministic value of 35 ft^3/s, and the model correction factor λ_C was not modelled.

The Gumbel distribution is also referred to as Extreme Value distribution Type 1, denoted by 'ExtValue 1' in Figure 3.11a and 3.11c. The last columns in Figure 3.11 show the values of $-n_i/\beta$, which are equal to the direction cosines (α_i) of the classical FORM procedure (Ang and Tang 1984, for example) when the random variables are independent.

Example 7: Reliability of a storm sewer system

Parts (a) and (b) of the question below are from Tung-Yen-Melching, 2006, Examples 4.10, and 4.11. Non-SI units are retained, for ease of comparison, and also because Manning's formula are unit-dependent.

The flow-carrying capacity Q_c of a storm sewer (ft³/s) is determined by Manning's formula, which, as given in Tung-Yen-Melching (2006), is :

$$Q_c = 0.463n^{-1}D^{8/3}S^{0.5}$$

in which n is the Manning roughness coefficient, D is the sewer diameter (in ft), and S is the pipe gradient (in ft/ft). Because roughness coefficient n, sewer diameter D, and sewer slope S in Manning's formula are subject to uncertainty owing to manufacturing imprecision and construction error, the sewer flow capacity would be subject to uncertainty.

Consider the case where the mean values and c.o.v. of n, D and S are 0.015, 3, 0.005 and 5%, 2% and 5%, respectively. Compute the reliability that the sewer can convey the inflow Q_L of 35 ft³/s, assuming:

(a) n, D, and S are Independent normal, lognormal, and Gumbel distribution, respectively.
(b) n, D, and S are all normally distributed, and n and D are negatively correlated with ρ = -0.75. (c)
(c) As in (a), but n and D are negatively correlated with ρ = -0.75.

Solution using spreadsheet FORM procedure: $x_i = F^{-1}\left[\Phi(n_i)\right]$

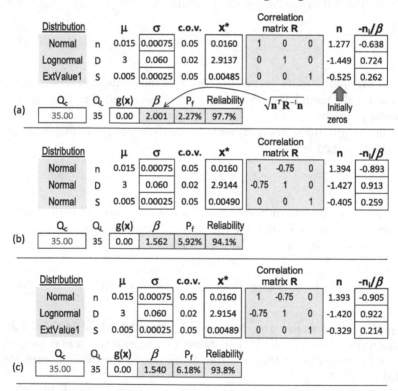

Figure 3.11 Reliability of a storm sewer system: (a) independent nonnormal variates, (b) correlated normal variaties, (c) correlated nonnormal variates.

Based on the descending order of the absolute values of the sensitivity indicators under the **n** columns in Figure 3.11, the variable D is a more sensitive parameter than Manning's roughness coefficient n, which is in turn more sensitive than sewer slope S, notwithstanding that the c.o.v. of D, at 0.02, is the smallest, while the c.o.v. of n and S, at 0.05, are equal. This interesting phenomenon can be understood as an interplay of all the factors which affect the location of the most probable failure point, which, as explained in Chapter 2, is the first point of contact between an expanding ellipsoid (or equivalent dispersion ellipsoid for nonnormal case) and the LSS defined by $g(x) = 0$. The factors at play are the relative magnitudes of mean values, the c.o.v. values, the influence of each parameter in the performance function and the correlation matrix. For example, the $g(x)$ Eq. 3.15 shows that D is raised to the power of 2.67 despite its smallest c.o.v. of the three. It is a significant merit of FORM that its most probable failure point reflects the sensitivities of the underlying parameters.

The first term $g(\mathbf{x})$ in Eq. 3.15 is a capacity entity, which increases with D and S but decreases with the roughness coefficient n. The most probable failure point is reached when the capacity is reduced to a magnitude of 35, equal to the deterministic inflow value. This failure condition is reached at the most probable failure point under the x^* column, with D^* and S^* values (2.9144 and 0.0049 in Figure 3.11b) at $1.427 \times \sigma_D$ and $0.405 \times \sigma_S$ *below* their respective mean values of 3.0 and 0.005, and with the roughness coefficient n^* (0.016) at $1.394 \times \sigma_n$ *above* its mean value of 0.015. These relationships between the normalized component distances from the mean-value point to the most probable failure point also apply to the nonnormal case of Figure 3.11a and 3.11c, in terms of *equivalent normal* mean values and *equivalent normal* standard deviations.

The failure probabilities P_f, based on $\Phi(-\beta)$, are 2.27%, 5.92% and 6.18% for the cases of independent nonnormal, correlated normal and correlated nonnormal in Figure 3.11a, 3.11b and 3.11c, respectively.

Monte Carlo simulations with 200,000 realizations using @RISK (www.palisade.com) yield P_f of 2.04%, 5.93% and 5.79%, respectively. (Note: The inputs for the @Risk ExtValue distribution are location parameter $a = 0.004887$, and scale parameter $b = 0.0001949$, and correlation matrix $\mathbf{R}$. The parameters a and b are related to the mean and standard deviation of the Extreme Value (Gumbel) distribution by: $\mu = a + 0.57721b$, and $\sigma = b\pi / \sqrt{6}$, as given in the Cambridge Dictionary of Statistics by Everitt and Skrondal (2010), for example.)

Tung et al. (2006, Examples 4.10 and 4.11) obtained failure probabilities of 2.02% and 5.5% for the cases of independent nonnormal and correlated normal, respectively. Correlated nonnormal was not conducted.

In principle, instead of the deterministic Q_L value of 35 ft³/s (1 m³/s) in Tung et al. (2006), one can model Q_L as an uncertain entity affected by the four random variables λ_L, C, i and A of Eq. 3.14, with correlations and non-Gaussian distributions if appropriate. This, and the correlated non-Gaussian case of Figure 3.11c, was not done in Tung et al. (2006, Examples 4.10 and 4.11), probably because the traditional FORM approach for correlated nonnormal cases in higher dimensions and rotated coordinate system could be too mathematical and lengthy to present in detail.

In comparison, it is straightforward to extend the spreadsheet FORM procedure of Figure 3.11c from the current three correlated variates to seven correlated non-Gaussian variates, or eight variates if a model correction factor λ_C is added to account for the model uncertainty of the sewer flow-carrying capacity Q_C,

Besides Eq. 3.15, Tung et al. (2006) mentioned that the performance function can also be expressed in two other ways, namely:

$$g_2(\mathbf{x}) = \frac{Q_c}{Q_L} - 1 = \frac{0.463}{n} \frac{\lambda_C D^{8/3} S^{0.5}}{\lambda_L C i A} - 1 \qquad (3.16)$$

and

$$g_3(\mathbf{x}) = \ln\left(\frac{Q_c}{Q_L}\right) \qquad (3.17)$$

The three performance functions of Eqs. 3.15–3.17 are mechanically equivalent at the LSS defined by $g(\mathbf{x}) = 0$. Any of the three performance functions can be used in the FORM approach without affecting the FORM results. Not so with the mean-value first-order second-moment (MVFOSM) method.

Second-order reliability methods (SORM), to improve the probability of failure predicted by FORM, were also discussed in Tung et al. (2006). Chan and Low (2012a) showed that SORM can be implemented in the spreadsheet platform, on the basis of FORM results.

3.9 EXAMPLE 8: RELIABILITY OF A BREAKWATER AT LA SPEZIA HARBOUR IN NORTHERN ITALY

This question is excerpted from Kottegoda and Rosso (2008, Example 9.12) but solved here using the spreadsheet FORM procedure, with discussions on the information attainable from the most probable failure point in FORM.

As explained in Figure 3.12a, the performance function against horizontal sliding of the breakwater at the La Spezia harbour in northern Italy is:

$$\begin{aligned} g(\mathbf{x}) &= c_f R'_v - Q_h \\ &= X_1(X_2 - 70X_3X_4) - X_3(17X_4^2 + 145X_4) \end{aligned} \qquad (3.18)$$

Example 8: Reliability of a harbor breakwater against sliding

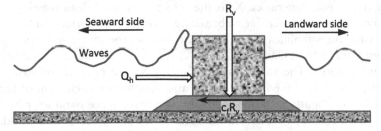

This question is excerpted from Kottegoda and Rosso, 2008, Example 9.12.
A harbor breakwater is constructed with massive concrete tanks filled with sand. It is necessary to evaluate the risk that the breakwater will slide under the lateral pressure of a large wave during a major storm. Stability against sliding exists when the total horizontal destabilizing force Q_h does not exceed the available horizontal frictional resistance $c_f R_v{}'$ along the the base, where c_f is the coefficient of friction. After considering the relevant forces acting, Kottegoda and Rosso expressed the ultimate limit state of sliding by the following performance function:

$$g(\mathbf{x}) = c_f R_v' - Q_h$$
$$= X_1\left(X_2 - a_1 X_3 X_4\right) - X_3\left(a_2 X_4^2 + a_3 X_4\right) = 0$$

in which X_1 represents the interface coefficient of friction at the base of the concrete tanks, X_2 is the effective vertical dead weight, $a_1 X_4$ is the dynamic uplift force due to breaking wave, X_3 is a correction factor to account for the simplifications adopted to model the dynamic vertical and horizontal wave forces, and $a_2 X_4{}^2 + a_3 X_4$ represent the horizontal wave force. The constants a_1, a_2 and a_3 depend on the geometry of the system.
   A unit width of a vertical wall located in La Spezia harbor is considered. All the four random variables are assumed independent. X_1, X_2 and X_3 are normal variates with mean μ_1, μ_2 and μ_3 of 0.64, 3400 kN/m and 1, with coefficient of variation of 15%, 5% and 20%. Frequency analysis of severe storms in the area suggests that X_4 has an extreme value type 1 distribution with mean $\mu_4 = 5.16$ m and standard deviation $\sigma_4 = 0.93$ m. Based on sea-bottom profile and the geometry of the breakwater wall, the constants a_i are $a_1 = 70$, $a_2 = 17$ m/kN, and $a_3 = 145$.

(a)

Solution using spreadsheet FORM procedure: $x_i = F^{-1}\left[\Phi\left(n_i\right)\right]$

			a₁	a₂	a₃	
			70	17	145	
		μ	σ	c.o.v.	n	x*
Normal	X₁	0.64	0.096	0.15	-0.5273	0.589
Normal	X₂	3400	170	0.05	-0.1905	3367.6
Normal	X₃	1	0.2	0.2	0.6664	1.133
ExtValue1	X₄	5.16	0.93	0.18	1.0339	6.056

$\sqrt{\sum n_i^2}$	g(x)	β	P_f	Reliability
	0.000	1.352	8.82%	91.2%

(b)

Figure 3.12 Reliability of a harbour breakwater with respect to the sliding limit state.

in which X_1 is the interface coefficient of friction c_f at the base of the sand-filled massive concrete tanks, X_2 is the effective vertical dead weight, $70X_4$ is the dynamic uplift force due to breaking wave, X_3 is a correction factor to account for the simplifications adopted to model the dynamic vertical and horizontal wave forces, and $a_2X_4^2 + a_3X_4$ represents the horizontal destabilizing wave force. The variables X_1, X_2 and X_3 are normal variates with mean μ_1, μ_2 and μ_3 of 0.64, 3400 kN/m and 1, and with coefficient of variation of 15%, 5% and 20%, respectively. The wave-force parameter X_4 has an extreme value type 1 distribution with mean $\mu_4 = 5.16$ m and standard deviation $\sigma_4 = 0.93$ m.

The spreadsheet FORM procedure is set up succinctly as shown in Figure 3.12b. Since all the four random variables are independent, there is no need to create a 4-by-4 identity matrix $\mathbf{R}$; the formula $\sqrt{\sum n_i^2}$ can be used in the β cell, entered as '=sqrt(sumsq(n)),' instead of $\sqrt{\mathbf{n}^T\mathbf{R}^{-1}\mathbf{n}}$. The g(x) cell refers to the values of the variables under the $\mathbf{x}^*$ column, which are computed via a short open-source VBA function created by Low and Tang (2007). Initially the $\mathbf{n}$ values were zeros. The Solver tool (Figure 3.1) was then invoked, to *minimize* the β cell, *by changing* the $\mathbf{n}$ values, *subject to* the constraint that the g(x) cell be ≤ 0.0. The solution was obtained promptly as shown in Figure 3.12b. The $\mathbf{x}^*$ solution is the most probable failure point where an expanding *equivalent* 4D normal ellipsoid is tangent to the LSS defined by g(x) = 0, and the sensitivity indicators under the $\mathbf{n}$ column are the normalized component distances between the mean-value point and the most probable failure point.

When the $\mathbf{n}$ column was initially zeros prior to invoking the Solver tool, the g(x) cell displayed a positive value. This means that the equivalent mean-value point is in the safe domain of the four dimensional space of X_1, X_2, X_3 and X_4. The FORM solution shows that the most probable failure point, at where the 4D expanding equivalent dispersion ellipsoid first touches the LSS defined by g(x) = 0, is at the point defined by $X_1^* = 0.589$, $X_2^* = 3367.6$, $X_3^* = 1.133$, $X_4^* = 6.056$.

By nature of their roles in the performance function (Eq. 3.18), X_1 and X_2, which represents the coefficient of base friction c_f and total vertical force R_v, are resistance parameters. The variable X_4 contributes to uplifting water force and horizontal water force in Eq. 3.18 and is an unfavorable load parameter. For this case with independent variables, the most probable failure values (0.589 and 3367.6) of resistance parameters X_1 and X_2 are below their respective mean values of 0.64 and 3400. In contrast, the most probable failure values of the wave-force load parameter X_4 and the wave-force correction factor X_3, at 6.056 and 1.133, are above their respective mean values of 5.16 and 1.0.

The location of the most probable failure point is affected by the LSS (defined by g(x) = 0) and the statistics and correlation matrix of the variables. Based on ranking by absolute values under the $\mathbf{n}$ column, X_4 (which appears three times and once as a square in the performance function) affects reliability most, followed by X_3, X_1 and X_2.

The reliability analysis using traditional FORM computational procedure is presented in detail by Kottegoda and Rosso (2008), obtaining $\beta = 1.352$, the same as the spreadsheet FORM solution of Figure 3.12b.

3.10 EXAMPLE 9: SPILLWAY CAPACITY

Ang and Tang (1984, Example 6.12) noted that inadequate spillway capacity to carry the inflow water during an extreme flood is a major cause of dam failure. A spillway with an uncontrolled overflow ogee crest was considered, and the following Bureau of Reclamation equation was used for discharge capacity Q_C:

$$Q_C = NCLH^{3/2} \tag{3.19}$$

where C is the discharge coefficient, L is the effective length of the crest, H is the total head on the crest and N is the correction for imperfections in the empirical formula.

The inflow rate at the spillway is denoted by Q_L (subscript L stands for 'load') and modelled by:

$$Q_L = RQ_I \tag{3.20}$$

where Q_I is the peak flow upstream to the reservoir and R is an attenuation factor.

The performance function is then given by:

$$g(\mathbf{x}) = Q_C - Q_L = NCLH^{3/2} - RQ_I \tag{3.21}$$

All the six variables in g($\mathbf{x}$) are regarded as uncertain and assumed to be independent, with their mean values, standard deviations and distribution types shown in Figure 3.13a. The probability of spillway discharge capacity smaller than the inflow from flood is required.

The solution using the spreadsheet FORM procedure is shown in Figure 3.13b. Since all the six variables are independent, there is no need to create a 6-by-6 identity matrix $\mathbf{R}$; the formula $\sqrt{\sum n_i^2}$ can be used in the β cell, entered as '=sqrt(sumsq(n)),' instead of $\sqrt{\mathbf{n}^T \mathbf{R}^{-1} \mathbf{n}}$. The g($\mathbf{x}$) cell displayed a positive value of 8914 when the $\mathbf{n}$ values were initially zeros; this means that the equivalent mean-value point is in the safe domain, and the reliability index value subsequently found by the *Solver* tool (2.020 in the β cell) can be regarded as positive reliability index. The probability of unsatisfactory

Example 9: Spillway capacity

Ang and Tang (1984, Example 6.12) considered the spillway of a dam with an uncontrolled overflow ogee crest, and used the following Bureau of Reclamation equation for spillway capacity:

$$Q_C = NCLH^{3/2}$$

where C is the discharge coefficient that depends on several factors, including depth of approach and upstream face slope; L is the effective length of the crest; H is the total head on the crest, including the velocity of approach head; and N is the correction for imperfections in the empirical formula.

During an extreme flood, the inflow rate at the spillway was modelled by :

$$Q_L = RQ_I$$

where Q_I is the peak flow upstream to the reservoir and R is the attenuation factor due to the volume effect of the reservoir. Ang and Tang assumed the following statistics:

	Mean	c.o.v	Distribution
N	1.0	0.20	Normal
C	3.85	0.07	Normal
L	93.4 ft (28.5 m)	0.06	Normal
H	12 ft (3.66 m)	0.06	Normal
R	0.7	0.14	Normal
Q_I	9146 cfs (259 m³/s)	0.35	Type 1 Largest

The engineer is required to determine the probability of unsatisfactory performance.
(a)

- - -- --- -- --- -- --- -- --- - --- -- --- -- --- -- --- -- --- --- - -- - - --

Solution using spreadsheet FORM procedure:

		x*	μ	σ	c.o.v.	n
Normal	N	0.780	1	0.2000	0.2	-1.101
Normal	C	3.7673	3.85	0.2695	0.07	-0.307
Normal	L	91.9	93.4	5.6040	0.06	-0.2616
Normal	H	11.715	12	0.7200	0.06	-0.3956
Normal	R	0.755	0.7	0.0980	0.14	0.5572
ExtValue1	Q_I	14351	9146	3201.1	0.35	1.4956

Type 1 Largest or Gumbel

$$\sqrt{\sqrt{\sum n_i^2}}$$

$Q_{capacity}$	Q_{inflow}	g(x)	β	P_f	Reliability
10829.5	10829.5	0.0	2.020	2.17%	97.8%

(b)

Figure 3.13 Reliability analysis of a spillway performance.

performance (i.e., $Q_C < Q_L$) is $P_f = \Phi(-\beta) = 2.17\%$. The corresponding reliability of adequate spillway capacity is 97.8%.

For comparison, the principles and sequence of the traditional FORM computational approach for this problem is well elucidated with calculation details in Ang and Tang (1984, pp. 374–378), involving iterative multiple partial derivatives, successive LSSs and tentative failure points. After three iterations, the solutions obtained were β = 2.00, P_f = 2.3%, and the most probable failure point values are N* = 0.78, C* = 3.77, L* = 92.0, H* = 11.72, R* = 0.75 and Q_I* = 14,416, which compare very well with the spreadsheet solution (Figure 3.13b) obtained by the Solver tool: β = 2.02, P_f = 2.17%, N* = 0.78, C* = 3.77, L* = 91.9, H* = 11.72, R* = 0.755 and Q_I* = 14,351.

3.11 EXAMPLE 10: TWO WAYS OF INCREASING SPILLWAY CAPACITY, AND RBD WITH COST CONSIDERATIONS

The previous spillway capacity has the following two extended investigations (a) and (b) in Ang and Tang (1984, Problem 6.13), as shown in Figure 3.14.

(a) The first extended investigation requires new FORM analysis on spillway capacity increased by the following two alternatives:
  (i) Extending the mean length L of the existing spillway (previous example) from 93.4 ft. (28.5 m) to 113.4 ft. (34.6 m).
  (ii) Lowering the crest of the existing spillway (previous example) so that the mean total head H on the crest increases from 12 ft. (3.66 m) to 13 ft. (3.96 m).

The spreadsheet FORM solutions for this first extended investigation are obtained readily:

  (i) Change the mean length of L from 93.4 to 113.4, in 'Solution (a), left' of Figure 3.14, and invoke the Solver tool to obtain β = 2.427, which is higher than β = 2.02 in the previous example.
  (ii) Change the mean total head H from 12 to 13 in 'Solution (a), right' of Figure 3.14, and invoke the Solver tool to obtain β = 2.273, which is higher than β = 2.02 in the previous example.
(b) The second extended investigation seeks design values of mean length L and mean total head H so that the probability of inadequate spillway capacity is smaller than 1%, which means the reliability against exceeding spillway capacity is 99%. The required reliability index is obtained using the Excel spreadsheet function NormSInv(0.99), which gives β_{target} = 2.326. Different mean values of L were tried in 'Solution (b), left' of Figure 3.14, and different mean values of total head H in 'Solution (b), right.' It was found that a mean L of 108.05 ft. (32.9 m) achieves the target β of 2.326 in Solution (b), left, and a mean total head of 13.223 ft. (4.03 m) achieves the target β of 2.326 in Solution (b), right. This means that the required change in mean effective crest length L is ΔL = 108.05–93.4 = 14.65 ft. (4.47 m). For increasing total head H, the required increase is ΔH = 13.223–12 = 1.223 ft. (0.373 m).

Example 10: Reliability analysis of increased spillway capacity, and cost

(a) Ang and Tang (1984, problem 6.13a) considered increasing the spillway capacity of the dam in the previous example by the following two alternatives:

(i) Extending the mean length of the existing spillway by an additional 20 ft (6.1 m).

(ii) Lowering the crest of the existing spillway so that the mean total head H on the crest is increased by 1 ft (0.305 m).

Assume the statistics of all the other parameters remain the same as those in the previous example. Determine the probability of unsatisfactory performance for each of the alternatives.

(b) Ang and Tang (1984, problem 6.13b) supposed that the probability of unsatisfactory performance of the spillway capacity should not exceed 0.01, which could be achieved either by lengthening the existing spillway or by lowering the crest of the existing spillway. Assume the cost (in $ million) associated with lengthening the spillway by ΔL feet is

$$C_1 = 0.5 + 0.08\Delta L$$

whereas that associated with lowering the spillway crest by ΔD feet is

$$C_2 = 0.8 + 0.7\Delta D + 0.4(\Delta D)^2$$

Determine the optimal plan for meeting the performance criterion.

Solution (a) using spreadsheet FORM procedure:

(a) Alternative I of increasing spillway capacity

		x*	μ	σ	c.o.v.	n
Normal	N	0.721	1	0.2000	0.2	-1.3936
Normal	C	3.7528	3.85	0.2695	0.07	-0.3607
Normal	L	111.3	113.4	6.8040	0.06	-0.307
Normal	H	11.665	12	0.7200	0.06	-0.4651
Normal	R	0.763	0.7	0.0980	0.14	0.645
ExtValue1	Q_i	15729	9146	3201.1	0.35	1.758

Type 1 Largest or Gumbel $\sqrt{\sum n_i^2}$

$Q_{capacity}$	Q_{inflow}	g(x)	β	P_f	Reliability
12004.2	12004.2	0.0	2.427	0.762%	99.24%

(a) Alternative II of increasing spillway capacity

		x*	μ	σ	c.o.v.	n
Normal	N	0.744	1	0.2000	0.2	-1.278
Normal	C	3.7581	3.85	0.2695	0.07	-0.341
Normal	L	91.8	93.4	5.6040	0.06	-0.29
Normal	H	12.657	13	0.7800	0.06	-0.44
Normal	R	0.760	0.7	0.0980	0.14	0.613
ExtValue1	Q_i	15209	9146	3201.1	0.35	1.6619

Type 1 Largest or Gumbel

$Q_{capacity}$	Q_{inflow}	g(x)	β	P_f	Reliability
11560	11560	0.0	2.273	1.152%	98.8%

Solution (b) using spreadsheet FORM procedure:

(b)(i) Lengthening L, Target β = 2.326

		x*	μ	σ	c.o.v.	n
Normal	N	0.737	1	0.2000	0.2	-1.3175
Normal	C	3.7562	3.85	0.2695	0.07	-0.3479
Normal	L	106.1	108.05	6.4830	0.06	-0.2962
Normal	H	11.677	12	0.7200	0.06	-0.4485
Normal	R	0.761	0.7	0.0980	0.14	0.6244
ExtValue1	Q_i	15391	9146	3201.1	0.35	1.696

$Q_{capacity}$	Q_{inflow}	g(x)	β	P_f	Reliability
11715.6	11715.6	0.0	2.326	1.00%	99.0%

Cost =	1.672		ΔL
	$ Million		14.65

(b)(ii) Lowering the crest to increase H, Target β = 2.326

		x*	μ	σ	c.o.v.	n
Normal	N	0.737	1	0.2000	0.2	-1.317
Normal	C	3.7562	3.85	0.2695	0.07	-0.348
Normal	L	91.7	93.4	5.6040	0.06	-0.296
Normal	H	12.867	13.223	0.7934	0.06	-0.448
Normal	R	0.761	0.7	0.0980	0.14	0.6243
ExtValue1	Q_i	15390	9146	3201.1	0.35	1.6958

$Q_{capacity}$	Q_{inflow}	g(x)	β	P_f	Reliability
11715	11715	0.0	2.326	1.00%	99.0%

Cost =	2.254		ΔH
	$ Million		1.223

Figure 3.14 Reliability analysis for increased spillway capacity; and optimal plan for increasing spillway capacity with cost considerations.

The required ΔH is brought about by lowering the crest of the existing spillway. The costs are obtained as follows:

$$C_1 = 0.5 + 0.08\Delta L = 0.5 + 0.08 \times 14.65 = \$1.672 \, \text{million}$$

and

$$C_2 = 0.8 + 0.7\Delta H + 0.4\Delta H^2 = 0.8 + 0.7 \times 1.223 + 0.4 \times 1.223^2 = \$2.254 \, \text{million}$$

Hence, the optimal (cheaper) plan for meeting the performance criterion of 99% reliability with respect to spillway capacity is to increase the effective length of the crest from the existing 93.4 ft. (28.5 m) to 108 ft. (32.9 m).

3.12 EXAMPLE 11: COLUMN SUBJECTED TO BIAXIAL BENDING MOMENTS AND AXIAL FORCE

Der Kiureghian (2005) conducted FORM analysis on a column subjected to biaxial bending moments m_1 and m_2 and axial force p. Assuming an elastic perfectly plastic material with yield stress y, the failure of the column is defined by the performance function

$$g(\mathbf{x}) = 1 - \frac{m_1}{S_1 y} - \frac{m_2}{S_2 y} - \left(\frac{p}{Ay}\right)^2 \tag{3.22}$$

where $\mathbf{x} = \{m_1, m_2, p, y\}^T$ denotes the vector of random variables, $A = 0.190 \, \text{m}^2$ is the cross-sectional area, and $S_1 = 0.030 \, \text{m}^3$ and $S_2 = 0.015 \, \text{m}^3$ are the flexural moduli of the fully plastic column section. The probability distributions, mean values, coefficients of variation (c.o.v.) and correlation coefficients of m_1, m_2, p and y are shown in Figure 3.15a.

The FORM and SORM spreadsheet solutions using Low and Tang (2007) FORM procedure and Chan and Low (2012a) SORM procedure are shown in Figure 3.15b. Using the Improved HL-RF algorithm, Der Kiureghian (2005) obtained (after nine steps) the following converged results:

$$\mathbf{x}* = \{341, 170, 3223, 31.8\}^T,$$

$$\alpha_i = [0.491, 0.283, 0.381, -0.731]$$

$$\beta = 2.467 \text{ and } p_f = \Phi(-\beta) = 0.00682$$

The failure probability, estimated from Monte Carlo simulations with 120,000 realizations, was 0.00931, with a coefficient of variation of 3%.

Example 11: Reliability analysis of a column: axially loaded and biaxial bended
Question based on Der Kiureghian (2005, 14.3.1.1)

Performance function $\quad g(\mathbf{x}) = 1 - \dfrac{m_1}{S_1 y} - \dfrac{m_2}{S_2 y} - \left(\dfrac{p}{Ay}\right)^{\theta}$

Variable	Distribution	Mean	c.o.v.	Correlation coefficient			
				m_1	m_2	p	y
m_1, kNm	Normal	250	0.3	1.0			
m_2, kNm	Normal	125	0.3	0.5	1.0		
p, kN	Gumbel	2500	0.2	0.3	0.3	1.0	
y, MPa	Weibull	40	0.1	0.0	0.0	0.0	1.0

(a)

Solution using spreadsheet FORM & SORM procedures:

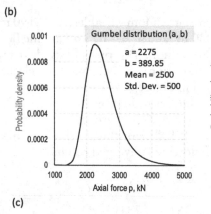

θ	A	S_1	S_2
2	0.19	0.03	0.015

		x*	Para1	Para2	n	Matrix R				n/β	u/β
Normal	m_1	340.3	250	75.00	1.204	1	0.5	0.3	0	0.487	0.487
Normal	m_2	170	125	37.5	1.204	0.5	1	0.3	0	0.487	0.281
ExtValue1	p	3215	2500	500.0	1.367	0.3	0.3	1	0	0.553	0.382
Weibull	y	31.71	12.15	41.72	-1.812	0	0	0	1	-0.733	-0.733

Gumbel

g(x)	β	P_f	Reliability
0.000	2.471	0.67%	99.3%

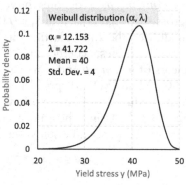

SORM formulas	$P_{f,SORM}$
Tvedt	0.907%
Breitung	0.882%
Hohenbichler_Rackwitz	0.920%
Koyluoglu_Nielsen	0.851%
Cai_Elishakoff	0.903%
Hong(P₃)	0.907%
Hong(P₄)	0.907%
AVERAGE	0.897%

(b)

Gumbel distribution (a, b)
a = 2275
b = 389.85
Mean = 2500
Std. Dev. = 500

Probability density / Axial force p, kN

Weibull distribution (α, λ)
α = 12.153
λ = 41.722
Mean = 40
Std. Dev. = 4

Probability density / Yield stress y (MPa)

(c)

Figure 3.15 (a) Performance function and statistics of a column subjected to axial loading and biaxial bending, (b) FORM and SORM solutions, (c) nonnormal probability density functions of axial load p and yield stress y.

The spreadsheet FORM solution is shown in Figure 3.15b, left. The results are:

$$\mathbf{x}* = \{340.3,\ 170,\ 3215,\ 31.7\}^{T},$$

$$\alpha = L^{-1}[n / \beta] = [0.487, 0.281, 0.382, -0.733]$$

$$\beta = 2.471 \text{ and } p_f = \Phi(-\beta) = 0.00674$$

in which L is the lower triangular matrix of the Cholesky decomposition of the **R** matrix.

The failure probabilities, estimated from three Monte Carlo simulations using the @RISK software, each with 500,000 realizations, are 0.918%, 0.901% and 0.909% (average 0.909%), with a coefficient of variation of 1.5%.

The @RISK Gumbel distribution uses location parameter a and shape parameter b, which can be obtained from the mean and standard deviation of Gumbel distribution as:

$$\text{Gumbel}(a, b): a = \mu - 0.45\sigma; \text{ and } b = 0.7797\sigma \qquad (3.23)$$

The @RISK Weibull distribution uses the shape parameter α and scale parameter λ, which are also used in the Low and Tang (2007) Weibull distributions. In this case $\alpha = 12.153$ and $\lambda = 41.722$.

Der Kiureghian (2005) mentioned that SORM can be conducted on FORM results, to improve the failure probability prediction based on FORM β. The SORM solution in Figure 3.15b, right, has been obtained, using the Chan and Low (2012a) spreadsheet SORM procedure, which created VBA functions for estimating component curvatures at the FORM design point. The FORM β value and the values of the component curvatures are then used to estimate the failure probability according to seven formulas suggested in the literature, yielding an average SORM P_f of 0.897%, which is in good agreement with the average P_f of 0.909% from Monte Carlo simulations using the @RISK software.

From the standpoint of failure probability, the 'exact' failure probability of 0.91% (from Monte Carlo simulations, or from SORM) is 1.35 times higher than the FORM P_f of 0.674%. However, from the standpoint of reliability, the FORM reliability value of 99.33% is only 1.002 times higher than the 'exact' reliability of 99.1%. One possible pragmatic standpoint to adopt, in cases where a target FORM β of 2.5 or 3.0 is typical in RBD involving ultimate limit states, is to regard the reliability estimate based on $\Phi(\beta)$ as sufficiently accurate for the given statistical inputs, probability distributions and correlation matrix.

Figure 3.15c shows the probability density curves of the Gumbel and Weibull distributions for the axial force p and the yield stress y of this case, respectively.

Der Kiureghian (2005, Section 14.3.3) discussed FORM importance and sensitivity measures. On the same theme, Chan et al. (2012) discussed probabilistic sensitivity analysis in the context of an underground rock excavation.

3.13 EXAMPLE 12: MOMENT CAPACITY OF A REINFORCED CONCRETE BEAM

Haldar and Mahadevan (2000, pp. 218–219) applied FORM to the following performance function involving the moment capacity of a singly reinforced rectangular prismatic concrete beam:

$$g(\mathbf{x}) = M_R - M = A_s f_y d \left(1 - \eta \frac{A_s \ f_y}{bd \ f_c'}\right) - M \qquad (3.24)$$

where M_R is the moment capacity or resistance, A_s is the area of the tension reinforcing bars, f_y is the yield stress of the reinforcing bars, d is the distance from the extreme compression fiber to the centroid of the tension reinforcing bars, η is the concrete stress block parameter, f_c' is the compressive strength of concrete, b is the width of the compression face of the member and M is the moment acting on the beam. Haldar and Mahadevan (2000) considered all seven variables in Eq. 3.24 as independent and random, with mean values and coefficients of variation as shown in Figure 3.16a. Four reliability index values were obtained, corresponding to four different probability distribution scenarios for the seven variables.

The spreadsheet FORM solutions for the four different distribution scenarios are shown in Figure 3.16b. Since all the seven random variables are assumed independent, there is no need to create a 7-by-7 identity matrix $\mathbf{R}$; the formula $\sqrt{\sum n_i^2}$ can be used in the β cell, entered as '=sqrt(sumsq(n)),' instead of $\sqrt{\mathbf{n}^T \mathbf{R}^{-1} \mathbf{n}}$.

The spreadsheet FORM solutions of the four β values in Figure 3.16b, namely 3.833, 3.761, 4.388 and 4.091, are the same as those obtained using a computer programme in Haldar and Mahadevan (2000, Table 7.6), which aptly noted that the distributions of random variables play a very important role in reliability index or failure probability estimation.

3.14 EXAMPLE 13: RELIABILITY ANALYSIS OF AN ASYMMETRICALLY LOADED BEAM ON WINKLER MEDIUM

This case is excerpted from Low and Tang (2004).

Beams on elastic foundations can be analyzed using various analytical and numerical methods. Figure 3.17 shows reliability analysis of a beam on Winkler medium, to test robustness and accuracy of the Low and Tang (2007) FORM approach when complicated nonlinear performance functions and correlated nonnormal distributions are involved. The beam has a length (l) of 15 m, a width (b) of 1 m. Six random variables are considered: the moment of inertia I_b (in m^4)) and Young's modulus E_b (MN/m^2) of the

Example 12: Reliability index of the moment capacity of a reinforced concrete beam

Question from Haldar and Mahadevan p218-219

The limit state equation is expressed by performance function g(x) = 0:

$$g(\mathbf{x}) = A_s f_y d\left(1 - \eta \frac{A_s \ f_y}{bd \ f_c'}\right) - M = 0$$

Uncertainty in the Design Parameters of a Reinforced Concrete Beam

Random variables	Mean	Coefficient of variation
A_s	1.56 in.² (10.1 cm²)	0.036
f_y	47.7 ksi (329 MPa)	0.15
f_c'	3.5 ksi (24 MPa)	0.21
b	8.0 in. (20.3 cm)	0.045
d	13.2 in. (33.5 cm)	0.086
η	0.59	0.05
M	326.25 kip-in. (36.86 kNm)	0.17

(a)

Solutions using spreadsheet FORM procedure, for different distributions:

		x*	μ	σ	n
Normal	A_s	1.537	1.56	0.0562	-0.4055
Normal	f_y	24.68	47.7	7.1550	-3.2177
Normal	f_c'	3.352	3.5	0.7350	-0.202
Normal	b	7.985	8	0.3600	-0.0415
Normal	d	11.79	13.2	1.1352	-1.2461
Normal	η	0.591	0.59	0.0295	0.0459
Normal	M	415.3	326.25	55.463	1.6051

g(x)	β	$\sqrt{\sum n_i^2}$
0.000	3.833	

		x*	μ	σ	n
Normal	A_s	1.538	1.56	0.0562	-0.3987
Normal	f_y	27.05	47.7	7.1550	-2.8859
Normal	f_c'	3.336	3.5	0.7350	-0.2237
Normal	b	7.984	8	0.3600	-0.0458
Normal	d	11.79	13.2	1.1352	-1.2463
Normal	η	0.591	0.59	0.0295	0.0506
Lognormal	M	451.8	326.25	55.463	2.0129

g(x)	β
0.000	3.761

		x*	μ	σ	n
Lognormal	A_s	1.521	1.56	0.0562	-0.6788
Lognormal	f_y	31.00	47.7	7.1550	-2.8136
Lognormal	f_c'	3.080	3.5	0.7350	-0.511
Lognormal	b	7.952	8	0.3600	-0.1106
Lognormal	d	11.04	13.2	1.1352	-2.0415
Lognormal	η	0.593	0.59	0.0295	0.1229
Normal	M	466.8	326.25	55.463	2.5335

g(x)	β
0.000	4.388

		x*	μ	σ	n
Lognormal	A_s	1.529	1.56	0.0562	-0.5442
Lognormal	f_y	33.69	47.7	7.1550	-2.2558
Lognormal	f_c'	3.133	3.5	0.7350	-0.4287
Lognormal	b	7.959	8	0.3600	-0.0928
Lognormal	d	11.41	13.2	1.1352	-1.6524
Lognormal	η	0.592	0.59	0.0295	0.1031
Lognormal	M	524.8	326.25	55.463	2.9009

g(x)	β
0.000	4.091

(b)

Figure 3.16 Reliability index β of a reinforced concrete beam under various probability distributions: (a) performance function and statistical inputs from cited source, (b) reliability index β under various probability distributions.

Example 13: Reliability index of an asymmetrically loaded beam on Winkler medium

This example from Low and Tang (2004, 2007).

Bending moment of finite beam on elastic foundation (free at ends A & B):

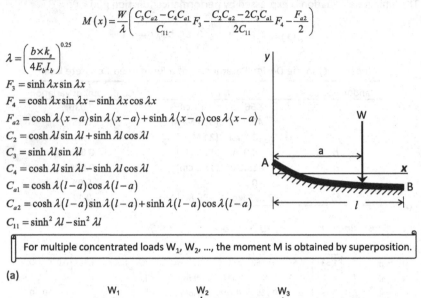

$$M(x)=\frac{W}{\lambda}\left(\frac{C_3 C_{a2}-C_4 C_{a1}}{C_{11}}F_3-\frac{C_2 C_{a2}-2C_3 C_{a1}}{2C_{11}}F_4-\frac{F_{a2}}{2}\right)$$

$\lambda=\left(\dfrac{b\times k_s}{4E_b I_b}\right)^{0.25}$

$F_3=\sinh\lambda x\sin\lambda x$

$F_4=\cosh\lambda x\sin\lambda x-\sinh\lambda x\cos\lambda x$

$F_{a2}=\cosh\lambda\langle x-a\rangle\sin\lambda\langle x-a\rangle+\sinh\lambda\langle x-a\rangle\cos\lambda\langle x-a\rangle$

$C_2=\cosh\lambda l\sin\lambda l+\sinh\lambda l\cos\lambda l$

$C_3=\sinh\lambda l\sin\lambda l$

$C_4=\cosh\lambda l\sin\lambda l-\sinh\lambda l\cos\lambda l$

$C_{a1}=\cosh\lambda(l-a)\cos\lambda(l-a)$

$C_{a2}=\cosh\lambda(l-a)\sin\lambda(l-a)+\sinh\lambda(l-a)\cos\lambda(l-a)$

$C_{11}=\sinh^2\lambda l-\sin^2\lambda l$

For multiple concentrated loads W_1, W_2, …, the moment M is obtained by superposition.

(a)

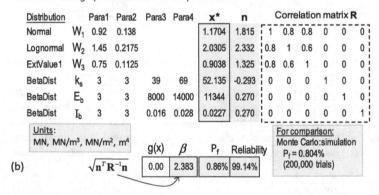

Solution using spreadsheet FORM procedure:

Distribution		Para1	Para2	Para3	Para4	x*	n	Correlation matrix R					
Normal	W_1	0.92	0.138			1.1704	1.815	1	0.8	0.8	0	0	0
Lognormal	W_2	1.45	0.2175			2.0305	2.332	0.8	1	0.6	0	0	0
ExtValue1	W_3	0.75	0.1125			0.9038	1.325	0.8	0.6	1	0	0	0
BetaDist	k_s	3	3	39	69	52.135	-0.293	0	0	0	1	0	0
BetaDist	E_b	3	3	8000	14000	11344	0.270	0	0	0	0	1	0
BetaDist	I_b	3	3	0.016	0.028	0.0227	0.270	0	0	0	0	0	1

Units:
MN, MN/m³, MN/m², m⁴

For comparison:
Monte Carlo:simulation
$P_f=0.804\%$
(200,000 trials)

	$g(x)$	β	P_f	Reliability
$\sqrt{n^T R^{-1} n}$	0.00	2.383	0.86%	99.14%

(b)

Figure 3.17 Reliability analysis of an asymmetrically loaded beam on Winkler medium.

beam, the subgrade modulus k_s (MN/m³) and the magnitudes (in MN) of three concentrated loads W_1, W_2 and W_3, acting at 1.5 m, 7.5 m and 13.5 m, respectively, from the left end of the beam. Failure is considered to have occurred if the bending moment reaches 1 MNm *anywhere* along the beam. The analytical solutions summarized in Young and Budynas (2002) and shown in Figure 3.17a are incorporated in a user-created function BMoment(...). The statistical inputs in Figure 3.17b are the three correlated loads W_1, W_2 and W_3, each with a coefficient of variation equal to 0.15, the subgrade modulus k_s, the Young's modulus E_b and the moment of inertia I_b of the beam. The bounded beta general distribution with four parameters $(a_1, a_2, \min, \max)$ is used for the variables k_s, E_b and I_b. In the four-parameter beta general distribution, the third and fourth parameters define the lower and upper limits of the range, while the first two parameters are shape parameters. If $a_1 = a_2$, as in this case, the beta distribution curve is non-skew. The mean and standard deviation of the three beta general distributions in Figure 3.17b are 54 and 5.67 for k_s, 11,000 and 1134 for E_b and 0.022 and 0.00227 for I_b. The assumed distribution types, statistical inputs and correlation matrix **R** are shown in Figure 3.17b.

Deterministic inputs include the x-coordinate values of the three loads (a_1 = 1.5 m, a_2 = 7.5 m, a_3 = 13.5 m), length l = 15 m and width b = 1 m.

The performance function is:

$$g(\mathbf{x}) = 1\,\text{MNm} - \max_{0<x<l}\left[\frac{W}{\lambda}\left(\frac{C_3 C_{a2} - C_4 C_{a1}}{C_{11}}F_3 - \frac{C_2 C_{a2} - 2C_3 C_{a1}}{2C_{11}}F_4 - \frac{F_{a2}}{2}\right)\right]$$

(3.25)

where λ depends on b, k_s, E_b and I_b, and the C_i and F_i terms inside the parenthesis are functions of sine, cosine, hyperbolic functions sinh and cosh, as defined in Figure 3.17b.

The computed β index using the spreadsheet *Solver* tool is 2.383. The reliability-based probability of failure is $P_f = \Phi(-\beta) = 0.86\%$, which compares well with $P_f = 0.804\%$ from Monte Carlo simulation using the @RISK software (http://www.palisade.com), with 200,000 realizations. In terms of reliability that the maximum bending moment along the beam will not exceed 1 MNm, it is Reliability = 99.14% based on FORM β, and Reliability = 99.2% based on Monte Carlo simulations.

3.15 EXAMPLE 14: A STRUT WITH COMPLEX SUPPORTS AND IMPLICIT PERFORMANCE FUNCTION

This example is excerpted from Low and Tang (2007), to illustrate FORM analysis of a case with implicit performance function that requires iterative numerical solution. The interesting sensitivity information at the most probable failure point is also discussed.

The deterministic analysis of a strut with complex supports was discussed in Coates et al. (1994). The member is initially straight (Figure 3.18) with a pin support at 2, an elastic support at 1 (rotational stiffness λ_1) which

Example 14: Reliability index of a strut with complex supports
This example from Low and Tang (2007).

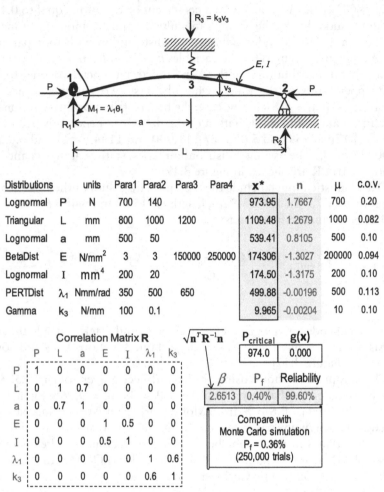

Distributions		units	Para1	Para2	Para3	Para4	x*	n	μ	c.o.v.
Lognormal	P	N	700	140			973.95	1.7667	700	0.20
Triangular	L	mm	800	1000	1200		1109.48	1.2679	1000	0.082
Lognormal	a	mm	500	50			539.41	0.8105	500	0.10
BetaDist	E	N/mm^2	3	3	150000	250000	174306	-1.3027	200000	0.094
Lognormal	I	mm^4	200	20			174.50	-1.3175	200	0.10
PERTDist	λ_1	Nmm/rad	350	500	650		499.88	-0.00196	500	0.113
Gamma	k_3	N/mm	100	0.1			9.965	-0.00204	10	0.10

Correlation Matrix **R**

	P	L	a	E	I	λ_1	k_3
P	1	0	0	0	0	0	0
L	0	1	0.7	0	0	0	0
a	0	0.7	1	0	0	0	0
E	0	0	0	1	0.5	0	0
I	0	0	0	0.5	1	0	0
λ_1	0	0	0	0	0	1	0.6
k_3	0	0	0	0	0	0.6	1

$\sqrt{\mathbf{n}^T \mathbf{R}^{-1} \mathbf{n}}$

$P_{critical}$	g(x)
974.0	0.000

β	P_f	Reliability
2.6513	0.40%	99.60%

Compare with
Monte Carlo simulation
P_f = 0.36%
(250,000 trials)

Figure 3.18 Reliability of a strut with complex supports. Performance function is implicit and requires numerical method using VBA function created in spreadsheet.

provides a restoring moment $M_1 = \lambda_1 \theta_1$, and a support at 3 (stiffness k_3) which provides a reaction force ($k_3 v_3$) proportional to the vertical displacement v_3. It is required to determine the smallest value of axial force P which will cause the strut to become elastically unstable. Coates et al. (1994) showed that the problem reduces to that of finding a value of P, referred to as $P_{critical}$, which would make the determinant of a 7-by-7 matrix (functions of P, L, a, E, I, λ_1 and k_3) vanish. It was also noted that a numerical procedure would be necessary to determine the determinant of the matrix by sequential variation of the axial load P until a zero determinant value is reached.

For the reliability analysis here, $P_{critical}$ can be obtained in a spreadsheet cell (Figure 3.18) that calls a user-defined Excel VBA function created in Low and Tang (2004), to compute (by an iterative algorithm) $P_{critical}$ as a function of L, a, E, I, λ_1 and k_3. The accuracy is verified by deterministic analysis with L, a, E, I, λ_1 and k_3 values of 1000, 500, 200,000, 125, 0, 0, obtaining $P_{critical}$ = 246.7 N, which agrees with the Euler load mentioned in Coates et al. (1994). If $k_3 = 4$, a $P_{critical}$ of 987 N was obtained, again in agreement with the deterministic case of Coates et al. (For the case in Figure 3.18, the mean value of I is 200 instead of 125 in Coates et al., to achieve a target β index greater than 2.5 in RBD.)

The performance function, g(x) in Figure 3.18, is '=$P_{critical}$ – P,' where P refers to the first value of the x* column. Hence, the single cell object 'g(x)' contains an implicit and iterative user-created programme code. The seven x_i* cells contain the open-source VBA function created in Low and Tang (2007), which automatically obtains x values from $x_i = F^{-1}[\Phi(n_i)]$ as the n values are varied by the Solver tool.

The seven n_i values were initially zeros. The Excel Solver tool was then invoked to minimize the β cell, by changing the n_i values, subject to the constraint that g(x) = 0. Solver found a β value of 2.6513 after eight trial solutions in 12 seconds. Solver's 'automatic scaling' option can be used, if desired.

The design point (x*) and the n values in Figure 3.18 indicate that the parameters λ_1 and k_3 hardly deviate from their mean values of 500 and 10, respectively. The implication is that the performance function and hence $P_{critical}$ is not sensitive to λ_1 and k_3 at the current mean values, standard deviations and correlations of P, L, a, E, I, λ_1 and k_3. These rather unexpected findings were further investigated in five analyses in which the initial values of n_1, n_2, n_3, n_4 and n_5 were zeros, but the initial values of n_6 and n_7 (pertaining to λ_1 and k_3) were randomized between –1 and + 1, followed by reliability analysis using the Solver tool. The β values from the five reliability analyses each with randomized initial n_6 and n_7 values were all equal to 2.6513, and their design point x* values were virtually identical, thus reinforcing the finding that $P_{critical}$ is insensitive to λ_1 and k_3. Coates et al. (1994) also investigated sensitivities, deterministically, by plotting curves of $P_{critical}$ versus k_3, for values of λ_1 ranging from 0 to 10^{20} Nmm/rad, and found that very large changes in λ_1 had only very small influence on $P_{critical}$. In comparison, for the mean values and standard deviations of Figure 3.18, reliability analysis reveals the insensitivities of $P_{critical}$ to λ_1 and k_3 with much less effort than deterministic contour plotting, besides providing additional information, including a good estimate of the probability of strut buckling.

Reliability analysis and RBD-via-FORM can reveal context-dependent parametric sensitivities in a way deterministic analysis and partial factor design cannot. This means that for the same problem, but with different input values (deterministic and statistical) for different actual cases, the sensitivities of the parameters could vary from case to case. It also means that a parameter (e.g. friction angle ϕ') can manifest different sensitivities in different problems (e.g. bearing capacity versus slope stability). It is therefore more meaningful to let FORM reveal context-dependent parametric sensitivities and to back-calculate partial factors from the design point of FORM for each case of the same problem type and for different problems, than to calibrate back-calculated partial factors when such calibration may not have general applicability under different scenarios of the same problem type and for different problems.

Chapter 4

Eurocode 7, LRFD and links with the first-order reliability method

4.1 INTRODUCTION

The design approach based on the overall factor of safety has long been used by geotechnical engineers. More recent alternatives are the partial factor design approach, including the load and resistance factor design (LRFD) method in North America, and the Eurocode 7 (EC7). Yet another approach can play at least a useful complementary role to LRFD and EC7, namely the design based on a target reliability index that explicitly reflects the uncertainty of the parameters and their correlation structure. Among the various versions of reliability indices, that based on the first-order reliability method (FORM) is most consistent.

In this book, the term *partial factor design approach* refers not just to the Eurocode 7 design method with its partial factors applied to characteristic values, but also to the LRFD method in North America with its load and resistance factors applied to nominal values. Both EC7 and LRFD are considered in this book as partial factor design methods, in the sense that both involve multiple safety factors in contrast to the traditional single global (lumped) factor of safety approach.

The sections below provide (i) an overview of the LRFD, and connections with RBD-via-FORM; (ii) an overview of Eurocode 7, and connections with RBD-via-FORM; (iii) an example of probability-based design of foundation settlement via FORM; (iv) probabilistic FORM settlement analysis of a Hong Kong land reclamation project involving the use of prefabricated vertical drains to accelerate soft clay consolidation and (v) a simple example illustrating the need to apply Monte Carlo simulations with understanding in order to avoid potential pitfalls.

4.2 LOAD AND RESISTANCE FACTOR DESIGN APPROACH, AND LINKS WITH FORM DESIGN POINT

The similarities and differences between the design point of the FORM and that of LRFD method used in North America (AASHTO 2020; CSA 2019) are summarized in Figure 4.1.

DOI: 10.1201/9781003112297-5

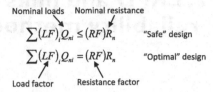

Design point in LRFD:

Nominal loads   Nominal resistance

$$\sum (LF)_i Q_{ni} \leq (RF) R_n \qquad \text{"Safe" design}$$

$$\sum (LF)_i Q_{ni} = (RF) R_n \qquad \text{"Optimal" design}$$

Load factor   Resistance factor

The design point in LRFD is obtained when the sum of amplified (factored) loads Q_1, Q_2, Q_3 ... is equal to the diminished (factored) resistance R

(a)

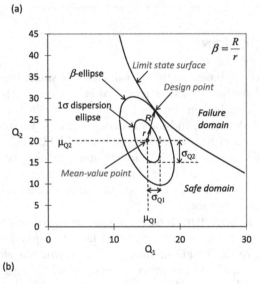

(b)

Figure 4.1 (a) Design point in LRFD, (b) FORM design point reflects parametric correlations and case-specific sensitivities.

The basic design criterion for LRFD is given as (e.g., Salgado and Kim 2014):

$$\sum (LF)_i Q_{ni} \leq (RF) R_n \quad \text{or} \quad \sum \gamma_i Q_{ni} \leq \varphi_R R_n \tag{4.1a}$$

where LF = γ = load factor, RF = φ_R = resistance factor, Q_{ni} = nominal loads and R_n = nominal resistance.

The load factors for unfavourable loads are typically greater than 1.0, and resistance factor is smaller than 1.0.

The most economical design is achieved when the above inequality sign becomes the equal sign:

$$\sum (LF)_i Q_{ni} = (RF) R_n \quad \text{or} \quad \sum \gamma_i Q_{ni} = \phi_R R_n \tag{4.1b}$$

which means:

$$\Sigma \text{ Factored loads} Q_i = \text{Factored resistance R} \qquad (4.1c)$$

The factored loads and factored resistance which satisfy Eq. 4.1c constitute the design point in LRFD. The design point is on the limit state surface (LSS), defined by the traditional global factor of safety $F_s = 1.0$, which separates safe combination of loads and resistance from unsafe combinations of the same. Although the design point is a failure combination of loads and resistance, the design has two tiers of safety built-in: conservative nominal values, and load and resistance factors applied on the nominal values.

Figure 4.1b shows the FORM design point on the LSS, in the two-dimensional space of loads Q_1 and Q_2, shown negatively correlated in this case. Visualizing the image of a FORM design point and a dispersion ellipsoid tangential to a LSS in the higher dimensional space of the examples in this book can be done in the mind's eye. For correlated nonnormal loads and resistance, the ellipsoid perspective is till applicable in terms of equivalent normal distributions, as explained in Chapter 2.

Examples on how RBD-via-FORM can complement LRFD are discussed in Low (2017, 2020a), and Parts II and III of this book.

The following may be noted concerning design points in LRFD and RBD, and design rationale:

(i) In LRFD, a design point is obtained when $\sum$Factored loads = Factored resistance, as conveyed in Eqs. 4.1b and c, and in Figure 4.1a for 'optimal' design. Two designers using the same load and resistance factors (as specified or recommended in design codes AASHTO 2020 or CSA 2019) may still end up with different design outcome if they use different nominal values.

(ii) In RBD, the LSS is the boundary between safe and unsafe domain in the coordinate space of resistance R and loads Q_i. One can imagine a dispersion ellipsoid centred at the mean values of the loads Q_i and resistance R. As the dispersion ellipsoid expands, the probability density decreases on the ellipsoidal contour, analogous to the one-dimensional bell-shaped curve where probability density decreases with distance from the mean value. The first contact point with the LSS is therefore the most probable failure point, also called the design point when there is a target reliability index, for example the case in Figure 2.8. This FORM design point on the LSS satisfies $R_D = \Sigma Q_{i,D}$, where subscript D denotes design value, and hence is similar to the LRFD design point. However, the design point in RBD reflects parametric uncertainties and sensitivities in a way that the LRFD design point (via Eq. 4.1b) cannot.

(iii) The RBD outcome corresponds to a target reliability index (and implied probability of failure P_f), whereas the LRFD outcome does not explicitly convey information on probability of failure, because its target reliability index values are not always known. In principle, even with code-prescribed load and resistance factors, a safer LRFD can be reached by using more conservative nominal values of loads and resistance, but 'how much safer' is not conveyed quantitatively.

(iv) RBD requires statistical inputs (mean values, standard deviations, correlations) and aims at a target reliability index, whereas LRFD requires conservative nominal values of loads and resistance, together with code-specified load and resistance factors. There is no logical way to judge how conservative the nominal values should be in order to reflect high and low consequences of failure and different load and resistance sensitivities from case to case. Because RBD-via-FORM is able to robustly and flexibly reflect case-specific input uncertainties and their sensitivities and does not involve judgemental decisions about nominal values or code-specified load and resistance factors, it can play a complementary role to partial factor design approach (e.g., LRFD, Eurocode 7), for example, by obviating difficult judgement on what nominal values and what partial factors to adopt in different cases.

(v) The results of RBD are only as good as the underlying statistical inputs. The often insufficient information for accurate determination of statistical inputs should not deter one from adopting RBD. One may note that the conservative nominal values and load and resistance factors in LRFD are also attempts to deal with input (and other) uncertainties in design, but in LRFD, the statistical and probabilistic considerations underlying the nominal values and load and resistance factors are hidden; in RBD, the statistical and probabilistic assumptions are laid bare and open to scrutiny and potential improvement in matching reality better. A necessary and pragmatic approach is to regard the probability of failure implied by the target reliability index in RBD as *aiming at a sufficiently safe design*. A target reliability index of 3.0 (or 3.5) in RBD corresponds to a probability of failure of about 0.13% (or 0.02%). This should not be regarded as the real probability of failure of the design, due to imprecise statistical inputs and other factors (model error, human error, unknown unknowns, for example). From a design perpective, a legitimate question to ask is whether the RBD outcome is still a sufficiently safe design even if the real probability of failure is several times higher, for example $2 \times 0.13\%$ and $2 \times 0.02\%$, or $3 \times 0.13\%$ and $3 \times 0.02\%$.

(vi) Most practitioners are unfamiliar with the concepts and procedure of RBD-via-FORM. In contrast, LRFD (or Eurocode 7) principles are easier to understand. Hence, RBD-via-FORM cannot replace LRFD (or Eurocode 7) at this stage. But RBD can play a complementary role to overcome some potential limitations and subtle inconsistencies of some designs based on LRFD and Eurocode 7, for example, the case

of retaining wall foundation in Section 5.9 with load-resistance duality. At the minimum, if statistical inputs are available or can be approximately estimated, it is instructive and even enlightening to do a FORM
reliability analysis of the design that results from LRFD or Eurocode.
Not only the implied notional probability of failure and the FORM
design point are revealed, but also the case-specific sensitivities of
loads and resistance. If desired, the implied load and resistance factors
can also be back-calculated from the FORM results. The examples in
this book illustrate these merits of FORM and RBD.

(vii) There are other reliability approaches in the literature, including the
mean-value first-order second moment method (MVFOSM), and the
Monte Carlo simulations, each with merits and limitations. Some of
the merits of FORM (e.g., information on design point and parametric
sensitivities) are not possessed by the MVFOSM.

4.3 EUROCODE 7 APPROACH, AND LINKS WITH THE DESIGN POINT OF RBD-VIA-FORM

Eurocode 7 is evolving, the latest version being EC7 (2020). However,
Clause 1.1 of the EC7 (2020) document 'EN 1997-3' states that it is intended
to be used in conjunction with EN 1997-1 (EC7 2004). Hence, the 2004
EC7 is referred to in this book.

In geotechnical design based on EC7 (2004), the characteristic values of
resistance parameters are divided by partial factors to obtain the design
values, while the characteristic values of action (load) parameters are multiplied by partial factors to obtain the design values. The resistance thus
diminished is required to be greater or equal to the amplified actions (loadings) as shown in Figure 4.2a, which also summarizes the three design
approaches (DA1, DA2 and DA3) of EC7. The EC7-DA1 is used in the
United Kingdom and Singapore, for example. The EC7-DA2 is similar in
principle to the LRFD of North America.

With respect to characteristic value, Clause 2.4.5.2(2)P EN 1997-1 defines
it as being 'selected as a cautious estimate of the value affecting the occurrence of the limit state,' and Clause 2.4.5.2(10) of EC7 states that statistical
methods may be used when selecting characteristic values of geotechnical
parameters, but they are not mandatory.

The ways in which RBD-via-FORM can complement Eurocode 7 are discussed in Low (2005, 2008a), Low and Phoon (2015b), Low et al. (2017),
and Parts II and III of this book.

One may note the following similarities and differences between a reliability-based design and one based on EC7:

(a) In EC7, a design point is obtained when Amplified actions = Diminished
resistance, as conveyed in Figure 4.2a. Two designers using the same

General concepts of ultimate limit state design in Eurocode 7:

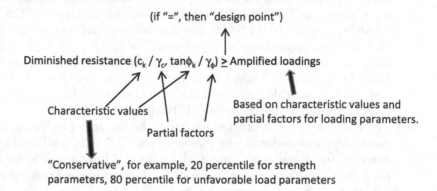

(if "=", then "design point")

Diminished resistance (c_k / γ_c, $\tan\phi_k / \gamma_\phi$) $\geq$ Amplified loadings

Characteristic values

Based on characteristic values and partial factors for loading parameters.

Partial factors

"Conservative", for example, 20 percentile for strength parameters, 80 percentile for unfavorable load parameters

The three sets of partial factors (on resistance, actions, and material properties) are not necessarily all applied at the same time.

In EC7, there are three possible design approaches:

● Design Approach 1 (DA1): (a) factoring actions only; (b) factoring materials only.

● Design Approach 2 (DA2): factoring actions and resistances (but not materials).

● Design Approach 3 (DA3): factoring structural actions only (geotechnical actions from the soil are unfactored) and materials.

(a)

$$\beta = R/r$$

Limit state surface: boundary between *safe* and *unsafe* domain

SAFE

σ_ϕ $\beta\sigma_\phi$

one-sigma dispersion ellipse

$\mu_{c'}$

β-ellipse

Cohesion, c'

$\sigma_{c'}$

UNSAFE

R

Design point

Mean-value point

$\mu_{\phi'}$

Friction angle, ϕ' (degrees)

(b)

Figure 4.2 (a) Design point in Eurocode 7 (2004), (b) FORM design point reflects parametric correlations and case-specific sensitivities.

partial factors recommended in EC7 may still end up with different design outcome if they use different characteristic values.

(b) In RBD-via-FORM, the design point is obtained as the most probable failure combination of parametric values, and the reliability index β, being the distance (in units of directional standard deviations) from the safe mean-value point to the nearest failure boundary (the LSS, Figure 4.2b), conveys information on the probability of failure. In Eurocode design or LRFD, there is no explicit information on the probability of failure.

(c) In reliability-based design one does not use partial factors. The design point in RBD-via-FORM is determined automatically by the spreadsheet Solver tool and reflects mean values, standard deviations, parametric sensitivities, parametric correlations and probability distributions in a way that prescribed partial factors and 'conservative' characteristic values cannot.

It will be clear from the RBD-via-FORM examples of this book that reliability-based design can provide insights and guidance to EC7 design or LRFD design under the following circumstances:

- When partial factors have yet to be proposed by Eurocode 7 to cover uncertainties of less common parameters, for example, *in situ* stress coefficient K in underground excavations in rocks, dip direction and dip angles of rock discontinuity planes and the smear effect of vertical drains installed in soft clay.

- When output has different sensitivity to an input parameter depending on the engineering problems (e.g., shear strength parameters in bearing capacity, retaining walls, slope stability).

- When input parameters are by their physical nature either positively or negatively correlated. One may note that EC7 does not mention use of different characteristic values and partial factors when some parameters are correlated. The same design is obtained in EC7 with or without modelling of correlations among parameters.

- When spatially autocorrelated soil properties need to be modelled. This will be illustrated in Chapter 8 on a Norwegian slope with spatially autocorrelated soil unit weight and undrained shear strength, and in Chapter 6 on vertically loaded and laterally loaded piles.

- When there is a target reliability index or probability of failure for the design in hand. The target reliability index or probability of failure can be set explicitly higher or lower in RBD depending on the importance of the structure, consequence of failure, or quality of inputs. In this regard, one may note that a design by EC7 or LRFD provides no explicit information on the reliability level or probability of failure, nor explicit guidance on how characteristic values or partial factors are to be modified when more stringent safety requirements are needed.

- When uncertainty in unit weight γ of soil needs to be modelled. This is illustrated in Chapter 8 on a Norwegian clay slope, and in Chapter 7 on the reliability-based design of an anchored sheet pile wall.

The insights and information at the design point of RBD-via-FORM and links with EC7 and LRFD are illustrated in the three examples below, on (i) probability-based design of consolidation settlement, (ii) reliability analysis of soft clay consolidation accelerated by prefabricated vertical drains and (iii) a simple example of RBD-via-FORM and comparison with Monte Carlo simulations, demonstrating the need to avoid potential pitfalls in Monte Carlo simulations.

4.4 EXAMPLE RBD OF FOUNDATION SETTLEMENT WITH BACK-CALCULATED PARTIAL FACTORS

This case is from Ang and Tang (1984, Example 6.24). It is required to determine the mean partial factors for the design of a foundation such that the reliability is $\beta = 2.0$ against a limiting consolidation settlement of 1 in. (2.54 cm) in an underlying clay layer, as shown in the soil profile of Figure 4.3a. Assume that the clay layer is normally consolidated and secondary consolidation settlement is negligible.

The performance function is:

$$g(\mathbf{x}) = s_L - N \frac{C_c}{1 + e_o} H \log \frac{\sigma_0' + \Delta\sigma}{\sigma_0'} \qquad (4.2)$$

where S_L is the limiting settlement, C_c is the compression index, e_o is the initial void ratio, H is the thickness of the clay layer, σ_0' is the original effective pressure at mid-depth of the clay layer and $\Delta\sigma$ is the increase in pressure at mid-depth. A correction factor N has been introduced to account for the idealizations underlying the analytical model.

For mathematical convenience of explaining the traditional FORM computational procedure step by step, Ang and Tang (1984) recast Eq. 4.2 as follows:

$$g(\mathbf{x}) = s_L - NY \log(1 + M) \qquad (4.3a)$$

in which

$$Y = \frac{C_c}{1 + e_o} H \qquad (4.3b)$$

and

$$M = \frac{\Delta\sigma}{\sigma_0'} \qquad (4.3c)$$

Example RBD of foundation for settlement. with back-calculated partial factors
This example compares Low and Tang (2007) FORM procedure with the
mathematical approach in Ang and Tang (1984, p428-431, Example 6.24)

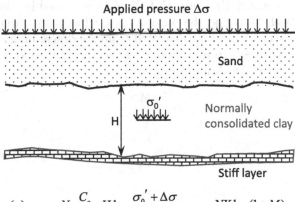

$$g(\mathbf{x}) = s_L - N \frac{C_c}{1+e_o} H \log \frac{\sigma_0' + \Delta\sigma}{\sigma_0'} = s_L - NY\log(1+M)$$

where S_L is the limiting settlement, N is a model error factor,

$$Y = \frac{C_c}{1+e_o} H \quad \text{and} \quad M = \frac{\Delta\sigma}{\sigma_0'}$$

(a)

Distribution		μ	σ	c.o.v.	x*	n	Partial factors	
Normal	N	1.000	0.100	0.10	1.070	0.701	1.070	$\gamma_{\mu N}$
Normal	Y	25.795	6.707	0.26	35.34	1.423	1.370	$\gamma_{\mu Y}$
Normal	M	0.050	0.0105	0.21	0.0628	1.217	1.256	$\gamma_{\mu M}$

	S_L	g(x)	β	P_f
	1	0.00	2.000	0.0228

μ_M	μ_Y	μ_N	$\gamma_{\mu M}$	$\gamma_{\mu Y}$	$\gamma_{\mu N}$
0.01	125.85	1.0	1.259	1.368	1.070
0.05	25.795	1.0	1.256	1.370	1.070
0.1	13.275	1.0	1.252	1.373	1.071
0.5	3.217	1.0	1.229	1.390	1.075

$\gamma_{\mu M}$, $\gamma_{\mu Y}$ and $\gamma_{\mu N}$ denotes the same as $\overline{\gamma}_M$, $\overline{\gamma}_Y$ and $\overline{\gamma}_N$ in Ang and Tang

(b)

Figure 4.3 (a) Example question from Ang and Tang (1984), (b) probability-based
design of foundation settlement via FORM.

The mean value of the model correction factor N is 1.0, with c.o.v. of 0.1. The c.o.v. of Y and M are 0.26 and 0.21, respectively. A range of mean values of M from 0.01 to 0.50 were investigated, and the traditional procedure involving iterative analytical partial derivatives was used in Ang and Tang 1984 (pp. 428–431) to find the mean Y that achieves the target β of 2.0. Partial factors were then back-calculated from the design point values M^*, Y^* and Y^* with respect to mean values (hence referred to as mean partial factors $\bar{\gamma}_M, \bar{\gamma}_Y$ and $\bar{\gamma}_N$), as follows:

$$\bar{\gamma}_M = \frac{M*}{\mu_M} \tag{4.4a}$$

$$\bar{\gamma}_Y = \frac{Y*}{\mu_Y} \tag{4.4b}$$

$$\bar{\gamma}_N = \frac{N*}{\mu_N} \tag{4.4c}$$

Ang and Tang (1984) aptly observed that the mean partial factors do not vary much over wide ranges of μ_M (0.01–0.5) and μ_Y (10–126), with $\bar{\gamma}_M \approx$ 1.26, $\bar{\gamma}_Y \approx 1.37$ and $\bar{\gamma}_N \approx 1.07$. The design requirement based on Eq. 4.3a was then derived as follows:

$$s_L - N^*Y^* \log\left(1 + M^*\right) \geq 0$$

For target β = 2.0, and s_L = 1", the above 4 equations lead to:

$$1 - 1.07 \times 1.37 \mu_Y \times \log\left(1 + 1.26\mu_M\right) = 0, \tag{4.5a}$$

Or

$$\mu_Y = \frac{0.682}{\log\left(1 + 1.26\mu_M\right)} \tag{4.5b}$$

Re-arranging, get:

$$\mu_M \approx \frac{\mu_{\Delta\sigma}}{\mu_{\sigma_b}} = 0.7937\left(10^{0.682/\mu_Y} - 1\right), \quad \text{for target } \beta = 2.0 \tag{4.5c}$$

This example of Ang and Tang (1984) elegantly demonstrates the connection between the design point of FORM (which yields the partial factors by Eqs. 4.4a–4.4c) and the partial factor design approach represented by Eq. 4.5b. For a given mean value of Y (defined by Eq. 4.3b), the designer can determine from Eq. 4.5c the upper bound on the mean value of M that will achieve the target β = 2.0.

However, Eq. 4.5 was derived for these conditions: (1) the c.o.v. values are 0.10, 0.26 and 0.21, for N, Y and M, respectively; (2) the recast three-parameter performance function of Eq. 4.3; and (3) the assumption of uncorrelated normal variates.

For mean M values of 0.01, 0.05, 0.10 and 0.50, the spreadsheet FORM procedure shown in Figure 4.3b efficiently obtains the four rows of mean Y and partial factors, which are virtually all identical to those obtained by the traditional FORM procedure in Ang and Tang (1984). Since this case involves uncorrelated normal variates, there is no need to set up the correlation matrix $\mathbf{R}$. The equation for reliability index reduces to Eq. 2.11, namely $\beta = \min_{x \in F} \sqrt{\sum n_i^2}$, which can be entered as '=sqrt(sumsq(n)),' where $\mathbf{n}$ is the block of three cells (initially zeros) to be changed by the Solver tool during minimization of the β cell. Also, for normal variates, VBA code is not necessary for the $\mathbf{x}^*$ cells, which contain the simple formulas '$=\mu_i + n_i\sigma_i$.'

It would be efficient and convenient to conduct RBD-via-FORM similar to Figure 4.3b when:

(a) Nonnormal distributions for N, Y and M are modelled instead of normal distributions.
(b) The c.o.v. of N, Y and M are different from the case in Figure 4.3.
(c) The original performance function (Eq. 4.2) with six random variables of (N, C_c, e_0, H, ...) is used instead of Eq. 4.3 with three random variables.
(d) Correlations among some of the parameters are warranted based on physical considerations.

Equations 2.9 and 2.10 are necessary for cases involving correlated nonnormal variates (e.g., Figures 2.7 and 2.15).

As nicely illustrated in this Ang and Tang (1984) example, RBD-via-FORM can account for uncertainties of parameters for which partial factors are not provided for in the EC7. In this case, the uncertain parameters are N, M and Y for performance function Eq. 4.3a, or N, C_c, e_0, H, σ_0' and $\Delta\sigma$ for performance function 4.2.

The next example deals with reliability analysis of the magnitude and rate of consolidation settlement in a Hong Kong land reclamation project involving use of vertical drains in marine clay. The deterministic model is described first, before extending the case into reliability analysis.

4.5 PROBABILISTIC SETTLEMENT-ANALYSIS OF HONG KONG LAND RECLAMATION TEST FILL

The Chek Lap Kok test reclamation fill on marine clay was part of a larger reclamation project for the construction of a Hong Kong replacement airport (which started operation in 1997). The objective of the test fill was to investigate the feasibility of reclamation over soft marine clay and the effectiveness of vertical drains in accelerating consolidation. The main test area, located about 200 m offshore and 100 m square in plan, was divided into quadrants: one was a control area, with no treatment of the marine clay, the remaining three quadrants were installed with vertical drains at different spacing through about 7-m thick marine clay. The offshore geotechnical investigations, the test fill and the instrumentation programme were described in substantial detail by Foott et al. (1987) and Koutsoftas et al. (1987) and further studied deterministically in Choa et al. (1990).

Low (2003b) presented a deterministic numerical method written in the VBA programming environment of Microsoft Excel spreadsheet for consolidation analysis involving vertical drains. The programme uses Barron's solution for equal vertical strain of consolidation due to radial drainage and Carillo's equation for combined radial and vertical drainage. The programme accounts—in an approximate manner—for stage loading, load reduction due to fill submergence, delayed vertical drain installation, changes in length of vertical drainage path with time and variation of soil stress history with depth. A practical algorithm for prediction of rate of settlement was adopted because, even in the relatively simple approach adopted, 15 or more values of individual input parameters were required. The uncertainties associated with some of these parameters will limit the accuracy of prediction even if sophisticated models are used. The computational algorithm in the programme is not fully rigorous, because idealizations and approximations have been made. Nevertheless, limited comparisons made by the author suggest that the degree of accuracy achieved is adequate for the purpose in hand. Its relative simplicity also gave rise to some insights on parametric relationships and sensitivities. The results of deterministic analysis using the programme compared well with the instrumented settlement records of the soft clay beneath the Chek Lap Kok test fills.

Figure 4.4 shows the results of a deterministic analysis using the programme, for one of the four test fills of the Chek Lap Kok land reclamation project where the prefabricated vertical band drains were installed at a spacing of 1.5 m. There was staged loading due to the increase in fill thickness with time as shown. The clay was overconsolidated since the initial effective vertical stress profile was less than the profile of maximum past pressures (preconsolidation pressures). The discrepancies between the computed curve and the measured range are partly attributable to the underestimation of final consolidation settlement by the programme. The final consolidation settlements are functions of compression ratio C_R, recompression ratio C_{RR}, stress history and applied loadings but are not affected by the values of the rate parameters c_v and c_h, which are the coefficients of consolidation for vertical and horizontal flow, respectively, nor by the idealizations and approximations in the modelling of excess pore pressure dissipation in the programme. The programme-computed results will be even closer to the observed settlement curves if different values of c_h and compression ratios

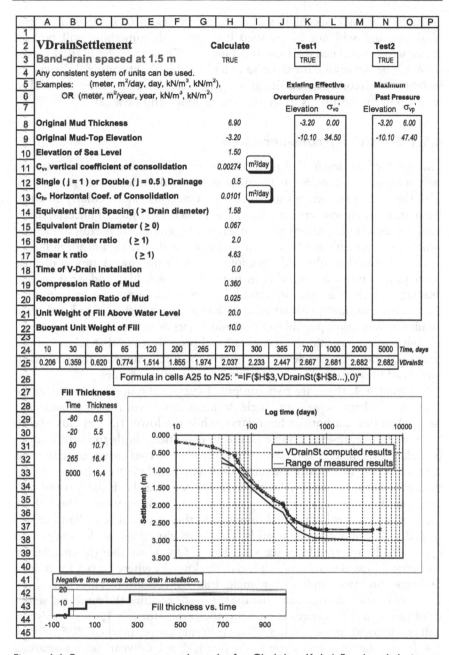

	A	B	C	D	E	F	G	H	I	J	K	L	M	N	O	P
1																
2	**VDrainSettlement**							Calculate			Test1			Test2		
3	Band-drain spaced at 1.5 m							TRUE			TRUE			TRUE		
4	Any consistent system of units can be used.															
5	Examples: (meter, m²/day, day, kN/m³, kN/m²),									Existing Effective			Maximum			
6	OR (meter, m²/year, year, kN/m³, kN/m²)									Overburden Pressure			Past Pressure			
7										Elevation σᵥₒ'			Elevation σᵥₚ'			
8	Original Mud Thickness							6.90		-3.20	0.00		-3.20	6.00		
9	Original Mud-Top Elevation							-3.20		-10.10	34.50		-10.10	47.40		
10	Elevation of Sea Level							1.50								
11	Cᵥ, vertical coefficient of consolidation							0.00274	m²/day							
12	Single (j = 1) or Double (j = 0.5) Drainage							0.5								
13	Cₕ, Horizontal Coef. of Consolidation							0.0101	m²/day							
14	Equivalent Drain Spacing (> Drain diameter)							1.58								
15	Equivalent Drain Diameter (≥ 0)							0.067								
16	Smear diameter ratio (≥ 1)							2.0								
17	Smear k ratio (≥ 1)							4.63								
18	Time of V-Drain Installation							0.0								
19	Compression Ratio of Mud							0.360								
20	Recompression Ratio of Mud							0.025								
21	Unit Weight of Fill Above Water Level							20.0								
22	Buoyant Unit Weight of Fill							10.0								
23																
24	10	30	60	65	120	200	265	270	300	365	700	1000	2000	5000	*Time, days*	
25	0.206	0.359	0.620	0.774	1.514	1.855	1.974	2.037	2.233	2.447	2.667	2.681	2.682	2.682	*VDrainSt*	
26				Formula in cells A25 to N25: "=IF($H$3,VDrainSt($H$8...),0)"												
27	**Fill Thickness**															
28	Time Thickness															
29	-80 0.5															
30	-20 5.5															
31	60 10.7															
32	265 16.4															
33	5000 16.4															
34																
35																
36																
37																
38																
39																
40																
41	*Negative time means before drain installation.*															
42																
43																
44																
45																

Figure 4.4 Programme-computed results for Chek Lap Kok 1.5 m band-drain test fill quadrant, and comparison with measured settlement range. The c_v and c_h values of 0.00274 and 0.0101 m²/day correspond to 1 m²/yr and 3.7 m²/yr, respectively.

are used, but such speculations belong to the realm of back-analysis and hindsight and will not be pursued here. Instead, something will be said about sensitivity. Then the Low and Tang (2007) spreadsheet FORM procedure will be illustrated for the case in hand to highlight some subtleties and insights of reliability-based design involving serviceability limit states. The material in this section is based on Low (2008b) and Low (2015).

4.5.1 Sensitivity considerations

The accuracy of design and prediction is often limited not so much by the lack of rigorousness in the analytical model as by the uncertainties associated with the input parameters used in the analysis. This is particularly so when the number of parameters is large, each with its own degree of uncertainty. In these circumstances, parametric and sensitivity studies are useful, to identify situations in which uncertainty in a parameter will have significant influence in the calculated results and situations in which the uncertainty is of little consequence. For instance, whether to use lab c_v or field c_v and what vertical drainage length to assume are certainly much more important questions for the control quadrant (without vertical drains) than for the 1.5 m band drain quadrant. Whenever possible, one should incorporate the uncertainty of the various parameters into the computational procedure, so that relationships/charts/equations can be developed to give the likely range of the settlement rate, in addition to the rate based on mean parametric values.

In Figure 4.4, there are two computed curves: the upper (slower) curve (with 'o' markers) represents single drainage condition (i.e., top boundary permeable, but not bottom boundary), while the lower (faster) curve (with '*' markers) represents double drainage condition (both top and bottom boundaries of the clay layer are permeable). These two curves of settlement versus time are practically the same. One may conclude that the rate of settlement is not sensitive to whether drainage occurs at both the top and bottom boundaries of the marine clay, or at the top boundary only. This reflects the predominant role of radial drainage in the quadrant with band drains spaced at 1.5 m in triangular grids. The sensitivity of the settlement rate to c_v can also be inferred from the same figure. One notes that the time factor for vertical flow depends on $c_v/(j \times \text{clay thickness})^2$, where $j = 0.5$ for double drainage condition, and 1.0 for single drainage condition. Therefore, if the values of c_v and j change such that the ratio c_v/j^2 remains the same, the settlement rate will not change. This means the results for ($j = 1$, $c_v = 4$ m²/year) will be identical to the ($j = 0.5$, $c_v = 1$ m²/year) curve (marked '*') in Figure 4.4; but the latter is very close to the ($j = 1$, $c_v = 1$ m²/year) points (marked 'o'). It follows that ($j = 1$, $c_v = 4$ m²/year) $\approx$ ($j = 1$, $c_v = 1$ m²/year). Hence, computed settlement curve is not sensitive to the uncertainty associated with the value of c_v in this quadrant, where the soft marine clay is about 7 m thick and the band drains are spaced at 1.5 m on triangular grids.

Similarly, if the values of c_h, equivalent drain spacing d_e, and drain diameter d_w all change but in such a way that the ratio $c_h/(d_e^2.F_s(n))$ remains constant, then the computed settlement versus time curves will be the same. These anticipations can be verified using the programme.

The computed settlement rate for the 3 m band-drain-spacing quadrant is by comparison more sensitive to variations in the ratio c_v/j^2. This is readily comprehensible because the $c_h/(d_e^2.F_s(n))$ value for this quadrant is only 22% of the $c_h/(d_e^2.F_s(n))$ value for the 1.5 m band-drain-spacing quadrant; the role of radial drainage is correspondingly less dominant in the 3-m band-drain-spacing quadrant. (Both quadrants have about the same thickness of marine clay.)

It follows that the sensitivities of the parameters which affect the rate of settlement (c_v, c_h, length of vertical drainage path H, equivalent drain spacing d_e and drain diameter d_w) can be studied via two terms: c_v/H^2, and $c_h/[d_e^2.F_s(n)]$, where $n = d_e/d_w$.

Assuming that the time for consolidation is the same for both vertical flow and horizontal flow (i.e., no time lag between the start of vertical flow and the start of radial flow), Low (2003b) obtained the chart shown in Figure 4.5 which displays the sensitivity of the average degree of consolidation for combined radial and vertical drainage (U_{vh}) to the λ values and hence to the $c_h H^2/c_v d_e^2 F(n)$ ratio:

$$\lambda = \frac{8}{F(n)} \frac{c_h / d_e^2}{c_v / H^2} \tag{4.6}$$

The chart is therefore convenient for assessing how deviations from the assumed values of the parameters will affect the calculated average degree of consolidation for simultaneous vertical and radial consolidation. Note that the uppermost curve, with $\lambda = 0$, is the case with no vertical drains.

Yet another alternative is to perform reliability analysis and reliability-based design accounting explicitly for the estimated uncertainties of the parameters. The classical method of computing reliability index using the FORM is intricate when correlation and nonnormal distributions are involved and when the performance function is complicated or implicit. These hitherto tedious problems can be solved, with relative ease and transparency, using spreadsheet-automated constrained optimization tool and the expanding equivalent ellipsoid perspective (Low and Tang 2004, 2007). By this perspective, the quadratic form defining the β index (Eq. 2.9) is visualized as a tilted multi-dimensional ellipsoid (centred at the mean μ or equivalent mean $μ^N$) in the original space of the random variables; there is no need to diagonalize the covariance or correlation matrix. The concepts of coordinate transformation and frame-of-reference rotation are not required. Iterative searching and partial derivatives are automatic using constrained

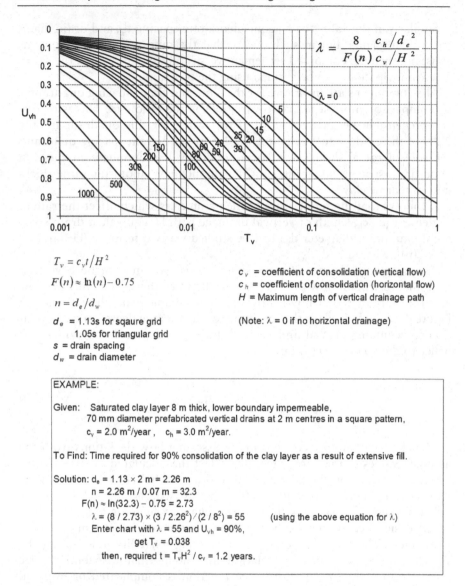

$$T_v = c_v t / H^2$$

$$F(n) \approx \ln(n) - 0.75$$

$$n = d_e / d_w$$

d_e = 1.13s for sqaure grid
        1.05s for triangular grid
s = drain spacing
d_w = drain diameter

c_v = coefficient of consolidation (vertical flow)
c_h = coefficient of consolidation (horizontal flow)
H = Maximum length of vertical drainage path

(Note: $\lambda = 0$ if no horizontal drainage)

EXAMPLE:

Given: Saturated clay layer 8 m thick, lower boundary impermeable,
        70 mm diameter prefabricated vertical drains at 2 m centres in a square pattern,
        c_v = 2.0 m^2/year , c_h = 3.0 m^2/year.

To Find: Time required for 90% consolidation of the clay layer as a result of extensive fill.

Solution: d_e = 1.13 × 2 m = 2.26 m
        n = 2.26 m / 0.07 m = 32.3
        F(n) ≈ ln(32.3) − 0.75 = 2.73
        λ = (8 / 2.73) × (3 / 2.26^2) / (2 / 8^2) = 55   (using the above equation for λ)
        Enter chart with λ = 55 and U_{vh} = 90%,
                get T_v = 0.038
        then, required t = $T_v H^2$ / c_v = 1.2 years.

Figure 4.5 Average degree of consolidation for combined radial and vertical drainage (U_{vh}) as a function of the lumped parameter λ.

optimization in the ubiquitous spreadsheet platform. The versatility of the spreadsheet constrained optimization approach is enhanced when used in combination with user-defined functions coded in the programming environment of the spreadsheet, for example, the Visual Basic (VBA) programming environment of the Microsoft Excel spreadsheet software. This means that the performance function can be implicit, iterative and based on numerical methods (Low et al. 2007).

4.5.2 Limit state surfaces and performance functions pertaining to magnitude and rate of soft clay settlement

The following three aspects are studied for the Hong Kong land reclamation test fills:

(i) The magnitude of the ultimate consolidation settlement s_{cf}.
(ii) The degree of consolidation U at time = 1 year.
(iii) The consolidation settlement remaining (s_r) at time = 1 year, where $s_r = s_{cf} - s_{1yr}$.

The performance functions are, respectively:

$$g_1(\mathbf{x}) = \text{Limiting } s_{cf} - s_{cf} \tag{4.7}$$

$$g_2(\mathbf{x}) = U - U_{required} \tag{4.8}$$

$$g_3(\mathbf{x}) = \text{Limiting } s_r - s_r \tag{4.9}$$

in which s_{cf}, U and s_r are functions of the various inputs $(\mathbf{x})$ shown in Figure 4.4, including parameters of compressibility, consolidation rate, staged loading and stress history.

The term 'limiting' connotes 'acceptable' or 'permissible.' Positive g(x) values correspond to safe domain, and negative g(x) values correspond to unsafe domain. Hence, for $g_1(\mathbf{x})$ and $g_3(\mathbf{x})$, by virtue of 'smaller settlement is safer,' safe domain is indicated when *Limiting s_{cf}* > s_{cf}, and *Limiting s_r* > s_r. In contrast, for $g_2(\mathbf{x})$, by virtue of 'larger degree of consolidation is safer,' safe domain is indicated when *U* > *Limiting U*. These are illustrated schematically in the plane in Figures 4.6 and 4.7. For all three performance functions, safe combinations of parametric values are separated from unsafe combinations by the LSS, defined by g($\mathbf{x}$) = 0.

4.5.3 Distinguishing positive and negative reliability indices

In Figure 4.6, when the limiting (i.e., permissible) ultimate consolidation settlement is s_{L1}, the settlement evaluated at mean-value point (s_μ) is already in the unsafe zone ($>s_{L1}$). Under this circumstance, the computed reliability index β must be regarded as negative: it is the minimum distance (in units of directional standard deviations) from the *unsafe mean-value point* to the *safe boundary* defined by limit state surface 1 (LSS1). On the other hand, if

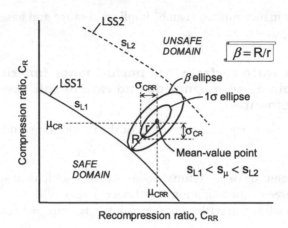

Figure 4.6 Illustration of reliability index in the plane. With respect to LSS1, reliability index β is negative; with respect to LSS2, β is positive.

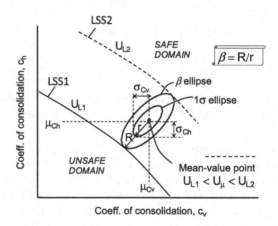

Figure 4.7 Illustration of reliability index in the plane. With respect to LSS1, reliability index β is positive; with respect to LSS2, β is negative.

a higher permissible settlement (s_{L2}) is specified, the mean-value point is in the safe zone, and the reliability index β is positive: it is the minimum distance (in units of directional standard deviations) from the *safe mean-value point* to the *unsafe boundary* defined by LSS2.

In contrast, in Figure 4.7, the reliability index β with respect to the limit state surface 1 (for which the limiting degree of consolidation is U_{L1}) is positive, while that with respect to limit state surface 2 (for which limiting $U = U_{L2}$) must be treated as negative, by virtue of $U_{L1} < U_\mu < U_{L2}$, where U_μ is the average value of the degree of consolidation evaluated using mean values (μ_{Cv}, μ_{Ch}) of the coefficients of consolidation c_v and c_h.

4.5.4 Reliability analysis for different limiting state surfaces

The Low and Tang (2007) procedure for FORM can deal with various correlated nonnormal distributions (lognormal, beta general, gamma, type 1 extreme, exponential, ...). For the case in hand, only correlated lognormals are illustrated. The values of compression ratio C_R, recompression ratio C_{RR} and coefficients of consolidation c_v and c_h in Figure 4.4 are taken to be the mean values in Figure 4.8. Assumed values of standard deviations are used,

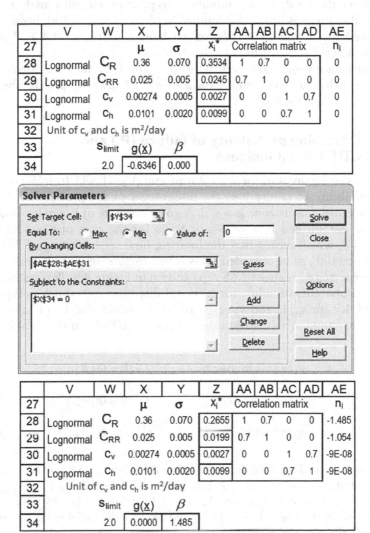

	V	W	X	Y	Z	AA	AB	AC	AD	AE
27			μ	σ	x_i^*	Correlation matrix				n_i
28	Lognormal	C_R	0.36	0.070	0.3534	1	0.7	0	0	0
29	Lognormal	C_{RR}	0.025	0.005	0.0245	0.7	1	0	0	0
30	Lognormal	c_v	0.00274	0.0005	0.0027	0	0	1	0.7	0
31	Lognormal	c_h	0.0101	0.0020	0.0099	0	0	0.7	1	0
32	Unit of c_v and c_h is m²/day									
33		s_{limit}	g(x)	β						
34		2.0	-0.6346	0.000						

Solver Parameters

Set Target Cell: `$Y$34` | Solve

Equal To: ⚬ Max ⦿ Min ⚬ Value of: `0` | Close
By Changing Cells:

`$AE$28:$AE$31` | Guess

Subject to the Constraints: | Options

`$X$34 = 0` | Add

Change

Reset All

Delete

Help

	V	W	X	Y	Z	AA	AB	AC	AD	AE
27			μ	σ	x_i^*	Correlation matrix				n_i
28	Lognormal	C_R	0.36	0.070	0.2655	1	0.7	0	0	-1.485
29	Lognormal	C_{RR}	0.025	0.005	0.0199	0.7	1	0	0	-1.054
30	Lognormal	c_v	0.00274	0.0005	0.0027	0	0	1	0.7	-9E-08
31	Lognormal	c_h	0.0101	0.0020	0.0099	0	0	0.7	1	-9E-08
32	Unit of c_v and c_h is m²/day									
33		s_{limit}	g(x)	β						
34		2.0	0.0000	1.485						

Figure 4.8 Initially the n_i values were zeros, and g(x) exhibits negative values. Hence, the computed β value must be treated as negative ($β = -1.485$).

for illustrative purpose. Positive correlations, logical between C_R and C_{RR} and between c_v and c_h, are modelled.

The deterministic setup of Figure 4.4 and the reliability analysis of Figure 4.8 are coupled easily by replacing the C_R, C_{RR}, c_v and c_h values (cells H19, H20, H11 and H13) of Figure 4.4 with the formulas '=Z28,' '=Z29,' '=Z30' and 'Z31' which refer to the x^* values in Figure 4.8. The performance function $g_1(x)$ is, by Eq. 4.7, '=W34–N25,' where cell W34 has value 2 m for this analysis. The computed β index is 1.485, treated as (–1.485) because the mean-value point is in the unsafe zone as indicated by the negative $g(x)$ value when the n_i values were initially zeros prior to reliability analysis. (The mean-value point is unsafe for limiting settlement $s_{limit} = 2$ m, because the computed mean final consolidation settlement in Figure 4.4 is 2.68 m, greater than the s_{limit} of 2 m in Figure 4.8.)

By varying the s_{limit} value (cell W34) between 1.2 and 4.8 at intervals of 0.2, and each time re-computing the β index, 19 values of β were obtained as shown in Figure 4.9.

4.5.5 Obtaining probability of failure (P_f) and CDF from β indices

Referring to Figure 4.6, for $s_{L1} = 2.0$ m and $\beta = -1.485$ from Figure 4.8, the probability of failure P_f is the integration of the probability density over the entire unsafe zone ($s > s_{L1}$). A good estimate of P_f can be obtained from the established $P_f \approx \Phi(-\beta)$, or 93% for the case of $\beta = -1.485$ in Figure 4.8, which means that the limiting final settlement s_{cf} of 2.0 m is almost certainly to be exceeded. This is not surprising given that the computed expected s_{cf} is about 2.68 m as shown in Figure 4.4. Besides estimating the probability of failure for serviceability limit states such as Eqs. 4.7, 4.8 and 4.9 for s_{cf}, U and s_r, respectively, it would also be of interest to obtain the cumulative distribution function (CDF) from the β indices, as follows:

$$g_1\left(x\right) = \text{Limiting } s_{cf} - s_{cf} : \quad CDF = \Phi\left(\beta\right) \tag{4.10}$$

$$g_2\left(x\right) = U - U_{required} : \quad CDF = \Phi\left(-\beta\right) \tag{4.11}$$

$$g_3\left(x\right) = \text{Limiting } s_r - s_r : \quad CDF = \Phi\left(\beta\right) \tag{4.12}$$

The reason for Eqs. 4.10 and 4.12 being different from Eq. 4.11 is readily appreciated if one notes that, for Eqs. 4.10 and 4.12, CDF = P[s < s_{limit}] = 1 – P[s > s_{limit}] = 1 – P_f. In contrast, for Eq. 4.11, CDF = P[U < U_{limit}] = P_f.

As shown in Figure 4.9, the CDF based on 19 values of β indices is practically indistinguishable from the CDF from 5000 realizations of Monte Carlo (MC) simulation using the software @RISK (http://www.palisade.com). For the assumed statistical inputs and correlation structure, the 90% confidence

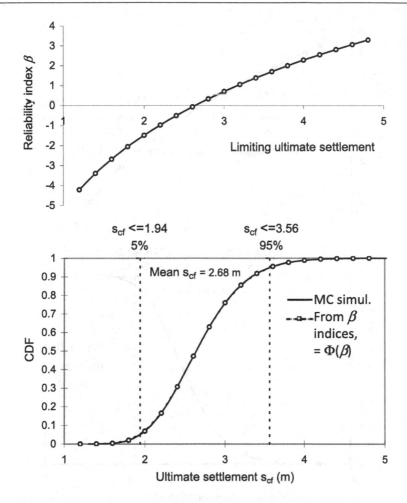

Figure 4.9 Reliability indices for different limiting ultimate settlements, and comparison of CDF based on β indices with CDF from Monte Carlo simulations.

interval of the ultimate consolidation settlement is (1.94 m, 3.56 m). The measured ultimate settlement shown in Figure 4.4 is within this interval. Other confidence intervals can also be read from Figure 4.9. Considerations such as this should be much more useful in the design stage than a deterministic analysis which yields a single ultimate settlement value with no indication at all of the effect of uncertainties (parametric, modelling and others) on the predicted ultimate settlement.

Figures 4.10 and 4.11 show the CDF curves obtained from reliability indices (Eqs. 4.11 and 4.12), for the degree of consolidation (U) and the consolidation settlement remaining (s_r), respectively, at time = 1 year. As in Figure 4.9, the two CDF curves in Figures 4.10 and 4.11, based on 17 and

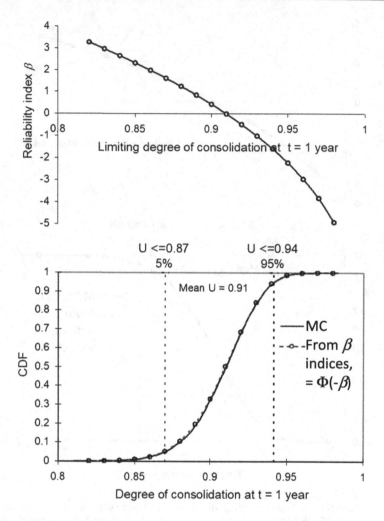

Figure 4.10 Reliability indices for different limiting U at **t** = I year, and comparison of CDF based on β indices with CDF from Monte Carlo simulations.

23 values of β indices, respectively, are practically the same as the CDF curves from 5000 realizations of Monte Carlo simulation.

It is of interest to note that the n_i values of c_v and c_h in cells AE30:AE31 of the lower Figure 4.8 are practically zeros. The implied insensitivities of the ultimate consolidation settlement to the coefficients of consolidation (c_v and c_h) are theoretically consistent (ultimate settlement not a function of rate parameters c_v and c_h) and are automatically revealed in a reliability analysis. In contrast, reliability analysis with respect to limiting degree of consolidation at t = 1 year and settlement remaining at t = 1 year will show

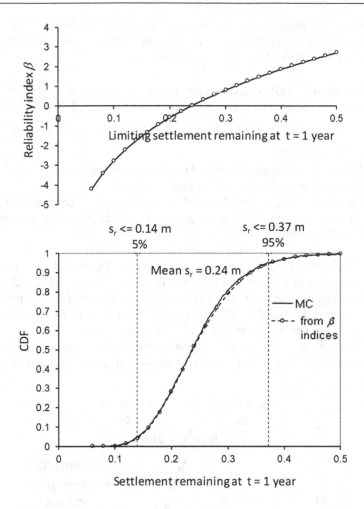

Figure 4.11 Reliability indices for different limiting settlement remaining at **t** = 1 year, and comparison of CDF based on *β* indices with CDF from Monte Carlo simulations.

non-zero values for the n_i values of the coefficients of consolidation c_v and c_h, both parameters having an effect on the rate of consolidation.

4.5.6 Obtaining PDF curves from β indices

The CDF curves shown in the lower plots of Figures 4.9, 4.10 and 4.11 were obtained easily—by Eqs. 4.10, 4.11 and 4.12—from the *β* values shown in the corresponding upper plots. In addition, it is simple to obtain the probability density function (PDF) of the respective outputs (s_{cf}, U and s_r), by applying cubic spline interpolation (e.g., Kreyszig 1988) to the CDF. This is accomplished easily in the Excel spreadsheet platform, as explained below with respect to the 19 CDF values of Figure 4.9:

(i) Autofill a 17-cell column vector $\mathbf{m}$ of m_i ($i = 2-18$) with the following formula, next to the column of the 19 CDF values of s_{cf}:
$m_i = \dfrac{3}{h}(CDF_{i+1} - CDF_{i-1})$, in which h is the s_{cf} interval of the CDF points (= 0.2 m).

(ii) A 17×17 tridiagonal matrix $\mathbf{D}$ is set up, with entries $d_{i,i} = 4$, $d_{i+1,i} = d_{i,i+1} = 1$ and all other entries equal to 0.

(iii) The 17 PDF values are obtained immediately and automatically upon entering '=mmult(minverse($\mathbf{D}$),$\mathbf{m}$)' as a spreadsheet array formula in a 17-cell column.

The PDF curve of s_{cf} thus obtained is shown in Figure 4.12. By the same procedure, 15 and 21 PDF values of the degree of consolidation (U) and of the settlement remaining (both at time = 1 year) were obtained easily from their respective 17 and 23 CDF values of Figures 4.10 and 4.11. These two PDF curves are shown in Figures 4.13 and 4.14.

The 5000 Monte Carlo (MC) realizations performed earlier for the plots of Figures 4.9, 4.10 and 4.11 can also be used to plot the outputs as PDF

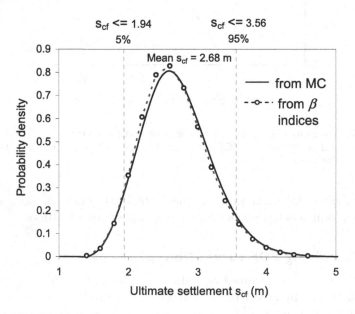

Figure 4.12 PDF of ultimate primary consolidation settlement s_{cf}, from Monte Carlo simulations with 5000 realizations, and from 19 values of reliability index.

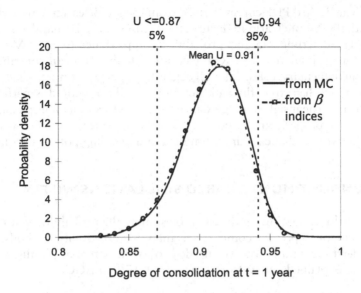

Figure 4.13 PDF of degree of consolidation (U) at time **t** = 1 year, from Monte Carlo simulations, and from 17 values of reliability index.

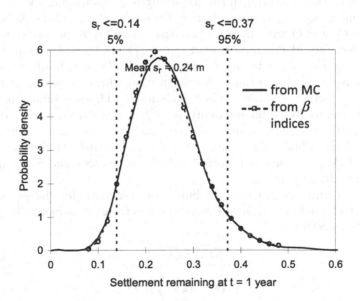

Figure 4.14 PDF of settlement remaining (s_r) at time **t** = 1 year, from Monte Carlo simulations, and from 23 values of reliability index.

curves. The dashed PDF curves derived from the β indices agree remarkably well with the Monte Carlo PDF curves in Figures 4.12, 4.13 and 4.14.

Different deterministic and probabilistic approaches (e.g., FEM, LEM, Monte Carlo simulations, FORM, etc.) are valuable and can contribute in their complementary roles. Like all scientific/engineering methods/models, each approach has strengths and limitations, and no method is perfect for all problems. The insights and guidance in the next section on implementing Monte Carlo properly are based on personal experience, and given with the sole objective of enhancing understanding and avoiding potential pitfalls.

4.6 CONDUCT MONTE CARLO SIMULATIONS WISELY

Monte Carlo simulation can be used to estimate the probabilities of failure of a system, or to provide comparative analysis to reliability methods. Care and understanding are however needed to avoid potential pitfalls, two of which are explained below, from Low (2010, 17th SEAGC).

4.6.1 Distortions caused by negative values of random numbers

Figure 4.15 shows a spring of rupture strength Q_u suspending a vertical load of magnitude Q, both in units of force. The mean values and standard deviations of Q_u and Q are shown for two cases, A and B, respectively, in which the uncertainty of the applied load Q is bigger in Case B than in Case A (hence larger standard deviations of Q in case B). The two random variables are assumed to be independent for illustration. One can use the first-order reliability method (FORM), which includes the Hasofer–Lind index as a special case, to obtain the reliability index β. Figure 4.15 uses the Low and Tang (2007) procedure of spreadsheet-automated constrained optimization approach that obtains the same design point and reliability index as the classical FORM procedure, but more intuitively and directly and on the ubiquitous spreadsheet platform.

The performance function (or limit state function) for the problem in Figure 4.15 can be expressed in two ways which are mathematically equivalent when $g(\mathbf{x}) = 0$:

$$g_1(\mathbf{x}) = Q_u - Q \tag{4.13a}$$

$$g_2(\mathbf{x}) = (Q_u / Q) - 1 \tag{4.13b}$$

Using FORM, one obtains identical value of reliability index β (2.236 for Case A and 1.562 for Case B) regardless of whether Eq. 4.13a or Eq. 4.13b is used. The corresponding probabilities of failure are 1.27% for Case A and 5.92% for Case B, respectively. Using Monte Carlo simulations with Latin

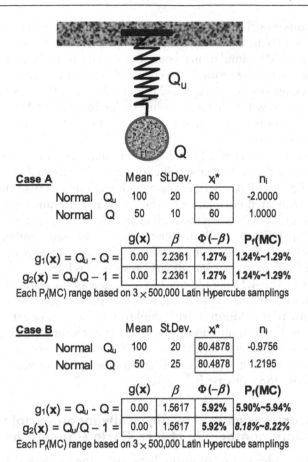

Case A		Mean	St.Dev.	x_i^*	n_i
Normal	Q_u	100	20	60	-2.0000
Normal	Q	50	10	60	1.0000

	g(x)	β	$\Phi(-\beta)$	P_f(MC)
$g_1(x) = Q_u - Q =$	0.00	2.2361	1.27%	1.24%~1.29%
$g_2(x) = Q_u/Q - 1 =$	0.00	2.2361	1.27%	1.24%~1.29%

Each P_f(MC) range based on 3 × 500,000 Latin Hypercube samplings

Case B		Mean	St.Dev.	x_i^*	n_i
Normal	Q_u	100	20	80.4878	-0.9756
Normal	Q	50	25	80.4878	1.2195

	g(x)	β	$\Phi(-\beta)$	P_f(MC)
$g_1(x) = Q_u - Q =$	0.00	1.5617	5.92%	5.90%~5.94%
$g_2(x) = Q_u/Q - 1 =$	0.00	1.5617	5.92%	8.18%~8.22%

Each P_f(MC) range based on 3 × 500,000 Latin Hypercube samplings

Figure 4.15 Monte Carlo simulations produce misleading results for Case B when performance function $g_2(x)$ of Eq. 4.13b is used.

Hypercube sampling (each 500,000 trials), the probabilities of failure (P_f) based on performance functions $g_1(x)$ of Eq. 4.13a and $g_2(x)$ of Eq. 4.13b are identical (in the range 1.24%~1.29%) and consistent with that based on FORM's $\Phi(-\beta)$ *only for Case A*. For case B, Monte Carlo simulations produce results (5.90%~5.94%) consistent with FORM's result of 5.92% only if the performance function $g_1(x)$ of Eq. 4.13a is used. When the performance function of $g_2(x)$ of Eq. 4.13b is used, Monte Carlo simulations yield a misleading P_f of 8.18%~8.22% versus the correct result of 5.92%, as summarized in the last column of Figure 4.15.

The wrong results of Monte Carlo for $g_2(x)$ of case B is due to the coefficient of variation of Q being 25/50 = 0.5, which means that about 2.275% of the 500,000 random sets in each simulation fall in the negative range of the lower tail (at greater than two standard deviations from the mean value of Q). These negative random numbers of Q do not distort $g_1(x)$ of Eq. 4.13a, which remains positive. However, negative random numbers of Q render $g_2(\mathbf{x})$ of Eq. 4.13b negative, since the Q_u in the numerator is virtually always positive during MC simulations, being five standard deviations away from the negative range. The result is: the correct 5.92% failure rate + a phantom 2.275% = erroneous 8.20%±. It helps to understand that a small positive load close to zero will yield large positive value for $g_2(\mathbf{x})$ of Eq. 4.13b, signifying safety; negative loads (which can happen in MC simulations) should logically be even safer.

The above inaccuracies in Monte Carlo simulations can be avoided easily by (i) expressing the performance functions in the form of $g_1(\mathbf{x})$ of Eq. 4.13a, or (ii) using probability distributions which exclude negative domain, e.g., lognormal, truncated normal, or the bounded four-parameter general beta distribution. Of course, when the coefficient of variation of the denominator in $g_2(\mathbf{x})$ of Eq. 4.13b is small such that negative random numbers are virtually impossible during MC simulations (as in Case A), performance functions in the form of Eq. 4.13b will yield results as reliable as Eq. 4.13a.

Case B in Figure 4.15 should not be construed as discouraging the use of normal distribution, merely that one needs to ponder the effect on the performance function if Monte Carlo simulation generates negative numbers. Unlike Monte Carlo simulations, the design point values of FORM (under the x_i^* columns in cases A and B) can be observed easily to check their legitimacy. There is nothing illegitimate about the design point values of 60 (for case A) and 80.488 (for case B).

4.6.2 Distortions caused by physically incompatible random numbers in Monte Carlo simulations

The admissible physical relationship between some random variables must not be violated if correct results are to be obtained from Monte Carlo simulations. Consider the case of a two-dimensional rock slope with a vertical tension crack of depth z, with height z_w of water in the tension crack. A correct treatment of the two random variables z and z_w is described first in the next paragraph, followed by incorrect treatment in the paragraph after.

The tension crack is of depth z, and filled with water to the depth z_w. From physical considerations, the minimum value for z_w/z is 0, when the tension crack is dry. The maximum value is 1, when the tension crack is completely filled with water. In the FORM reliability analysis and Monte Carlo simulations of Low (2007), both z and z_w will change from their mean values, but both must be restricted to the domain $0 \le z_w/z \le 1.0$. Normal distribution was assumed for z, and truncated exponential (trimmed_exp) probability distribution was assumed for z_w/z. For the case without reinforcement (T = 0), the computed reliability index β was 1.556, corresponding to a failure probability of about 6.0%, obtained from $P_f = \Phi(-\beta)$. This compares reasonably well with the failure rates of 6.8%, 6.2%, 6.5%, 6.5%, 6.6%, 6.4% from six Monte Carlo simulations (each with a sample size of 5000 based on Latin Hypercube sampling using the commercial software @RISK (http://www.palisade.com). The difference between the reliability-index-inferred probability of failure of 6% and the average of 6.5% from Monte Carlo simulations is due to the approximate nature of the equivalent normal transformation in FORM when nonnormals are involved, and possible nonlinearity in the limit state function $g(\mathbf{x}) = [F_s(\phi, c, z, z_w/z, \alpha) - 1]$, which render the equation $P_f = \Phi(-\beta)$ approximate. (The equation is exact when the random variables are normally distributed and the LSS is planar, as in the cases of Figure 4.15.)

Instead of modelling z and z_w/z as two random variables (with the mean value of the latter equal to 0.5), one could conceivably treat z and z_w as two random variables, with the mean value of z equal to 14 m, and that of z_w equal to 7 m. However, doing so leads to paradoxical values of probability of failure from Monte Carlo simulations which are several times greater than the average of 6.5% (also from Monte Carlo simulations) mentioned in the preceding paragraph when z and z_w/z were modelled. Examination of the random numbers of z and z_w generated in MC simulations (when z and z_w were modelled separately) revealed that there were many failure cases (performance function $g(\mathbf{x}) < 0$) involving generated z_w values greater than z values, which are physically inadmissible, but happened in the random numbers generated during MC simulations.

One can think of two other random variables whose physical dependency on each other should not be violated in Monte Carlo simulations if correct results are to be obtained: the friction angle ϕ' of retained backfill and the interface friction angle δ between the retained fill and the retaining wall. Physical considerations require $\delta \leq \phi'$. If δ values greater than ϕ' values are generated in MC simulations together with positive performance function values, the probability of failure from MC simulation could be underestimated.

Part II

Reliability-based design applied to soil engineering

Chapter 5

Spread foundations

5.1 INTRODUCTION

The design approach based on overall factor of safety (or lumped factor of safety) has long been used by geotechnical engineers. More recent alternatives are the characteristic values and partial factors used in the limit state design approach in Eurocode 7 (EC7), and the load and resistance factor design (LRFD) approach in North America. Yet another approach can play a useful complementary role to EC7 and LRFD, namely reliability-based design (RBD) based on a target reliability index β that explicitly reflects the uncertainty of the parameters and their correlation structure. Among the various versions of reliability indices, that based on the first-order reliability method (FORM) for correlated nonnormals is more consistent and has merits not found in other probabilistic methods. A special case of FORM is the earlier Hasofer–Lind index (1974) for correlated normal random variables. These reliability methods are described in Ditlevsen (1981), Ang and Tang (1984), Madsen et al. (1986), Haldar and Mahadevan (2000), Melchers (1999), Baecher and Christian (2003), for example. In addition, Low and Tang (2004, 2007) presented spreadsheet-automated practical procedures for FORM reliability-based analysis and design. Geotechnical examples based on the FORM procedures of Low and Tang (2004, 2007) are presented succinctly in Low (2005, 2008a, 2015, 2017, 2020a) and Low et al. (2017, Chapter 4 of the TC205/TC304 Joint Report).

This chapter illustrates RBD and FORM analysis in the context of spread foundations, which are also referred to as footing foundations, spread footings or shallow foundations. The analyses and discussions here are deeper than previously published by the author and also contain novel investigations and cases not found in the author's earlier writings. Focus is on information and insights from RBD-via-FORM which can enhance partial factor design methods like EC7 and LRFD. Ultimate limit states (ULS) will be considered first, followed by serviceability limit states (SLS).

An intuitive perspective is given next for the Hasofer–Lind reliability index β for correlated normal random variables, and its use in the RBD of a simple spread foundation. This will be followed by (i) reliability analysis

DOI: 10.1201/9781003112297-7

for estimating failure probability of a strip footing originally designed according to EC7, and RBD of the footing width for a target reliability level; (ii) a deterministic example of retaining wall foundation design from Tomlinson (2001) involving eccentric and inclined loading, extended into RBD in this chapter to demonstrate the merits of RBD in reflecting context-dependent sensitivities and resolving automatically the stabilizing–destabilizing duality of the vertical load when horizontal load and overturning moment are also acting; (iii) probabilistic extension of the Burland and Burbidge settlement estimation method applied to the foundation of a water storage tank, including statistical analysis of SPT test data extracted from Terzaghi et al. (1996).

RBD-via-FORM for correlated normal and nonnormal random variables is illustrated, with increasing complexity and more dimensions. The chapter will end with a lucid illustration of the difference between negative and positive reliability indices, and a simple way to distinguish them when doing RBD.

5.2 HASOFER–LIND INDEX APPLIED TO THE RBD OF A SPREAD FOUNDATION WITH TWO RANDOM VARIABLES

The 1974 Hasofer–Lind reliability index β for correlated *normal* random variables can be regarded as a special case of the subsequent FORM for correlated *non-normal* (non-Gaussian) random variables. The RBD of a spread foundation with only two normally distributed random variables is presented in this section to facilitate graphical visualization and an intuitive perspective of the Hasofer–Lind index. The same perspective for higher dimensions in the later sections can then be readily imagined in the mind's eye.

The example concerns the bearing capacity of a strip footing sustaining a vertical load Q_v, Figure 5.1. With respect to bearing capacity failure, the performance function (*PerFn*) for a strip footing is:

$$PerFn = g(\mathbf{x}) = q_u - q \qquad (5.1a)$$

where

$$q_u = c'N_c + p_o N_q + \frac{1}{2}\gamma BN_\gamma \qquad (5.1b)$$

in which $\mathbf{x}$ denotes the set of random variables affecting performance, q_u is the ultimate bearing capacity, q the applied bearing pressure (= Q_v/B), c' the effective cohesion of soil below foundation level, p_o the effective overburden pressure at foundation level, B the foundation width, γ the unit weight of

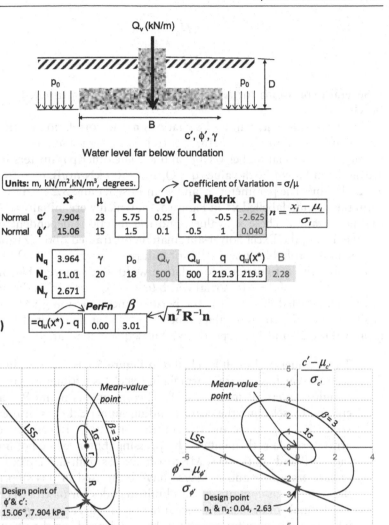

Figure 5.1 (a) Reliability-based design of footing width B for a target β = 3.0, high c' and low ϕ', (b) design point and β ellipse in original space and (c) normalized space.

soil below the base of foundation and N_c, N_q and N_γ are bearing capacity factors, which are functions of the angle of shearing resistance (ϕ') of soil:

$$N_q = e^{\pi \tan\phi'} \tan^2\left(45 + \frac{\phi'}{2}\right) \tag{5.2a}$$

$$N_c = (N_q - 1)\cot \phi' \tag{5.2b}$$

$$N_\gamma = 2(N_q + 1)\tan \phi' \tag{5.2c}$$

Several expressions for N_γ exist. The above N_γ is attributed to Vesic in Bowles (1996).

The mean values $\boldsymbol{\mu}$, standard deviations $\boldsymbol{\sigma}$ and correlation matrix $\mathbf{R}$ of c' and ϕ' are shown in Figure 5.1a. A correlation coefficient ρ of -0.5 models the negative correlation between c' and ϕ'. The other parameters in Eq. 5.1 are assumed known with values q = Q_v/B = (500 kN/m)/B, p_o = 18 kPa and γ = 20 kN/m³. The parameters c' and ϕ' in Eqs. 5.1 and 5.2 read their values from the column labelled x* in Figure 5.1a, which were initially set equal to the mean values. These x* values, and the functions dependent on them, change during the Excel Solver-automated constrained optimization search for the most probable point (MPP) of failure.

The objective is to design a foundation width B for a target Hasofer–Lind reliability index of β = 3.0. A trial width (e.g., 2.5 m or 3 m) is the first input in the cell labelled B, such that the performance function g(x*) = q_u – q is positive when the column values labelled x* are at their initial values (i.e., mean values, 23 and 15, respectively). Subsequent steps are:

1. The formula in the cell labelled β in Figure 5.1 is Eq. 5.3b in the next section, which is an array formula in Excel: '=sqrt(mmult(transpose(n), mmult(minverse(R matrix), n))).' The arguments n and R matrix are entered by selecting the corresponding numerical cells of the column vector n = $(x_i - \mu_i)/\sigma_i$ and the correlation matrix R, respectively. This array formula is then entered by pressing 'Enter' while holding down the 'Ctrl' and 'Shift' keys. Microsoft Excel's built-in matrix functions *mmult*, *transpose* and *minverse* have been used in this step. Each of these functions contains Excel's built-in codes for matrix operations.

2. The formula of the performance function in cell 'PerFn' is g(x*) = q_u – q, where the equation for q_u is Eq. 5.1b and depends on the x* values.

3. Microsoft Excel's built-in constrained optimization programme *Solver* is invoked, *To Minimize* the cell β, *By Changing* the x* values, *Subject To* cell 'PerFn' = 0. (The solving method selected is *GRG nonlinear*. The 'Use automatic scaling' option of Solver is also ticked, but not the 'Make unconstrained variables non-negative' option. If used for the first time, Solver needs to be activated once via File\Options\Add-ins\ Excel Add-ins\Solver Add-in.)

4. Repeat step 3 with a different trial value of width B until the target reliability index of β = 3.0 is achieved.

After a few trial values of width B, it is found that a foundation width B = 2.28 m achieves the target β value of 3.0. For comparison, Figure 5.2 shows a different case, with higher mean angle of friction ϕ' and lower c'

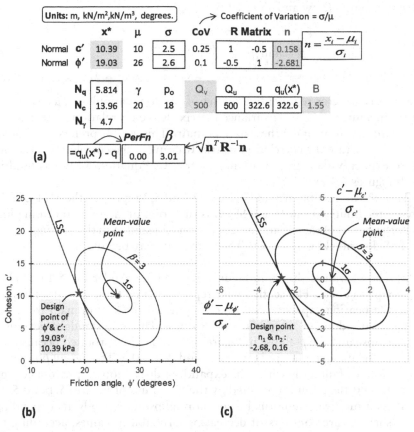

Figure 5.2 (a) Reliability-based design of footing width B for a target $\beta = 3.0$, low c' and high ϕ', (b) design point and β ellipse in original space and (c) normalized space.

than the case in Figure 5.1, which requires a width B of 1.55 m to achieve the target β value of 3.0. The RBD outcomes and ellipsoidal plots of these two cases will be further discussed shortly. (Excel files of RBD examples of this book can be downloaded freely at the book's website.)

The meaning of Hasofer–Lind index is explained next, together with the insights and information afforded by RBD-via-FORM.

5.3 HASOFER–LIND INDEX REINTERPRETED VIA EXPANDING ELLIPSOID PERSPECTIVE

The matrix formulation of the Hasofer–Lind (1974) reliability index β is:

$$\beta = \min_{\mathbf{x} \in F} \sqrt{(\mathbf{x} - \boldsymbol{\mu})^T \mathbf{C}^{-1} (\mathbf{x} - \boldsymbol{\mu})} \qquad (5.3a)$$

or, equivalently (Low and Tang 1997):

$$\beta = \min_{x \in F} \sqrt{\left[\frac{x_i - \mu_i}{\sigma_i}\right]^T \mathbf{R}^{-1} \left[\frac{x_i - \mu_i}{\sigma_i}\right]} \qquad (5.3b)$$

where $\mathbf{x}$ is a vector representing the set of random variables x_i, $\boldsymbol{\mu}$ the vector of mean values μ_i, $\mathbf{C}$ the covariance matrix, $\mathbf{R}$ the correlation matrix, σ_i the standard deviation and F the failure domain. This author prefers Eq. 5.3b to Eq. 5.3a because the correlation matrix $\mathbf{R}$ conveys the correlation structure more explicitly than the covariance matrix $\mathbf{C}$. Equation 5.3b was used in both Figures 5.1 and 5.2.

The 'x*' values obtained in Figures 5.1 and 5.2 represent the MPP of failure on the limit state surface (LSS). It is the point of tangency (bottom plots in both figures) of the expanding dispersion ellipsoid with the bearing capacity LSS described by $q_u - q = 0$. The following may be noted:

(a) The x* values shown in Figures 5.1 and 5.2 render Eq. 5.1a (*PerFn*) equal to zero. Hence, the point represented by these x* values lies on the bearing capacity LSS, which separates the safe domain from the unsafe domain of parametric values. The one-standard-deviation ellipse and the β-ellipse in Figures 5.1 and 5.2 are rotated because the correlation coefficient between c' and ϕ' is -0.5. The *design point* (the MPP of failure) is where the expanding dispersion ellipse touches the LSS, at the point represented by the x* values of Figures 5.1 and 5.2.

(b) As a multivariate normal dispersion ellipsoid expands, its expanding surfaces are contours of decreasing probability values, according to the established probability density function of the multivariate normal distribution:

$$f(\underline{x}) = \frac{1}{(2\pi)^{\frac{n}{2}} |\mathbf{C}|^{0.5}} \exp\left[-\frac{1}{2}(\mathbf{x} - \boldsymbol{\mu})^T \mathbf{C}^{-1} (\mathbf{x} - \boldsymbol{\mu})\right] \qquad (5.4a)$$

$$= \frac{1}{(2\pi)^{\frac{n}{2}} |\mathbf{C}|^{0.5}} \exp\left[-\frac{1}{2}\beta^2\right] \qquad (5.4b)$$

where β is defined by Eq. 5.3a or Eq. 5.3b, without the 'min.' Hence, to minimize β (or β^2 in the above multivariate normal distribution) is to maximize the value of the multivariate normal probability density function, and to find the smallest ellipsoid tangent to the LSS is equivalent to finding the MPP of failure. If this MPP of failure is based on a target reliability index β, one may also call it the *design point*. This intuitive and visual understanding of the *design point* is consistent with the more mathematical approach in Shinozuka (1983), in which correlated random variables were transformed into their rotated u-space together with the limit state equation. The Low and Tang (2004, 2007) n-space FORM approach does not rotate the frame of reference.

(c) The design point, being the first point of contact between the expanding ellipsoid and the LSS in Figures 5.1 and 5.2, is the MPP of failure with respect to the safe mean-value point at the centre of the expanding ellipsoid. The reliability index β is the axis ratio (R/r) of the ellipse that touches the LSS and the 1σ dispersion ellipse/ellipsoid. By geometry, this co-directional axis ratio is the same along any 'radial' direction. Two perspectives on the LSS, the β ellipse and the design point are plotted in Figures 5.1 and 5.2, one (bottom left) based on the original random variables ϕ' and c', with design points (the x^* values) located by Excel Solver: '15.06°, 7.904 kPa' and '19.03°, 10.39 kPa,' in Figures 5.1b and 5.2b, respectively, the other (bottom right) based on normalized random variables $n_1 = (\phi'^* - \mu_{\phi'})/\sigma_{\phi'}$ and $n_2 = (c'^* - \mu_{c'})/\sigma_{c'}$ with *dimensionless* design points (n_1, n_2) equal to '0.04, –2.63' and '–2.68, 0.16,' respectively. A negative n value of –2.63 means that the value of the parameter at the MPP of failure (the design point) is 2.63 times its standard deviation *below* its mean value, while a positive n value of 0.04 means that its design value is 0.04 times its standard deviation *above* its mean value.

(d) While the design point in terms of original random variables—ϕ' and c' for the case in hand, Figures 5.1b and 5.2b—facilitates direct comprehension, the design point in terms of normalized dimensionless n_1, n_2, ... (Figure 5.1c and 5.2c), portrays the sensitivity of LSS and the aspect ratios of rotated dispersion ellipses/ellipsoids more objectively. For example, Figure 5.1c shows a case with the LSS more sensitive to the normalized c' than normalized ϕ', while Figure 5.2c shows a case with the LSS more sensitive to the normalized ϕ' than normalized c'. Also, the normalized dispersion ellipses are always bounded vertically and horizontally from β to $-\beta$, regardless of the value of the correlation coefficient (which governs the rotation of the ellipses).

(e) Context-dependent sensitivities are revealed automatically in RBD-via-FORM in a way EC7 cannot. For example, based on the values of the dimensionless sensitivity indicators n_1 and n_2, c' is more sensitive than ϕ' in Figure 5.1, but vice versa in Figure 5.2, by virtue of higher $\mu_{c'}$ and lower $\mu_{\phi'}$ in Figure 5.1 than Figure 5.2. The design value of the less sensitive parameter (ϕ' in Figure 5.1, and c' in Figure 5.2) can be slightly larger than its mean value due to its negative correlation ($\rho = -0.5$) with the other more-sensitive resistance parameter. Note that the sensitivities of the dimensionless indicators n_i are with respect to not only the LSS (defined by $g(\mathbf{x}) = q_u - q = 0$), but also the size and orientation of the equivalent dispersion ellipse/ellipsoid, as defined by the statistics of the random variables in the quadratic form in the equation of β (Eq. 5.3). This can be readily appreciated from the fact the MPP of failure (design point) depends on the interplay of the expanding dispersion ellipsoid and the LSS. Context-dependent sensitivities will be further illustrated in the case of a retaining wall foundation subjected to both inclined load and overturning moment later.

(f) Monte Carlo simulation software may provide probability of failure and sensitivity analysis based on regression or correlation techniques. Such sensitivities are with respect to LSS only, which are not the same as the sensitivities implied by the design point of RBD-via-FORM, which reflects the interplay of both the LSS and the expanding dispersion ellipsoid.

(g) For each parameter, the ratio of the mean value to the x^* value is similar in nature to the partial factors in limit state design (e.g., EC7). However, in a RBD, one does not specify the partial factors. The design point values (x^*) are determined automatically and reflect sensitivities, standard deviations, correlation structure and probability distributions in a way that prescribed partial factors and judgemental characteristic values cannot.

(h) In Figures 5.1 and 5.2, the respective mean value points, at '15°, 23 kPa' and '26°, 10 kPa,' are safe against bearing capacity failure; but bearing capacity failure occurs when the ϕ' and c' values are decreased to the values shown: '15.06°, 7.904 kPa,' and '19.03°, 10.39 kPa,' or, in terms of the dimensionless n_1 and n_2 values, '0.04, –2.63' and '–2.68, 0.16.' The distance from the safe mean-value point to this most probable failure combination of parametric values, in units of directional standard deviations, is the reliability index β, equal to 3.0 in these cases.

(i) The probability of failure (P_f) can be estimated from the reliability index β. Microsoft Excel's built-in function NormSDist(.) can be used to compute $\Phi(.)$ and hence P_f. Thus, for the bearing capacity problem of Figures 5.1 and 5.2, P_f = NormSDist(–3.0) = 0.13%. This value compares remarkably well with the range of P_f values 0.129% ~ 0.140% (for the case in Figure 5.1) and 0.130% ~ 0.136% (for the case in Figure 5.2), obtained from three Monte Carlo simulations each with 500,000 trials using the commercial Monte Carlo simulation software @RISK (http://www.palisade.com). The correlation matrix was accounted for in the simulation. The excellent agreement between 0.13% from reliability index and that from Monte Carlo simulation is hardly surprising given the almost linear LSS and normal variates shown in Figures 5.1 and 5.2. However, for the retaining wall foundation shown later, where five random variables are involved and non-normal distributions are used, the five-dimensional equivalent hyperellispoid and the limit state hypersurface can only be perceived in the mind's eye. Nevertheless, the probabilities of failure inferred from reliability indices are again in close agreement with Monte Carlo simulations. Computing the reliability index and $P_f = \Phi(-\beta)$ by the present approach takes only a few seconds. In contrast, the time needed to obtain the probability of failure by Monte Carlo simulation is several orders of magnitude longer, particularly when the probability of failure is small.

(j) The sensitivity indicator values for n_1 and n_2, '0.04, −2.63' in Figure 5.1 and '−2.68, 0.16' in Figures 5.1 and 5.2, suggest that the design point outcome is much more sensitive to c' than ϕ' in Figure 5.1, and vice versa in Figure 5.2, due to different relative magnitudes of c' and ϕ' in the two cases. These context-dependent sensitivities of the random variables are automatically reflected in the design point of RBD-via-FORM but will be difficult to deal with in EC7 and LRFD. This will be clear in the retaining wall foundation of Sections 5.9 and 5.10.

The phenomenon of correlated sensitivities is demonstrated in Low (2020b), and presented in Chapter 2.

5.4 EFFECT OF PARAMETRIC CORRELATIONS ON FAILURE PROBABILITY

A negative correlation coefficient of value $\rho = -0.5$ between c' and ϕ' was modelled in the previous section, in line with well-established experimental findings in the literature. One may understand this phenomenon from the effective stress Mohr-Coulomb failure envelope which is concave downwards. The linearized failure criterion defined by c' and ϕ' tends to be less steep with increasing horizontal coordinate values. This means that higher c' intercepts on the y axis tend to occur with smaller inclination angle ϕ' values of the linearized failure envelope, and vice versa, hence negatively correlated.

Figure 5.3 shows that ignoring the negative correlation between resistance parameters c' and ϕ' will result in smaller computed reliability indices: $\beta = 2.34$ and 2.37 for the foundation width determined in Figures 5.1 and 5.2, respectively, compared with $\beta = 3.0$ in the two figures when negative correlation between the resistance parameters is modelled. This could have been anticipated from the shape and orientation of the expanding canonical dispersion ellipses (which reduces to circles in the dimensionless n_1 and n_2 space when $\rho = 0$) in Figure 5.3 compared with the elongated ellipses of Figures 5.1c and 5.2c which are rotated such that the major axes of the ellipses tend to be in a similar direction as the LSS. Similar graphical visualization in the mind's eye is instructive for other scenarios.

Figure 5.3 also demonstrates the MPP of failure as the outcome of the interplay between dispersion ellipse/ellipsoid and LSS: despite the LSS of Figure 5.3a and 5.3b being the same as Figures 5.1c and 5.2c, respectively,

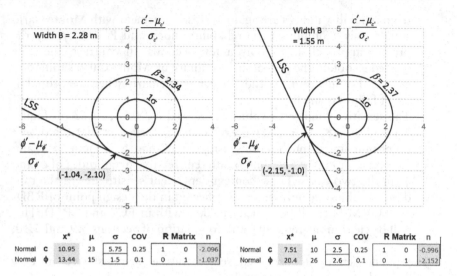

Figure 5.3 Smaller reliability index for the cases in Figures 5.1 and 5.2 if negative correlation between resistance parameters c' and ϕ' is not modelled.

the MPP of failure (in n_1 and n_2) are different from those in Figure 5.1c and 5.2c, being '–1.04, –2.10' and '–2.15, –1.00,' respectively, compared with '0.04, –2.63' and '–2.68, 0.16' in the earlier figures. One may also note that when uncorrelated, the two resistance parameters will have their MPP of failure values lower than their respective mean values.

Figure 5.4 shows the rotation of the 1σ dispersion ellipse for different values of correlation coefficient, in the normalized dimensionless n_1 and n_2 space.

Table 5.1 summarizes the underestimation and overestimation of reliability index—which means overestimation and underestimation of failure probability—for different combinations of resistance and unfavourable load when positive or negative parametric correlations are justified (based on physical considerations) but not modelled. Other examples in this book will illustrate cases where parametric correlations need not be considered, and cases where parametric correlations are warranted.

5.5 FIRST-ORDER RELIABILITY METHOD (FORM)

The examples in the previous sections deal with random variables obeying the normal (Gaussian) distribution, for which the Hasofer–Lind reliability method (Eq. 5.3) for *correlated normals* was used. There are situations where the uncertainty of a random variable is more appropriately modelled by a nonnormal distribution, for example, lognormal distribution, exponential distribution, beta distribution and so on. The established first-order

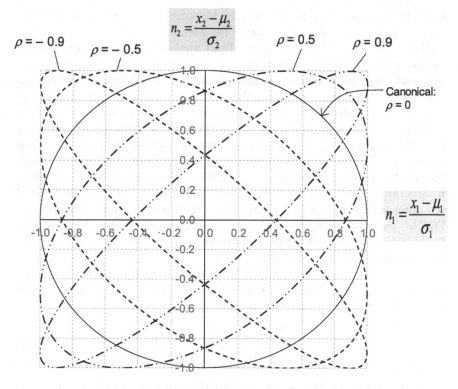

$$n_2 = \frac{x_2 - \mu_2}{\sigma_2}$$

$\rho = -0.9$ $\rho = -0.5$ $\rho = 0.5$ $\rho = 0.9$

Canonical:
$\rho = 0$

$$n_1 = \frac{x_1 - \mu_1}{\sigma_1}$$

Figure 5.4 The 1σ dispersion ellipse rotates with correlation coefficient ρ: anti-clockwise for negative ρ, and clockwise for positive ρ.

Table 5.1 Overestimation and underestimation of reliability index if parametric correlations are warranted but not modelled

	Resistance parameter (or favourable load)	Unfavourable load parameter
Resistance parameter (or Favourable load)	• Overestimate β if ignore positive correlation. • Underestimate β if ignore negative correlation (e.g., Case in Figure 5.3).	• Underestimate β if ignore positive correlation (e.g., Case in Figure 2.9). • Overestimate β if ignore negative correlation.
Unfavourable load parameter	(same as upper right)	• Overestimate β if ignore positive correlation. • Underestimate β if ignore negative correlation.

reliability method (FORM) extends the Hasofer–Lind method to deal with *correlated non-normal* random variables. Hence, FORM is a more general method which includes the Hasofer–Lind method as a special case. When parametric correlations exist, the classical FORM procedure (the u-space approach) requires rotating the frame of reference prior to normalization, as described in the references cited at the *Introduction* of this chapter. This section will briefly describe the similarities and differences between the Low and Tang (2004, 2007) n-space FORM procedure with the classical u-space FORM procedure. This will be followed by sections on (i) a design example of footing width B based on Eurcode 7 (EC7), (ii) FORM analysis to assess the reliability and failure probability of the EC7 design, (iii) RBD of width B for a target β, to compare with the width B from EC7 design, (iv) RBD of retaining wall foundation revealing the subtle load-resistance duality, (v) FORM analysis of SLS accounting for uncertainty in the Burland and Burbidge method of estimating settlement and (vi) distinguishing positive from negative reliability indices.

In FORM, one can rewrite Eq. 5.3b as follows and regard the computation of β involving nonnormal random variables as that of finding the smallest *equivalent* ellipsoid that is tangent to the LSS:

$$\beta = \min_{x \in F} \sqrt{\left[\frac{x_i - \mu_i^N}{\sigma_i^N} \right]^T \mathbf{R}^{-1} \left[\frac{x_i - \mu_i^N}{\sigma_i^N} \right]} \tag{5.5}$$

where μ_i^N and σ_i^N are the *equivalent normal mean* and *equivalent normal SD* of the nonnormal random variables. These μ_i^N and σ_i^N can be calculated by the Rackwitz and Fiessler (1978) transformation. Hence, for correlated non-normals, the ellipsoid perspective of Figures 5.1 and 5.2 still applies in the original coordinate system or the normalized but unrotated n space, except that the nonnormal distributions are replaced by an *equivalent normal ellipsoid*, centred not at the original mean values of the nonnormal distributions, but at the equivalent normal mean μ^N.

Equation 5.5 and the Rackwitz–Fiessler equations for μ_i^N and σ_i^N were used in the spreadsheet-automated constrained optimization FORM computational approach in Low and Tang (2004). An alternative to the 2004 FORM procedure is given in Low and Tang (2007), which uses the following equation for the reliability index β:

$$\beta = \min_{x \in F} \sqrt{\mathbf{n}^T \mathbf{R}^{-1} \mathbf{n}} \tag{5.6a}$$

where

$$n = \frac{x_i - \mu_i^N}{\sigma_i^N}$$

$\left(\text{The 2007 procedure obviates the computation of } \mu_i^N \text{ and } \sigma_i^N \right)$

$$\tag{5.6b}$$

Low and Tang 2004 FORM procedure: minimize β by directly varying **x**

$$\beta = \min_{x \in F} \sqrt{\left[\frac{x_i - \mu_i^N}{\sigma_i^N}\right]^T \mathbf{R}^{-1} \left[\frac{x_i - \mu_i^N}{\sigma_i^N}\right]}$$

$$\sigma^N = \frac{\phi\{\Phi^{-1}[F(x)]\}}{f(x)}$$

$$\mu^N = x - \sigma^N \times \Phi^{-1}[F(x)]$$

Use Excel's *Solver* to change the **x** vector.
Subject to g(**x**) = 0

Low and Tang 2007 FORM procedure: minimize β by varying **n**, on which **x** depends

$$\beta = \min_{x \in F} \sqrt{\mathbf{n}^T \mathbf{R}^{-1} \mathbf{n}}$$

Use Excel's *Solver* to change the **n** vector.

Subject to g(**x**) = 0

For each trial **n**, get $x_i = F^{-1}[\Phi(n_i)]$

Third spreadsheet-based FORM procedure:
minimize β by varying **u**, from which **n** and **x** are readily obtainable

$$\beta = \min_{x \in F} \sqrt{\mathbf{u}^T \mathbf{u}}$$ Use Excel's *Solver* to change the **u** vector, subject to g(**x**) = 0

For each automated trial **u**, get $\mathbf{n} = \mathbf{Lu}$, and $x_i = F^{-1}[\Phi(n_i)]$

Figure 5.5 Comparison of the two FORM computational approaches of Low and Tang (2004, 2007), and the additional **u-to-n-to-x** approach. All three procedures use the optimization routine Solver resident in the Microsoft Excel spreadsheet.

The computational approaches of Eqs. 5.5 and 5.6 and associated equivalent ellipsoidal perspective yield identical results as the classical rotated u-space computational approach, and, the n-space ellipsoidal perspective being more intuitive, may help reduce the conceptual and language barriers of FORM.

The two Low and Tang (2004, 2007) spreadsheet-based computational approaches of FORM are compared in Figure 5.5. Either method can be used as an alternative to the classical u-space FORM procedure. A third alternative (illustrated in Lü and Low 2011) is also shown in Figure 5.5, for which the Microsoft Excel's built-in constrained optimization routine (*Solver*) is invoked to automatically vary the u vector so that β and the design point are obtained. This requires only adding one u column to the 2007 procedure, and expressing the unrotated n vector in terms of u, where u is the uncorrelated standard equivalent normal vector in the rotated space of the classical mathematical u-space approach of FORM. The vectors n and u can be obtained from one another, n = Lu and u = L⁻¹n, derived as follows (e.g., Low et al., 2011):

$$\beta = \min_{x \in F} \sqrt{\mathbf{n}^T \mathbf{R}^{-1} \mathbf{n}} = \min_{x \in F} \sqrt{\mathbf{n}^T (\mathbf{LU})^{-1} \mathbf{n}} = \min_{x \in F} \sqrt{(\mathbf{L}^{-1}\mathbf{n})^T (\mathbf{L}^{-1}\mathbf{n})} \quad (5.7a)$$

i.e.,

$$\beta = \min_{x \in F} \sqrt{\mathbf{u}^T \mathbf{u}}, \quad \text{where } \mathbf{u} = \mathbf{L}^{-1}\mathbf{n}, \text{ and } \mathbf{n} = \mathbf{Lu} \quad (5.7b)$$

in which **L** is the lower triangular matrix of Cholesky decomposition of **R**. When the random variables are uncorrelated, **u** = **n** by Eq. 5.7b, because then $L^{-1} = L = I$ (the identity matrix).

A widely used nonnormal distribution is the lognormal distribution. The mean and SD of a lognormally distributed random variable x can be given in two ways, either the mean and SD of ln$x$ (i.e., $\mu_{\ln x}$ and $\sigma_{\ln x}$), or the mean and SD of x (i.e., μ_x and σ_x). The two statistical functions LOGNORM.DIST and LOGNORM.INV in Microsoft Excel require inputs in $\mu_{\ln x}$ and $\sigma_{\ln x}$, whereas some statistical example problems (including the RBD examples in this book) and statistical software (e.g., @RISK) use μ_x and σ_x as inputs of the lognormal distribution. One can obtain μ_x and σ_x from $\mu_{\ln x}$ and $\sigma_{\ln x}$, or $\mu_{\ln x}$ and $\sigma_{\ln x}$ from μ_x and σ_x, using the following established statistical relationships (e.g., Ang and Tang (1976):

$$\mu_{\ln x} = \ln \mu_x - 0.5\left[\ln\left(1+\left(\sigma_x/\mu_x\right)^2\right)\right] \tag{5.8a}$$

$$\sigma_{\ln x} = \sqrt{\ln\left(1+\left(\sigma_x/\mu_x\right)^2\right)} \tag{5.8b}$$

$$\mu_x = \exp\left(\mu_{\ln x} + 0.5\sigma_{\ln x}^2\right) \tag{5.9a}$$

$$\sigma_x = \sqrt{\left[\exp\left(\sigma_{\ln x}^2\right)-1\right]\exp\left(2\mu_{\ln x}+\sigma_{\ln x}^2\right)} \tag{5.9b}$$

Apart from normal and lognormal distributions, the Low and Tang (2004, 2007) Excel FORM procedures allow beta distribution, exponential distribution and also Weibull, gamma, extreme value type 1, triangular, uniform and PERT distributions.

Next, an EC7 design of a strip footing foundation from Knappett and Craig (2019) is presented in an Excel worksheet. Then, FORM analysis with lognormal random variables will be conducted on the EC7 design example to obtain information on reliability index and failure probability, followed by RBD-via-FORM of foundation width for a target reliability index, for comparison with the foundation width based on EC7.

5.6 AN EC7 DESIGN OF STRIP FOOTING WIDTH BASED ON CHARACTERISTIC VALUES AND PARTIAL FACTORS

Knappett and Craig (2019, Example 8.4) shows the solution for the design of the width B of a spread foundation against bearing capacity ultimate limit state (ULS) using EC7. The thickness of the concrete strip footing is 0.7 m. The characteristic values of vertical loads to be supported is 500 kN/m for dead load and 300 kN/m for live load, at a depth of 0.7 m in a gravelly sand.

The characteristic values of the shear strength parameters are $c' = 0$ and $\phi' = 40°$. It is assumed that the water table may rise up to foundation level. The unit weight of the sand above the water table is 17 kN/m³, and below the water table the saturated unit weight is 20 kN/m³. The bulk unit weight of the concrete is 24 kN/m³. The foundation width is yet unsized and will apply an additional action (permanent, unfavourable) of 24 kN/m³ × 0.7 m × B = 16.8B kN/m, resulting in a total dead load of 500 + 16.8B kN/m. The limit state formulation is the same as Eqs. 5.1 and 5.2, except that $N_\gamma = 2(N_q - 1)\tan\phi$ is used instead of Eq. 5.2c. All the three design approaches—DA1, DA2 and DA3, as summarized in Figure 5.6—were checked. More details and guidance on EC7 are in Frank et al. (2005), Bond et al. (2013), Bond et al. (2015), Orr (2017), Simpson (2017) and EC7 (2020).

The EC7 design outcomes from Knappett and Craig (2019, Example 8.4) are summarized in the Excel worksheet in Figure 5.7. It can be seen that DA1b or DA3 are most critical in this case, with a required foundation width of B ≥ 1.46 m.

The next two sections illustrate how FORM can be used in two ways to complement EC7, by giving useful information and insights not found in EC7. The similarities and differences between RBD-via-FORM and EC7 will also become clear.

General concepts of ultimate limit state design in Eurocode 7:

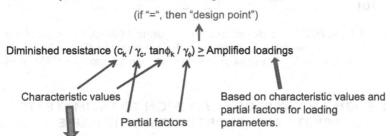

(if "=", then "design point")

Diminished resistance (c_k / γ_c, $\tan\phi_k / \gamma_\phi$) ≥ Amplified loadings

Characteristic values

Based on characteristic values and partial factors for loading parameters.

Partial factors

"Conservative", for example, 10 percentile for strength parameters, 90 percentile for loading parameters

The three sets of partial factors (on resistance, actions, and material properties) are not necessarily all applied at the same time.

In EC7, there are three possible design approaches:

• Design Approach 1 (DA1): (a) factoring actions only; (b) factoring materials only.

• Design Approach 2 (DA2): factoring actions and resistances (but not materials).

• Design Approach 3 (DA3): factoring structural actions only (geotechnical actions from the soil are unfactored) and materials.

(The draft "Eurocode 7 (2020): Geotechnical design - Part 3" states that it is meant to be used in conjunction with the Parts 1 and 2 of earlier versions)

Figure 5.6 Characteristic values, partial factors, design point and design approaches (DA) in Eurocode 7.

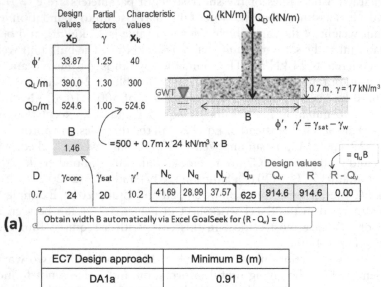

(a)

EC7 Design approach	Minimum B (m)
DA1a	0.91
DA1b	1.46 (shown above)
DA2	1.16
DA3	1.46

(b)

Figure 5.7 (a) EC7 DA1b design of a strip footing width B, based on characteristic values from Example 8.4 of Knappett and Craig (2019), (b) summary of EC7 designs by different design approaches (DA) of EC7.

5.7 FORM ANALYSIS OF EC7 DESIGN TO ESTIMATE ITS RELIABILITY AND PROBABILITY OF FAILURE

In geotechnical design based on EC7, the characteristic values of resistance parameters are divided by partial factors to obtain their design values, while the characteristic values of unfavourable action (load) parameters are multiplied by partial factors to obtain their design values. The resistance thus diminished is required to be greater or equal to the amplified actions (loadings), Figure 5.6, which also summarize the three design approaches DA1, DA2 and DA3. The DA2 is akin in principle to the LRFD of North America.

With respect to characteristic values, Clause 2.4.5.2(2)P EN 1997-1 defines it as being 'selected as a cautious estimate of the value affecting the occurrence of the limit state,' and Clause 2.4.5.2(10) of EC7 states that statistical methods may be used when selecting characteristic values of geotechnical parameters, but they are not mandatory.

RBD-via-FORM does not need characteristic values and partial factors but instead requires probability distribution and statistics—mean values, standard deviations and parametric correlations (if any)—of random variables. For a meaningful comparison of FORM analysis of the EC7 design outcome of Figure 5.7, it is necessary to relate the characteristic values of the EC7 example to the mean values and standard deviations required in FORM analysis.

This section conducts *reliability analysis* on a design outcome from EC7, hence the term '*FORM analysis*,' which means computing the reliability index (and associated P_f) of a design using FORM, in contrast with the term '*RBD-via-FORM*' in the next section, which means conducting *reliability-based design* for a *target reliability index* (e.g., $\beta = 3.0$).

The selection of characteristic values is described only briefly in the prevailing version of EC7. Some guidance is given in Orr (2017). Most EC7 design examples, for example, Tomlinson (2001), Knappett and Craig (2019), Smith (2014) and Frank et al. (2005), mention inputs in characteristic values, without explaining the deliberations leading to those characteristic values. This means that in practice the outcomes of EC7 design based on code-specified partial factors may not be unique if different engineers adopt different levels of conservatism in estimating characteristic values, for example, 5/95 percentiles (fractiles), or the less conservative 30/70 percentiles suggested by Schneider and Schneider (2013). For a normally distributed random variable, the 5/95 percentiles correspond to a characteristic value at 1.64σ below and above the mean μ, for a resistance parameter and an unfavourable action (load), respectively, and the 30/70 percentiles correspond to 0.52σ below and above the mean value μ.

We will assume $\mu_x = x_k \pm \sigma_x$ for the purpose of estimating mean value μ_x from characteristic value x_k, to strike a balance between the conservative $\mu_x = x_k \pm 1.64\sigma_x$, corresponding to 5/95 percentiles of normal distribution, and the less stringent $\mu_x = x_k \pm 0.52\sigma_x$, corresponding to 30/70 percentiles of normal distribution. If desired, the Microsoft Excel's built-in function LOGNORM.INV and Eq. 5.8 can be used for more precise percentiles of characteristic values when dealing with lognormals.

In Figure 5.8, the mean values (μ) and SD values (σ) correspond to coefficient of variation (σ/μ) values of 0.11, 0.2 and 0.1 for ϕ', Q_L and Q_D, respectively. Also, the mean values and SD values satisfy the assumed $\mu_x = x_k \pm \sigma_x$, namely mean $\pm$ SD = characteristic value:

$$\mu_\phi - \sigma_\phi = 45° - 5° = 40° = \phi_k \quad \text{of Figure 5.7} \quad (5.10a)$$

$$\mu_{QL} + \sigma_{QL} = 250 + 50 = 300 = \left(Q_L\right)_k \quad \text{of Figure 5.7} \quad (5.10b)$$

$$\mu_{QD} + \sigma_{QD} = 476.9 + 47.7 = 500 + 0.7m \times 1.46m \times 24kN/m^3 = \left(Q_D\right)_k \quad (5.10c)$$

The FORM analysis in Figure 5.8a shows that, for the lognormal random variables of ϕ', Q_L and Q_D and a EC7 design width of B = 1.46 m (from the EC7 design outcome in Figure 5.7), the reliability index computed using the

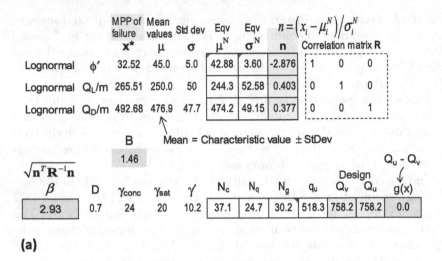

(a)

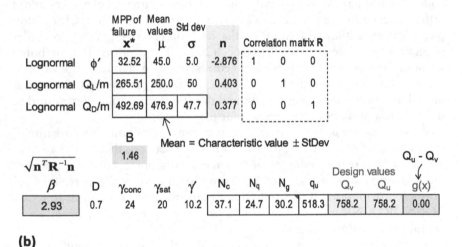

(b)

Figure 5.8 FORM analysis for reliability index β of the EC7 design of width B = 1.46 m, using the Excel-automated methods of (a) Low and Tang (2004), and (b) Low and Tang (2007).

Low and Tang (2004) FORM procedure (Eq. 5.5) is $\beta = 2.93$, or a probability of failure against bearing capacity is $P_f = \Phi(-\beta) = 0.17\%$. Figure 5.8b shows the reliability analysis using the Low and Tang (2007) FORM procedure (Eq. 5.6), which obviates calculation of equivalent normal mean (μ^N) and equivalent normal SD (σ^N) shown in the upper figure. The values of the MPP of failure ($\mathbf{x}^*$), the dimensionless sensitivity indicators ($\mathbf{n}$) and the computed reliability index (β) in Figure 5.8a are the same as Figure 5.8b. In Figure 5.8a, Excel Solver was invoked to minimize the β cell (Eq. 5.5), by changing the x^* column, subject to the constraint that the performance function $g(\mathbf{x}) = Q_u - Q_v = 0$. In Figure 5.8b, Excel Solver was invoked to minimize the β cell (Eq. 5.6), by changing the n column, subject to the constraint that the performance function $g(\mathbf{x}) = Q_u - Q_v = 0$. Figure 5.8 thereby provides an illustration of the two procedures summarized earlier in Figure 5.5. The Hasofer–Lind method (Figure 5.1 and Eq. 5.3) as a special case of FORM is also better understood by comparing Figure 5.1 with Figure 5.8, and Eq. 5.3 with Eqs. 5.5 and 5.6. The dispersion ellipsoid perspective in Figures 5.1 and 5.2 also applies to the nonnormal case of Figure 5.8, when visualized as an equivalent ellipsoid (centred at μ^N, and with equivalent normal standard deviations σ^N) tangent to the 3D LSS.

The FORM analysis in Figure 5.8 provides information on reliability index and the probability of failure. In addition, column n of FORM analysis in Figure 5.8a and 5.8b suggests that ϕ' is a much more sensitive parameter than Q_L and Q_D, for the case in hand. Researchers using other probabilistic approach (e.g., Monte Carlo simulation) may obtain the same information on probability of failure, but not the valuable and insightful information at the MPP of failure, as elaborated in the RBD-via-FORM of retaining wall foundation later. The case-dependent sensitivities and the subtle load-resistance duality will be better appreciated then.

The above is FORM analysis for reliability index of a design outcome from EC7.

Next we consider RBD-via-FORM of width B for a target reliability index β, to compare with the EC7 design width B of Figure 5.7. Comparisons between RBD-via-FORM and EC7 for other geotechnical problems are given in Low and Phoon (2015b).

5.8 RBD-VIA-FORM FOR TARGET RELIABILITY INDEX, TO COMPARE WITH EC7 FOOTING WIDTH DESIGN

Instead of the FORM analysis to compute reliability index β (=2.93) for the EC7 design width of B = 1.46 m in the previous section, one can use the same statistics of Figure 5.8 to obtain an RBD-via-FORM width B, for a target reliability index β (e.g., 2.5 or 3.0), for comparison with the width B = 1.46 m based on EC7 partial factors and characteristic values. This is shown in Figure 5.9. A slightly larger width of B = 1.50 m is needed to achieve β = 3.0, Figure 5.9a, and a smaller width of B = 1.25 m for a target β = 2.50, Figure 5.9b, both using the Low and Tang (2007) FORM procedure. The values under the x^* column may now be referred to as *design values* because there is a target reliability index β, instead of *MPP of failure* in the previous figure.

For lower consequence of failure, a design width B = 1.25 m (with $\beta = 2.5$) could be adequate. For higher consequence of failure, B = 1.5 m (with $\beta = 3.0$) can be used.

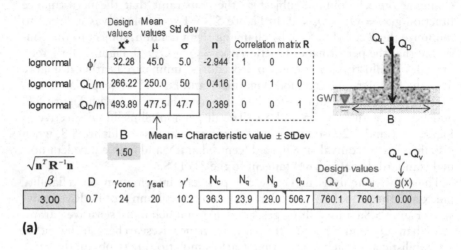

(a)

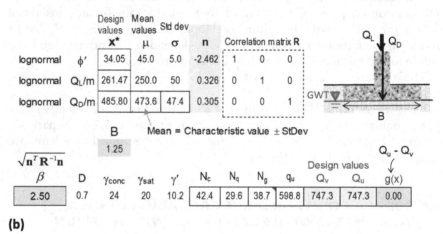

(b)

Figure 5.9 RBD-via-FORM of width B, for a target reliability index β of 3.0 in (a), and β of 2.5 in (b), using the Low and Tang (2007) Excel procedure for FORM.

The case-dependent sensitivity of ϕ' is obvious if one notes the different n values of ϕ' in Figures 5.1, 5.2 and 5.9a, where $n_{\phi'}$ = 0.04, –2.681, –2.944, respectively, all based on a target reliability index β = 3.0. Comparison of the n values of ϕ' in Figure 5.9a and 5.9b, which are $n_{\phi'}$ = –2.94 and –2.46, respectively, indicates that sensitivity indicator **n** depends on target β value of reliability index. For cases with different β values, one can compare n/β instead.

RBD-via-FORM automatically reflects context-dependent and reliability-level-dependent sensitivities at its design point in a way partial factor design approach cannot. It also resolves the subtle load-resistance duality as shown in the next section involving five nonnormal random variables.

5.9 RBD-VIA-FORM RESOLVES LOAD-RESISTANCE DUALITY OF A RETAINING WALL FOUNDATION

Tomlinson (2001, Example 2.2) analyzed deterministically the factor of safety (F_s) against bearing capacity failure of a retaining wall foundation carrying both vertical and horizontal loads and computed a lumped $F_s = 3.0$ against general shear failure of the soil below the base of the wall. The wall footing had base width B = 5 m and length L = 25 m. Figure 5.10a revisits the Tomlinson example but extends it to RBD-via-FORM.

The ultimate bearing capacity q_u of the retaining wall footing is given by:

$$q_u = c'N_c s_c d_c i_c + p_o N_q s_q d_q i_q + 0.5 B' \gamma N_\gamma s_\gamma d_\gamma i_\gamma \qquad (5.11)$$

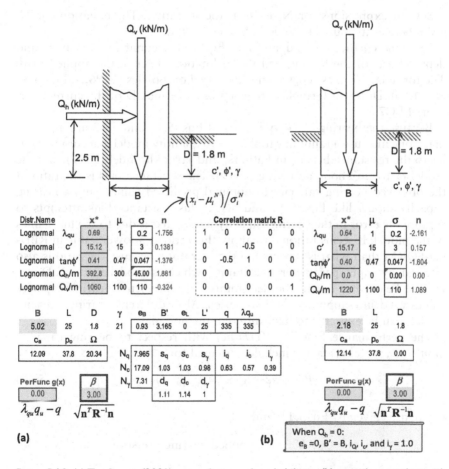

Figure 5.10 (a) Tomlinson (2001) example turned probabilistic, (b) a similar case but with vertical load only, to demonstrate load-resistance duality and context-sensitivity of case (a). The σ values are for coefficients of variation 0.1, 0.2, 0.1, 0.15 and 0.1, respectively.

in which c' is the effective cohesion of soil, p_o is the effective overburden pressure at foundation level, $B' = B - 2e_B$, where e_B is the eccentricity of the resultant inclined load, γ is the unit weight of the foundation soil and N_c, N_q and N_γ are bearing capacity factors, which are functions of the friction angle (ϕ') of soil:

$$N_c = (N_q - 1)\cot(\phi') \tag{5.12a}$$

$$N_q = e^{\pi \tan\phi} \tan^2\left(45 + \frac{\phi'}{2}\right) \tag{5.12b}$$

$$N_\gamma = 2(N_q + 1)\tan\phi' \tag{5.12c}$$

Several expressions for N_γ exist in the literature. The equation for N_γ given here is attributed to Vesic in Bowles (1996).

The nine coefficients s_j, d_j and i_j in Eq. 5.11 account for the shape and depth effects of the footing and the inclination effect of the applied load. The formulas for these coefficients are based on Bowles (1996, Tables 4.5a and 4.5b). Different formulas are given in the Annex D of the current version of EC7.

Because the bearing capacity factors in Eqs. 5.11 and 5.12 are approximate, one may use a random variable λ_{qu} to quantify model inaccuracy, similar to the resistance bias λ_R in Bathurst and Javankhoshdel (2017), and the model factor in Phoon and Tang (2019). The resistance bias is the ratio of the observed bearing capacity to nominal predicted value using a bearing capacity model like Eqs. 5.11 and 5.12. This resistance bias attempts to capture the error in the bearing capacity model that is due to the mechanics of the model. It seems rational to apply this ratio at a lower value in RBD-via-FORM than what is reported in the literature, because the ratio is also affected by input parameter uncertainty, which is modelled in RBD-via-FORM below but not in the deterministically predicted bearing capacity values used in computing the bias ratios. Also, different bearing capacity models can be expected to have different method bias.

The performance function (*PerFunc*) with respect to bearing capacity limit state, with bias factor included, is equal to:

$$PerFunc: \quad g(\mathbf{x}) = \lambda_{qu}q_u - q \tag{5.13}$$

where q_u is given by Eq. 5.11, and

$$q = Q_v / B' \quad \text{is the applied bearing pressure} \tag{5.13a}$$

The statistical values and assumed lognormal distribution in Figure 5.10 are illustrative. The insights on load-resistance duality and context sensitivity that are the main focus here are not affected by the assumed statistical values or the choice of bearing capacity formulation adopted in this example.

The mean values are: $c' = 15$ kPa, $\tan\phi' = 0.47$ (corresponding to $\phi' = 25°$), $\gamma = 21$ kN/m^3, $Q_h = 300$ kN/m and $Q_v = 1100$ kN/m, based on the deterministic values in the Tomlinson example. The mean value of λ_{qu} is assumed to be 1.0. The coefficient of variation of λ_{qu}, c', $\tan\phi'$, Q_h and Q_v are 0.2, 0.2, 0.1, 0.15 and 0.1, respectively, yielding the five standard deviation values under the column labelled σ in Figure 5.10a. Negative correlation between c' and $\tan\phi'$ is modelled by $\rho = -0.5$ in the correlation matrix, as explained in Section 5.4. For comparison, Figure 5.10b shows the RBD of the wall base with vertical load only.

The RBD for the case in hand aims to find a foundation width B for a target reliability index β value of 3.0, or an estimated probability of failure $P_f = \Phi(-\beta) = 0.13\%$. The five values under the column labelled n (initially zeros) and the corresponding five design values under the column labelled x* in Figure 5.10a were obtained automatically using the Excel Solver constrained optimization procedure (Eq. 5.6a) of Low and Tang (2007). Other software packages (e.g., MATLAB, R) can also be used to implement FORM. A few values of the width B were tried before B = 5.02 m was found to achieve the target of $\beta = 3$ against bearing capacity failure.

5.9.1 Information and insights at the design point of RBD-via-FORM

Each of the n values in Figure 5.10a represents the normalized distance from the equivalent normal mean-value μ_i^N of a random variable to the tangent point at the LSS (Figure 5.1b and 5.1c, but in 5D space) where the performance function $g(\mathbf{x}) = 0$. The five x* values represent the design point (MPP of failure), where the expanding five-dimensional dispersion ellipsoid first touches the 5D LSS. The FORM reliability index β is the shortest distance from the equivalent normal mean-value point to the 5D LSS, in units of directional equivalent normal standard deviations, that is, the distance R/r in Figure 5.1b, which can be visualized in the mind's eye in 5D space. It can be appreciated intuitively that each value of n, equal to $\left(x_i - \mu_i^N\right)/\sigma_i^N$ as defined in Eq. 5.6b, can be regarded as a sensitivity indicator: greater absolute n value indicates higher sensitivity. Positive n value means that the design value is bigger than the equivalent normal mean value by n times the equivalent normal standard deviation, and negative n value means that the design value is smaller than the equivalent normal mean value by n times the equivalent normal standard deviation.

The slightly positive n value of c' in Figure 5.10a, 0.138, is due to negative correlation coefficient of −0.5 between c' and $\tan\phi'$. For the case in hand, the design is much more sensitive to Q_h than to Q_v, with n values 1.881

versus −0.324, and much more sensitive to λ_{qu} and $\tan\phi'$ than c', with n values −1.756 and −1.376 versus 0.138. One may note in Eq. 5.11 that the ultimate bearing capacity q_u consists of three components, with c' affecting the first component only, but with ϕ' affecting all the three components through terms N_c, N_q, N_γ, s_q, i_q, i_γ, d_q. Hence, it is not surprising that, for the case in hand, the sensitivity to $\tan\phi'$ is much more than the sensitivity to c'. This sensitivity inference should not be generalized, because the μ and σ values of c' and $\tan\phi'$ (and their correlation) in Figure 5.10 are similar to Figure 5.2. A different combination of the statistics of c' and $\tan\phi'$ (e.g., higher $\mu_{c'}$ and lower $\mu_{\phi'}$, as in Figure 5.1) can reverse the sensitivities of c' and $\tan\phi'$, even for the same bearing capacity problem. Regardless of circumstances, it is an important merit of RBD-via-FORM that it reflects case-specific sensitivity automatically. The practical implication of sensitivity indicator values in RBD-via-FORM is that the designer can focus attention on the choice of the magnitude of input parameters that most influence the probability of failure. In this case, the choice of Q_h, λ_{qu} and friction angle ϕ' has the greatest influence on the level of safety for a trial footing of width B.

When $Q_h = 0$, the vertical load Q_v is an unfavourable load without ambiguity. However, when Q_h acts together with Q_v, the latter possesses load-resistance duality, because load inclination and eccentricity decrease with increasing Q_v. RBD automatically takes this load-resistance duality into account when locating the design point. Interestingly, RBD reveals that the design value of Q_v (1060 kN/m) in Figure 5.10a is lower than its mean value of 1100 kN/m, in contrast to the Q_v^* value of 1220 kN/m (> mean Q_v) in Figure 5.10b without horizontal load, thereby revealing the load-resistance duality of Q_v in Figure 5.10a. It is difficult for partial factor and LRFD design approaches (e.g., EC7 (2020) and AASHTO (2020)) to deal with a parameter that possesses load-resistance duality, such as the vertical load Q_v in the presence of horizontal load Q_h.

Figure 5.11 reveals the load-resistance duality of the vertical load Q_v when the mean value of the horizontal load Q_h increases from 0 to 300 kN/m. The y axis of Figure 5.11 is the normalized equivalent design value, $n_i = \left(x_i^* - \mu_i^N\right)/\sigma_i^N$, as defined in Eq. 5.6b. The n_{Qv} curve labelled '$\beta = 3.0$' starts on the left with a value of 1.089, which is the last value under the column labelled n in Figure 5.10b when $\mu_{Qh} = 0$. This means that the design value of Q_v at the MPP of failure is at 1.089 times the equivalent standard deviation (σ^N) of Q_v *above* the equivalent mean (μ^N) of Q_v, thereby revealing the unambiguous destabilizing nature of Q_v when there is no horizontal load. However, the value of n_{Qv} decreases with increasing mean horizontal load μ_{Qh}, reaching a value of $n_{Qv} = -0.324$ (the last value under the column labelled n in Figure 5.10a), which means that the design value of Q_v at the MPP of failure is at 0.324 times the equivalent standard deviation (σ^N) of Q_v *below* the equivalent mean (μ^N) of Q_v when the mean horizontal load Q_h is 300 kN/m and the overturning moment is 750 kN/m, thereby revealing the

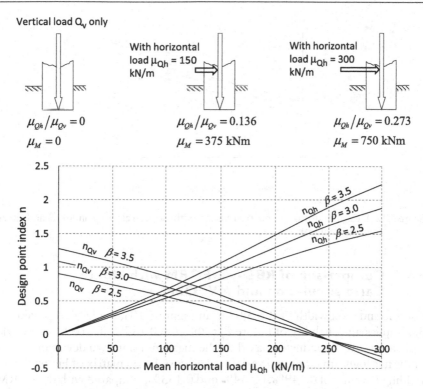

Figure 5.11 The load-resistance duality (or stabilizing–destabilizing duality, or favourable-unfavourable duality) of vertical loads Q_V is revealed by the decreasing values of n_{QV} with increasing mean horizontal load μ_{QH}.

load-resistance duality of Q_V in the face of horizontal load. One may also note in Figure 5.11 that the high n_{QV} values (and zero n_{Qh} value) on the left of the curves gradually transform to negative n_{QV} values (and high n_{Qh} values) on the right, due to increasing load-resistance duality. The values of n_{QV} and n_{Qh} are equal only when $\mu_{Qh} \approx 110$ kN/m and $\mu_{Qh} = 103$ kN/m for β equal to 3.0 and 2.5, respectively. Load-resistance duality was referred to as favourable–unfavourable duality and stabilizing–destabilizing duality by Low (2017). These three duality terms can be used interchangeably.

Figure 5.12 shows that the required foundation width B increases with mean horizontal load μ_{Qh} and target reliability index β. For the case with concentric vertical load Q_V only (Figure 5.10b), the design width B is about 1.9 m, 2.2 m and 2.5 m for reliability index of 2.5, 3.0 and 3.5, respectively, and increases to about 4.6 m, 5.0 m and 5.5 m when the mean horizontal load and overturning moment are 300 kN/m and 750 kNm, respectively. The reliability index values of 2.5, 3.0 and 3.5 are typical target β values for ultimate limit states, corresponding to P_f of 0.62%, 0.13% and 0.023%, respectively.

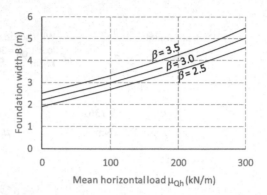

Figure 5.12 Design width of foundation varies with target reliability index β and mean horizontal load.

5.9.2 Comparison of RBD-via-FORM with Monte Carlo simulation and MVFOSM

For a foundation width of B = 5.02 m in Figure 5.10a, the FORM β value of 3.0 implies a probability of failure $P_f \approx \Phi(-\beta) = 0.135\%$. Three Monte Carlo simulations were conducted based on the mean and standard deviation values of the five lognormal random variables and correlation matrix of Figure 5.10a, yielding P_f values of 0.149%, 0.144% and 0.153%, compared with the FORM P_f of 0.135%. The agreement is more than adequate from a pragmatic standpoint. This is summarized in the second and third columns of Figure 5.13.

A simpler but less consistent reliability method is the mean-value first-order second moment method (MVFOSM). The last two columns of

	FORM	Monte Carlo	MVFOSM	
Performance function $g_1(\mathbf{x})$ or $g_2(\mathbf{x})$?	Same answer for $g_1(\mathbf{x})$ or $g_2(\mathbf{x})$	Same answer for $g_1(\mathbf{x})$ or $g_2(\mathbf{x})$	$g_1(\mathbf{x}) = \lambda_{qu} q_u - q$	$g_2(\mathbf{x}) = \dfrac{\lambda_{qu} q_u}{q} - 1$
β index and/or P_f	$\beta = 3.0$ $P_f = \Phi(-\beta) = 0.13\%$	$P_f = 0.149\%$, 0.144%, 0.153%, (Average = 0.15%), in 3 simulations each with 300,000 iterations	$\beta = \dfrac{\mu_{g1(\mathbf{x})}}{\sigma_{g1(\mathbf{x})}}$ $= \dfrac{525.8}{266.5} = 1.97$	$\beta = \dfrac{\mu_{g2(\mathbf{x})}}{\sigma_{g2(\mathbf{x})}}$ $= \dfrac{1.763}{0.927} = 1.90$
MPP of failure (i.e. Design point) available?	Yes, design point x* and sensitivity index n_i are part of the FORM outcome	Not part of the outcome of Monte Carlo simulation	No	No
Guidance/caveats for EC7 and LRFD available?	Yes, FORM reveals load-resistance duality and context-sensitivity	Not part of the outcome of Monte Carlo simulation	No	No

Figure 5.13 Comparison of FORM, Monte Carlo and MVFOSM for the case in Figure 5.10a.

Figure 5.13 show that two different values of reliability index (1.97 and 1.90) are computed, based on MVFOSM method, for two seemingly different but mathematically equivalent requirements on the performance functions, namely $g_1(\mathbf{x}) = \lambda_{qu} q_u - q = 0$ and $g_2(\mathbf{x}) = \dfrac{\lambda_{qu} q_u}{q} - 1 = 0$. This inconsistency of MVFOSM is in line with established knowledge (e.g., Haldar and Mahadevan 2000; Wu 2008, p. 420) that the reliability index computed by the MVFOSM method can vary according to different (but mathematically equivalent) ways of defining the LSS. In contrast, FORM and Monte Carlo simulation do not have this shortcoming.

The bottom two rows of Figure 5.13 emphasize that, besides probability of failure, FORM also provides insights and information at its MPP of failure (the design point represented by x* values in Figure 5.10) and the sensitivity indicators n_i.

5.10 RBD-VIA-FORM INSIGHTS FOR LRFD

RBD of the case in Figure 5.10 aims at achieving a sufficiently safe design, in terms of a target reliability index β, say 2.5, 3.0 or 3.5, and associated probability of failure P_f, without relying on nominal values and load and resistance factors. Nevertheless, to appreciate the similarities and differences between RBD-via-FORM and LRFD, the latter's load and resistance factors (LF and RF, or γ_i and φ_R) can be back-calculated from the results of RBD-via-FORM.

Figure 5.14a shows the main concepts of LRFD. The LF and RF are applied to nominal loads Q_{ni}, which (if destabilizing) are higher than mean loads μ_{Qi}, and nominal resistance R_n, which is lower than mean resistance μ_R. For the purpose of back-calculating LF and RF, the nominal values of loads and resistance are taken to be at $\mu \pm \sigma$, which means taking the nominal Q_v at about 84 percentile of the lognormal distribution. Comparisons will be made later with nominal values taken to be at *mean values*.

5.10.1 Back-calculations of LF and RF from the design point of RBD-via-FORM

For the case in Figure 5.10a, the vertical load has mean value μ_{Qv} = 1100 kN/m, and standard deviation σ_{Qv} = 110kN/m. Hence, nominal vertical load $Q_{v,n}$ is 1100 + 110 = 1210 kN/m, as shown in the 5th column of Figure 5.14b. The resistance R is $\lambda_{qu} Q_u = \lambda_{qu} \times q_u \times B'$. The mean value and standard deviation of resistance R shown under the columns labelled μ_R and σ_R were obtained from Monte Carlo simulations (with correlated coefficient $\rho_{c\phi}$ and other statistical inputs as in Figure 5.10). Hence, nominal resistance R_n is $\mu_R - \sigma_R$, shown in the column labelled $R_{nominal}$.

Case-specific LF and RF can be back-calculated from the design point of RBD-via-FORM. For example, the x* column of Figure 5.10a indicates that

Load and Resistance Factor Design (LRFD)

$$\sum (LF)_i Q_{ni} \le (RF) R_n$$

$$\text{OR} \quad \sum \gamma_i Q_{ni} \le \varphi_R R_n$$

where LF = γ = load factor,
RF = φ = resistance factor,
Q_{ni} = nominal load,
R_n = nominal resistance.

The most economical design is achieved
when " $\le$ " becomes " = ":

$$\sum \gamma_i Q_{ni} = \varphi_R R_n$$

i.e.: ΣDesign (factored) loads Q_i
= Design (factored) resistance R

(a)

	kN/m	kN/m	m		kN/m	kN/m		kN/m	Nominal = $\mu \pm \sigma$			
	μ_{Qh}	σ_{Qh}	B	$Q_{v,nominal}$	μ_R	σ_R	$R_{nominal}$	$Q_v^*=R^*$	LF$_{Qv}$	RF	LF$_{Qh}$	Q_h^*
$\beta = 2.5$	1	0.15	1.91	1210	2355	677	1678	1198	0.99	0.71	0.86	1
	150	22.5	3.1	1210	2358	733	1625	1134	0.94	0.70	0.97	168
	300	45	4.59	1210	2600	912	1688	1066	0.88	0.63	1.08	374
$\beta = 3.0$	1	0.15	2.19	1210	2713	791	1922	1220	1.01	0.63	0.86	1
	150	22.5	3.41	1210	2713	846	1867	1147	0.95	0.61	0.99	171
	300	45	5.02	1210	3039	1055	1984	1060	0.88	0.53	1.14	393
$\beta = 3.5$	1	0.15	2.51	1210	3138	922	2215.6	1243	1.03	0.56	0.86	1
	150	22.5	3.75	1210	3120	975	2145	1161	0.96	0.54	1.01	174
	300	45	5.48	1210	3539	1223	2316	1052	0.87	0.45	1.20	414

(b)

Figure 5.14 (a) The load and resistance factor design (LRFD) method use nominal values, load factor LF and resistance factors RF, (b) backcalculated LF and RF for the retaining wall foundation, assuming nominal value = mean ± SD.

the design value of Q_v is 1060. The LF is therefore $Q_v^*/Q_{v,\text{ nominal}}$ = 1060/1210 = 0.88, shown in the sixth row under the column labelled LF_{Qv}. The design vertical load Q_v^*, 1060 kN/m, is equal to the design resistance $R^* = \lambda_{qu} \times q_u \times B' = 1060$. On the other hand, the mean and standard deviation of resistance R are (from Monte Carlo simulation) $\mu_R = 3039$ kN/m and $\sigma_R = 1055$ kN/m, hence $R_{\text{nominal}} = \mu_R - \sigma_R = 3039{-}1055 = 1984$ kN/m. The ratio of R_{Design} to R_{nominal}, 1060/1984, is the resistance factor of 0.53, shown in the sixth row of the *RF* column, next to the LF_{Qh} value of '1.14,' which is computed from $LF_{Qh} = Q_h{}^*/Q_{h,nominal} = 392.8/(300 + 45) = 1.14$. All the other values of LF_{Qv}, RF and LF_{Qh} in Figure 5.14b were calculated in like manner and plotted as dashed curves in Figure 5.15 for mean horizontal load varying from 0 to 300 kN/m, and reliability index from 2.5 to 3.5. Note that the values of LF and RF back-calculated from the FORM design point are not intrinsic but are functions of the assumed nominal values of loads and resistance. Had 70 percentile been used for the nominal loads of Q_v and Q_h and 30 percentile for the resistance R, the back-calculated LF_{Qv} and LF_{Qh} would be higher and the back-calculated RF smaller than the corresponding values in Figure 5.14b which are based on (more conservatively) nominal loads at 84 percentile and nominal resistance at 16 percentile. In other words, assuming less conservative nominal values will result in more conservative back-calculated LF and RF, and vice versa. With this in mind, the solid curves in Figure 5.15 are presented as alternative back-calculated values of LF and

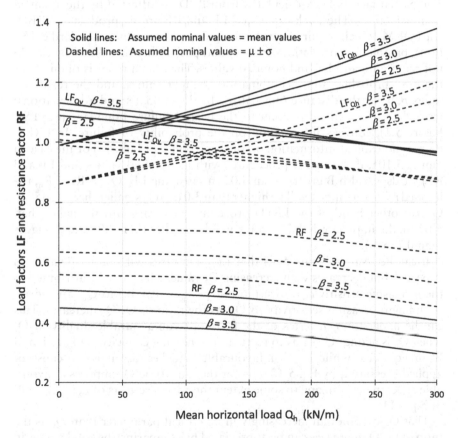

Figure 5.15 Context-dependent nature of back-calculated load factor LF and resistance factor RF. The load-resistance duality of Q_v is also revealed with increasing Q_h.

RF, perhaps less affected by different levels of conservatism regarding nominal values. These solid curves assume nominal values at mean values and hence are simply $LF_{Qv} = Q_v^*/\mu_{Qv}$, $LF_{Qh} = Q_h^*/\mu_{Qh}$ and $RF = R^*/\mu_R$.

In LRFD, two engineers using the same sets of LF and RF values may still end up producing different designs if they adopt different nominal values of loads and resistance. Nominal values are not inputs in RBD, hence the design outcome of RBD is not affected by one's assumption of nominal values or LF and RF. The above and Figure 5.15b attempt to show the non-intrinsic nature and case-dependent variability of back-calculated LF and RF. Figure 5.15b also reveals load-resistance duality, like Figure 5.11.

5.10.2 Load-resistance duality revealed in load factors back-calculated from RBD-via-FORM

The partial factors back-calculated from RBD are affected by the assumed nominal values. The back-calculated LF and RF are by-products of RBD-via-FORM, which requires statistical inputs of mean values, standard deviations, parametric correlations (if any) and probability distributions, but do not require LF and RF and nominal values. The design point is obtained as the most probable failure combination of input values, and the reliability index β, being the distance (normalized by directional standard deviations) from the safe mean-value point to the nearest failure boundary (the LSS, Figure 5.1), conveys information on the probability of failure. In LRFD, there is no explicit information on the probability of failure. For the case in Figure 5.10a, if the LRFD procedure (with its nominal values and LF and RF) yields a width B greater than 5.02 m satisfying $LF_{qv}Q_{v,n} = RF \times R_n$, the implied reliability index will be higher than 3.0 (and P_f smaller than 0.135%). On the other hand, if the LRFD procedure yields a width B smaller than 5.02 m, the implied reliability index will be smaller than 3.0 (and P_f bigger than 0.135%).

It can be seen in Figure 5.15 that LF_{Qv} increases with target β value, as expected, but decreases with increasing Q_h value, which can be attributed to the increasing sensitivity of resistance R ($= \lambda_{qu} \times q_u \times B'$) to Q_h (and corresponding decreasing sensitivity to Q_v) as the ratio of Q_h/Q_v increases. This can be understood in terms of the favourable/unfavourable duality of Q_v when Q_h is acting: Q_h leads to reduction in bearing capacity via i_c, i_q, i_γ and B' in Eq. 5.11, while Q_v is unfavourable load because it is the cause of applied pressure q by Eq. 5.13a, but at the same time Q_v imposes a favourable effect in neutralizing to some extent the adverse effects of i_c, i_q, i_γ and B' in Eq. 5.11.

That Q_h becomes an increasingly more critical parameter than Q_v as the ratio of Q_h/Q_v increases can be appreciated by comparing the solid curves in Figure 5.15 labelled LF_{Qv} and LF_{Qh}, which shows that as LF_{Qv} decreases, LF_{Qh} increases. These solid curves of LF_{Qv} and LF_{Qh}, based on nominal values = mean values, are simply the ratios of Q_v^*/μ_{Qv}, and Q_h^*/μ_{Qh}, where '*' indicates design value.

The dashed curves of LF and RF in Figure 5.15 were computed from $LF = Q_{Design}/Q_{Nominal} = Q_{Design}/(\mu_Q + \lambda\sigma_Q)$, where $\lambda = 1$ has been assumed for the dashed curves. Had less conservative nominal loads been assumed, the back-calculated dashed curves of LF_{Qv} and LF_{Qh} would have been higher. LRFD design codes typically assume $LF > 1$, but there is no theoretical reason for this. For cases where a load has stabilizing effect, or possesses stabilizing–destabilizing duality, LF can logically be smaller than 1.0. The entanglement of nominal loads with LF (and nominal resistance with RF) and their effects on LRFD design will be examined further in an example on design of later-ally loaded piles in another chapter.

5.11 RELIABILITY ANALYSIS OF SERVICEABILITY LIMIT STATE INVOLVING THE BURLAND AND BURBIDGE METHOD

Tomlinson (2001, Example 2.1) checked the ULS and SLS of a water tank found on sand, based on EC7. The standard penetration test (SPT) N values were used to estimate the average settlement using the Burland and Burbidge (1985) method. This section presents (i) a summary of EC7 deliberations with respect to the ULS and SLS of the foundation of the water tank, (ii) statistical analysis of the Burland and Burbidge SPT N data, (iii) FORM analysis of settlement accounting for the input uncertainty underlying the Burland and Burbidge method.

5.11.1 Deliberations by Tomlinson (2001) on the ULS and SLS of a spread foundation

In Tomlinson (2001, Example 2.1), a water tank 5 m wide by 20 m long is to be constructed at a depth of 0.8 m below ground level. The depth of water in the tank is 7.5 m. Borings showed a loose, becoming medium-dense, normally consolidated sand with the water table 2 m below ground level. Standard penetration tests made in the boreholes are plotted in Figure 5.16a. It is required to check that the ultimate limit and serviceability limit states (ULS and SLS) are not exceeded.

Tomlinson (2001) deliberations on the ULS of bearing capacity are summarized in Figure 5.16b. There is no problem at all with respect to the ULS of bearing capacity.

Reliability analysis using FORM with respect to ULS will not be a problem and is omitted here.

With respect to SLS, the *average* settlement of the tank foundation was calculated using the Burland and Burbidge (1985) empirical relationship for settlements of foundations on sands and gravels, based on the average

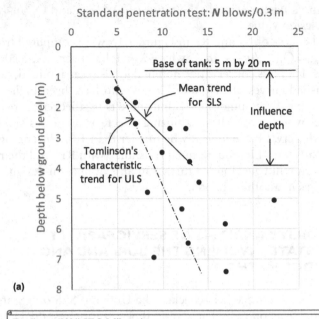

Standard penetration test: *N* blows/0.3 m

(a)

Tomlinson (2001) ULS deliberations:
•Self weight of storage tank= 4190 kN
•Weight of stored water = 7500 kN. Partial factor = 1.3
•Total design value of actions: V_d = 4190 kN + 1.3 × 7500 kN= 13,940 kN.
•Characteristic value of N = 6 within the influence depth. (dashed line above)
  Characteristic value of angle of shearing resistance is ϕ_k = 29°, from chart of N vs ϕ'
  Design value of ϕ' = $\tan^{-1}[(\tan29°)/1.25]$=23.9°
  Design depth of foundation: 0.5 m (instead of 0.8 m), to allow for possible regrading

The design value of vertical resistance against bearing capacity failure is 42,500 kN,
which is bigger than net V_d = 13,940 – 17 kN/m³ × 0.5 m × 20 m × 5 m = 13090 kN.
Hence the ULS against bearing capacity failure is not exceeded.

(b)

Figure 5.16 Tomlinson (2001) Example 2.1: the base of a water tank 5 m wide by
20 m long is at a depth of 0.8 m below ground level. (a) SPT *N* versus
depth, (b) checking for the ULS against bearing capacity failure.

standard penetration test (SPT) $\bar{N}$ value within the influence depth z_I. The
Burland and Burbidge (1985) procedure was described in Tomlinson (2001,
pp. 67–69), similar to the Burland and Burbidge procedure described in
Terzaghi et al. (1996, pp. 395–398) and Knappett and Craig (2019,
pp. 383–385). The procedure can deal with overconsolidated sand, but for
the normally consolidated sand underlying the water tank of the example
here, the equation reduces to:

$$\text{Settlement} = s = f_s f_I f_t \times q'_n B^{0.7} m_v \quad \text{(for NC sand)} \tag{5.14}$$

where

$$m_v = \text{compressibility} = \frac{1.7}{\left(\bar{N}_{60}\right)^{1.4}} \quad \left(\text{same as compressibility index } I_c\right) \quad (5.14a)$$

(Unit of m_v is in MPa^{-1}, that is, m^2/MN. If q_n in kPa and B in m, will get s in mm)

$$f_s = \text{shape factor} = \left(\frac{1.25L/B}{L/B + 0.25}\right)^2 \quad\quad (5.14b)$$

$$f_I = \frac{H}{z_I}\left(2 - \frac{H}{z_I}\right) \quad\text{, if sand thickness } H < z_I, \text{ where } z_I = B^{0.7} \quad (5.14c)$$

(f_I is a correction factor if sand thickness H below foundation level is less than z_I)

$$f_t = \left[1 + R_3 + R_t \log\left(\frac{t}{3}\right)\right] \quad\quad (5.14d)$$

where R_3 is the time-dependent settlement, as a proportion of s, occurring during the first three years after construction, and R_t the settlement occurring during each log cycle of time in excess of 3 years.

The influence depth is $z_I = B^{0.75}$ in Terzaghi et al. (1996), slightly different from $z_I = B^{0.7}$ in Tomlinson (2001) calculations and Knappett and Craig (2019).

For checking SLS, an average of SPT $\bar{N}$ value (unfactored) over the influence depth (Figure 5.16a) was estimated to be 8 from the mean trend in Figure 5.16, yielding an average compressibility index I_c (same meaning as m_v in Terzaghi et al. 1996) of 0.092 MPa^{-1} by Eq. 5.14a or Figure 5.17. EC7 recommends all partial factors to be 1.0 for checking SLS, hence net applied pressure is:

$$q'_n = \frac{4190\,\text{kN} + 7500\,\text{kN}}{5\,\text{m} \times 20\,\text{m}} - 17\,\text{kN}/\text{m}^3 \times 0.8\,\text{m} = 103\,\text{kPa}$$

Shape factor $= f_s = \left(\dfrac{1.25L/B}{L/B + 0.25}\right)^2 = \left(\dfrac{1.25 \times 4}{4 + 0.25}\right)^2 \approx 1.4$

Sand thickness factor $= f_I = 1$

Time factor $= f_t = \left[1 + R_3 + R_t \log\left(\dfrac{t}{3}\right)\right] = \left[1 + 0.3 + 0.2\log\left(\dfrac{30}{3}\right)\right] = 1.5$ for 30-year period

Average settlement at 30 years $= 1.4 \times 1 \times 1.5 \times 103$ kPa $\times$ (5 m)$^{0.7} \times 0.092$ MPa^{-1} = **61.4 mm**

Short-term (immediate) settlement $= 1.4 \times 1 \times 1 \times 103$ kPa $\times$ (5 m)$^{0.7} \times 0.092$ MPa = **41 mm**

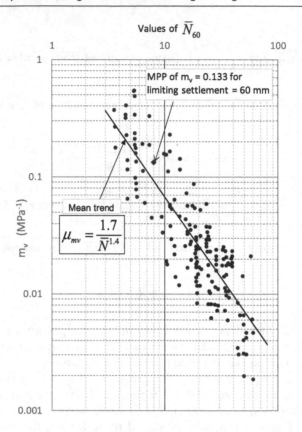

Figure 5.17 Relation between average N values within influence depth and compressibility m_v of sand (data from Burland and Burbidge 1985, in Figure 50.3 of Terzaghi et al. 1996).

The predicted short-term settlement of 41 mm and long-term settlement of 61 mm may be regarded as the *average (or most likely)* short-term and long-term settlements, respectively, because they are based on the mean compressibility value of $\mu_{mv} = 1.7 / \bar{N}^{1.4}$ from Eq. 5.14a or Figure 5.17, and also using average values of loads with all partial factors set to 1.0. Although the mean value of m_v (for $\bar{N} = 8$) is 0.092 MPa^{-1} by Eq. 5.14a or Figure 5.17, Tomlinson (2001) also gave minimum and maximum estimates of compressibility m_v as 0.04 MPa^{-1} and 0.30 MPa^{-1}, respectively, giving a possible range of 30-year settlement from 27 to 200 mm. Similarly, a short-term settlement range based on the Burland and Burbidge data is from 18 to 133 mm. Table 5.2 summarizes the possible values of calculated settlements.

In EC7, SLS is considered not exceeded if the estimated settlement is smaller than a *limiting* settlement. Tomlinson (2001) mentioned that the limiting short-term settlement for raft foundations on sand is from 40 to 65 mm.

Table 5.2 Summary of Tomlinson's calculated settlements of water tank on sand based on Burland and Burbidge (1985) method

	Compressibility m_v (MPa^{-1})	Short-term settlement (calculated)	Long-term settlement (calculated)
Average	0.092	41 mm	61 mm
Lower limit	0.04	18 mm	27 mm
Upper limit	0.30	133 mm	200 mm

The predicted *average* short-term settlement of 40 mm provides no information on the chance of exceeding the limiting settlement of 40 mm ~ 65 mm, because the uncertainty and scatter of m_v in Figure 5.17 has not been modelled in the settlement calculations.

It is shown in the next section that uncertainty analysis of the data in Figure 5.17 can be conducted to yield the standard deviation of compressibility m_v, followed by FORM analysis to obtain reliability indices and probabilities of exceeding SLS for different values of limiting settlement.

5.11.2 Statistical properties of compressibility m_v in Burland and Burbidge data

Information on the probabilities of exceeding different limiting settlements can be obtained from FORM analysis. But first one needs to estimate the statistical properties which quantify the uncertainty of the input variables in the Burland and Burbidge method. For the case in hand, the short-term settlement is calculated from the following equation (a reduced version of Eq. 5.14, with f_l and f_t both equal to 1.0):

$$\text{Short-term settlement: } s = \left(\frac{1.25 L/B}{L/B + 0.25} \right)^2 \times q'_n B^{0.7} m_v \qquad (5.15)$$

$$\text{OR} \quad s \approx 1.4 q'_n B^{0.7} m_v \qquad (5.15a)$$

$$\text{in which} \quad q'_n = \frac{Q_D + Q_L}{5\,\text{m} \times 20\,\text{m}} - 17\,\text{kN/m}^3 \times 0.8\,\text{m} \qquad (5.15b)$$

The three random variables to be modelled are dead load Q_D, live load Q_L and compressibility m_v. The mean values μ of Q_D and Q_L are taken to be 4190 kN and 7500 kN, as in the previous Tomlinson (2001) EC7 design calculations, and their CoV (i.e., σ/μ) are assumed to be 0.10 and 0.15, respectively, yielding σ values of 419 kN and 1125 kN, respectively.

The mean m_v value corresponding to the mean of STP N values ($\bar{N} = 8$) within the influence depth is $\mu_{mv} = 0.092$ MPa^{-1}, by Eq. 5.14a or Figure 5.17.

For the purpose of estimating the statistical properties of compressibility m_v, the x and y axes of Figure 5.17 are linearized as $\log(m_v)$ and $\log(\bar{N})$ in Figure 5.18, followed by using the *Add Linear Trendline* option in Microsoft Excel. The displayed equation on chart is practically the same as the following:

$$\text{Mean trend:} \quad \mu_{\log(mv)} = \log(1.7) - 1.4\log(\bar{N}) = \log\frac{1.7}{\bar{N}^{1.4}} \quad (5.16a)$$

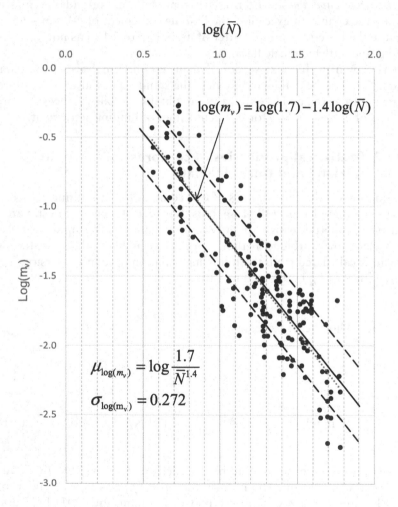

Figure 5.18 Replotting the Burland and Burbidge (1985) data in log(average N) versus log(m_v), with statistical analysis for the mean linear trend and standard deviation of log(m_v).

For the case in hand, $\bar{N}$ is equal to 8 within the influence depth, hence:

$$\mu_{\log(mv)} = \log\frac{1.7}{\bar{N}^{1.4}} = \log(0.0925) = -1.034 \qquad (5.16b)$$

Further analysis of the data in Figure 5.18 then yields:

$$\sigma_{\log(mv)} = 0.272 \qquad (5.16c)$$

The $\mu_{\log(mv)} \pm 1\sigma_{\log(mv)}$ dashed lines are shown in Figure 5.18, above and below the mean trend solid line.

5.11.3 FORM analysis for probability of settlement exceedance

FORM analysis can now be conducted, as shown in Figure 5.19a. The three random variables Q_D, Q_L and $\log(m_v)$ are assumed to be normally distributed. The input values shown for μ and σ of Q_D, Q_L and $\log(m_v)$ have already been explained previously. The performance function in the cell 'g(x*) immediate' at the bottom left of Figure 5.19a is:

$$g(x*) = \text{Limiting settlement } s_0 - \text{Settlement calculated by Eq. 5.15a} \qquad (5.17)$$

Initially, the values under the column labelled n were zeros. Excel Solver was then invoked to obtain the final solutions of x* values and n values, using the Low and Tang (2007) FORM procedure.

For a short-term limiting settlement of 60 mm, the computed reliability index β is 0.59, corresponding to a probability of 27.7% that the short-term settlement will exceed 60 mm.

Higher values of limiting settlement result in higher reliability index (Figure 5.20), and lower probability of exceedance (Figure 5.21a, where $P_f = \Phi(-\beta)$ is practically identical to the P_f from Monte Carlo simulation with 100,000 trials. The FORM reliability and outcome probability information conveyed in Figures 5.20 and 5.21a are valuable for decision-making regarding SLS and also reduce the ambiguity and subjectivity in the selection of limiting settlement. The target reliability index with respect to SLS may logically be lower than that for ULS (for which the target β is about 2.5–3.5 depending on the consequence of failure), because exceeding SLS does not immediately result in failure.

Although the mean short-term settlement is about 41 mm, its relatively high probability (27.7%) of exceeding 60 mm is not surprising, because of the large scatter in the compressibility m_v, Figures 5.17 and 5.18. This large uncertainty is also testified by the large discrepancies between predicted settlement and measured settlement, in Figure 5.21b. For the case in hand where the predicted settlement is 41 mm, past records in Figure 5.21b indicate measured settlements from 12 to 90 mm.

		x*	μ	σ	n	R matrix		
Normal	Q_D	4205.7	4190	419	0.038	1	0	0
Normal	Q_L	7613.3	7500	1125	0.101	0	1	0
Normal	$\log(m_v)$	-0.8767	-1.034	0.27	0.582	0	0	1

$$z_I = B^{0.7}$$

	f_s	f_I	f_t	m_v	q_n	B	z_I
Limiting s_0	1.4	1	1	0.1328	104.6	5	3.09
60							

g(x*) immediate		β	P_f
0.0		0.59	0.277

(a)

		x*	μ	σ	n	R matrix		
Normal	Q_D	4205.7	4190	419	0.038	1	0	0
Normal	Q_L	7613.3	7500	1125	0.101	0	1	0
Lognormal	m_v	0.1328	0.1122	0.077	0.582	0	0	1

I_c

$$z_I = B^{0.7}$$

	f_s	f_I	f_t	q_n	B	z_I
Limiting s_0	1.4	1	1	104.6	5	3.09
60						

g(x*) immediate		β	P_f
0.0		0.59	0.277

(b)

Figure 5.19 FORM analysis of the serviceability limit state (SLS) of water tank settlement, accounting for input uncertainties in the Burland and Burbidge method, (a) based on normally distributed $\log(m_v)$, and (b) based on lognormally distributed m_v.

5.11.4 From normally distributed $\log(m_v)$ to lognormally distributed m_v for FORM analysis

It is desirable to render the third random variable in Figure 5.19a, $\log(m_v)$, directly in terms of m_v. The transformation is possible through the theoretical relationship that a variable X is lognormally distributed when the natural logarithm of X is normally distributed. First we need to get the mean and SD of $\ln(m_v)$ from those of $\log(m_v)$, as follows:

Since:

$$\ln(m_v) = \frac{\log(m_v)}{\log(2.71828)} = 2.3026\log(m_v) \qquad (5.18a)$$

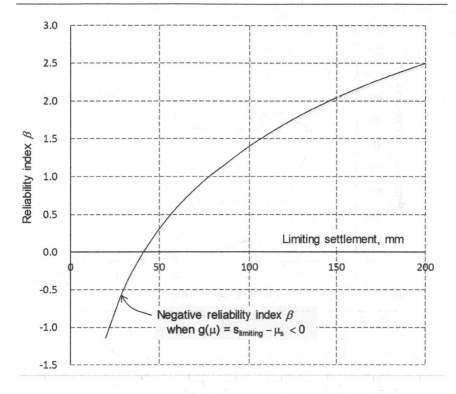

Figure 5.20 Reliability against exceeding limiting settlement, accounting for the scatter of compressibility m$_v$ in Burland and Burbidge method.

Hence:

$$\mu_{\ln(mv)} = 2.3026\mu_{\log(mv)} = 2.3026 \times (-1.034) = -2.381 \quad (5.18b)$$

$$\sigma_{\ln(mv)} = 2.3026\sigma_{\log(mv)} = 2.3026 \times 0.27 = 0.6217 \quad (5.18c)$$

Next, the mean and SD of the lognormally distributed m$_v$ are obtained from Eq. 5.9a and 5.9b:

$$\mu_{mv} = \exp\left(\mu_{\ln mv} + 0.5\sigma_{\ln mv}^2\right) = \exp\left(-2.381 + 0.5 \times 0.6217^2\right) = 0.1122 \ (5.19a)$$

$$\sigma_{mv} = \sqrt{\left[\exp\left(\sigma_{\ln(mv)}^2\right) - 1\right]\exp\left(2\mu_{\ln(mv)} + \sigma_{\ln(mv)}^2\right)} = 0.0771 \quad (5.19b)$$

The lognormally distributed m$_v$, with $\mu_{mv} = 0.1122$ and $\sigma_{mv} = 0.077$, is used in the FORM analysis of Figure 5.19b, yielding the same reliability index β, the same sensitivity indicator n values, and the same MPP values (4206, 7613, 0.1328) of Q$_D$, Q$_L$ and m$_v$, as Figure 5.19a. This verifies the

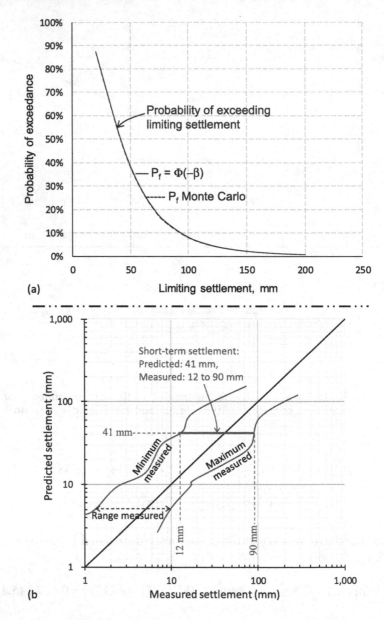

Figure 5.21 (a) Probability of exceeding limiting settlement by FORM analysis, (b) comparison of calculated settlement by Burland and Burbidge method versus measured settlement at the end of construction (Figure 5.21b based on Figure 50.6 of Terzaghi et al. 1996).

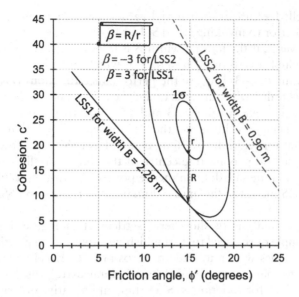

Figure 5.22 Distinguishing negative from positive reliability index: reliability index β is 3 for LSS1, and negative 3 for LSS2.

correct transformation of the *normally distributed random variable log(m_v)* of Figure 5.19a to the *lognormally distributed random variable m_v* of Figure 5.19b, using Eqs. 5.18 and 5.19.

The MPP value of $m_v^* = 0.133$ (for a limiting settlement of 60 mm) is well within the scattered data of Figure 5.17, hence the relatively high probability (27.7%) of exceeding 60 mm. The exceedance probability decreases as the limiting settlement becomes larger (Figure 5.21a). The three n values (0.038, 0.101, 0.582) in Figure 5.19 reveal that the calculated settlement is more sensitive to m_v than to Q_D and Q_L for the case in hand.

5.12 DISTINGUISHING POSITIVE FROM NEGATIVE RELIABILITY INDICES

In FORM analysis, one must distinguish negative from positive reliability indices. The computed β index can be regarded as positive only if the mean-value point is in the safe domain. Although the discussions in the next paragraph assume normally distributed random variables, they are equally valid for the equivalent normals of nonnormal random variables in FORM.

For the design footing width B = 2.28 m shown in Figure 5.1, the mean-value point is in the safe domain (the upper right domain above the LSS in Figure 5.1b and 5.1c). This can be confirmed by noting that the cell of the performance function, $g(x) = q_u(x^*) - q$, displays a positive value when the column labelled x^* are at their initial values (i.e., mean values, 23 and 15, respectively) prior to invoking Excel Solver to obtain the solution $\beta = 3.0$ by Eq. 5.3. However, if the footing width B is 0.96 m (instead of 2.28 m), the Solver-computed β is also 3.0 by Eq. 5.3, which must be given a negative sign (to become $\beta = -3$) because the performance function (*PerFn*) cell displays negative value when the cells under the column labelled x^* are at their initial values (i.e., mean values, 23 and 15, respectively) prior to invoking Excel Solver. The $\beta = 3$ ellipse and the LSS for width B = 2.28 m, denoted as LSS1, and that for width B = 0.96 m, denoted as LSS2, are shown in Figure 5.22. The mean-value point (the centre of the ellipse, or ellipsoid in higher dimensions) is in the safe domain (hence positive β) with respect to LSS1 for width B = 2.28 m, but in the unsafe domain with respect to LSS2 for width B = 0.96 m.

The FORM analysis of short-term settlement in Figure 5.19 provides another example of the need to distinguish negative β from positive β values. When the n values were initialized to zeros prior to Excel Solver optimization, the performance function cell ('g(x*) immediate') displays a positive value by Eq. 5.17 for *Limitimg* $s_0 > 41$ mm, and negative otherwise. Hence, the β values computed by Eq. 5.5 or 5.6 must be given a negative sign (resulting in the lower left portion of the curve in Figure 5.20 with negative β values) if the performance function is negative at the mean-value point.

The above leads to a very simple but necessary check: If the performance function g(x) is positive when the n values are initialized to zeros, treat the computed β as positive. If the performance function g(x) is negative when the n values are initialized to zeros, a negative sign must be given to the computed β.

Chapter 6

Pile foundations

6.1 INTRODUCTION

This chapter will discuss RBD-via-FORM of the ultimate limit state (ULS) and serviceability limit state (SLS) of axially loaded piles and laterally loaded piles, with modelling of uncertain inputs, spatially autocorrelated shear strength and model inaccuracy. It will be shown that RBD-via-FORM can reflect different parametric sensitivities from case to case and provide valuable insights to enhance the partial factor design approach.

We will begin with a case of *deterministic* pile length design based on average value of undrained shear strength and adhesion factor, before extending it statistically and probabilistically into RBD-via-FORM. This will be followed, in the section after next, by RBD of the allowable design load of a driven pile in sand. The effect of autocorrelated shear strength is investigated. Subsequent sections deal with FORM analysis of the serviceability limit state (SLS) of pile settlement using the Randolph and Wroth method, and probabilistic analysis and design with respect to the ULS and SLS of laterally loaded single piles obeying depth-dependent nonlinear and strain softening Matlock p-y curves.

Focus is on the information and insights afforded by the most probable point (MPP) of failure in FORM. Comparisons will be made with Eurocode 7 (EC7) and the load and resistance factor design (LRFD) method.

6.2 A DRIVEN PILE IN STIFF CLAY BELOW A JETTY

A deterministic design from Tomlinson (2001, Example 7.4) is summarized, followed by new extensions involving uncertainty quantification and RBD-via-FORM. Spatial autocorrelation will be considered.

6.2.1 Deterministic design of the length of a jetty pile

The soil profile below seabed consists of 4.5 m of very soft claycy silt overlying stiff to very stiff clay, Figure 6.1. Undrained shear strength tests on undisturbed samples gave the data shown. Steel tube piles of diameter 0.5 m

Deterministic analysis in Tomlinson (2001, Example 7.4): A jetty is sited on 4.5 m of very soft clayey silt overlying stiff to very stiff clay. Tests on undisturbed samples gave the shear strength-depth relationship shown in the figure below. For structural reasons it is desired to use 500 mm steel tube piles. Determine the required penetration to carry an allowable load of 400 kN.
In view of lateral movements in the jetty structure due to berthing forces, wind, and wave action, the skin friction in the very soft clayey silt was ignored.
Consider a penetration length of 13 m below sea-bed level

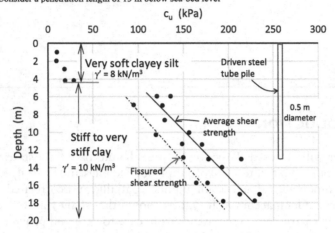

From the figure above, the average shear strength from 4.5 to 13 m ≈ 140 kN/m²
Fissured shear strength at 13 m = 145 kN/m²
At 4.5 m below seabed, stiff clay c_u = 100 kPa. Allowing for 1 m erosion, σ_{vo}' = 28 kPa
At 13 m below seabed, stiff clay c_u = 180 kPa. Allowing for 1 m erosion, σ_{vo}' = 113 kPa
Based on c_u/σ_{vo}' values, and charts in Tomlinson Fig.7.13, $F\alpha_p$ = 1×0.5 along entire shaft.
Hence total shaft resistance in stiff clay is:

$$Q_s = F\alpha_p \mu_{cu} A_s = 1\times0.5\times140 \text{ kPa} \times (\pi \times 0.5 \text{ m} \times 8.5 \text{ m}) = \textbf{935 kN}$$

The end resistance in *fissured* stiff clay is:

$$Q_b = q_b A_b = N_c c_{ub} A_b = 9\times145\times\frac{1}{4}\pi\times0.5^2 = \textbf{256 kN}$$

Therefore total pile resistance = $Q_s + Q_b$ = 935 + 256 = **1191 kN**

Figure 6.1 Deterministic analysis based on mean values of shear strength and sustained load.

will be driven through the 4.5 m thick very soft clayey silt (buoyant unit weight γ' = 8 kN/m³) to the underlying stiff to very stiff clay (buoyant unit weight γ' = 10 kN/m³). Each pile will carry a vertical load of 400 kN.

The shaft resistance for the driven pile is estimated by Tomlinson as follows, for a trial length of 13 m in soil below seabed:

$$R_s = F\alpha_p \mu_{cu} A_s \tag{6.1}$$

$$= 1\times0.5\times140 \text{ kPa} \times (\pi \times 0.5 \text{ m} \times 8.5 \text{ m}) = 935 \text{ kN}$$

where the length factor F and peak adhesion factor α_p are read from charts, and μ_{cu} is the average undrained shear strength of 140 kPa, taken at the mid-length of the pile in the stiff clay layer, which is at a depth of (L + 4.5)/2, or 8.75 m below seabed when an initial pile length of 13 m below seabed is tried. The skin friction in the very soft clayey silt was ignored in view of lateral movements in the jetty structure due to berthing forces, wind and wave action. The shaft resistance is assumed to act along the entire embedded length of 8.5 m (=13 m – 4.5 m) in the stiff clay layer. When calculating effective vertical stress σ_{v0}', needed for c_u/σ_{v0}', allowance is made for 1 m erosion below seabed, so that the overburden thickness of the top layer of very soft clayey silt is 3.5 m instead of 4.5 m.

The base bearing resistance by Tomlinson (2001) is:

$$R_b = q_b A_b = N_c c_{ub} A_b = 9 \times 145 \times \frac{1}{4} \pi \times 0.5^2 = 256 \, \text{kN} \tag{6.2}$$

where c_{ub} is based on the fissured shear strength at pile toe level, which is assumed to be representative of the average c_{ub} of the soil volume near pile toe relevant to end bearing failure. The dashed line for the fissured shear strength, from Tomlinson (2001), indicates a c_{ub} of 145 kPa at pile toe when the trial length is 13 m.

The total vertical resistance R of the pile is:

$$R = R_s + R_b = 935 \, \text{kN} + 256 \, \text{kN} = 1191 \, \text{kN} \tag{6.3}$$

The lumped factor of safety, Fs = R/Q = 1191/400 ≈ 3, was deemed to be high, because a safety factor of 2.5 would be acceptable.

An alternative deterministic approach was also used to compute the allowable load Q_a:

$$Q_a = R_s/1.5 + R_b/3 \tag{6.4}$$

Even with a conservative Q_s value of 734 kN, based on adopting a c_u value of 110 kPa (at the top of the stiff clay layer) for the entire 8.5 m shaft length in the stiff clay layer, an allowable Q_a of 575 kN was obtained, well in excess of the required safe working load of 400 kN. Tomlinson suggested that the pile length below seabed could perhaps be reduced from 13 to 12 m.

The lumped factor of safety (e.g., F_s = 2.5), or the resistance factors 1.5 and 3 to shaft resistance and base resistance, respectively, were meant to provide safety margins against failure arising from uncertainties and scatter in the input values of α, c_u, $c_{u,fissured}$ and applied load Q, and from the approximate nature of calculating pile bearing resistance.

An alternative is to extend the above deterministic approach into RBD-via-FORM, which accounts explicitly for parametric uncertainties and scatter. This is done next.

6.2.2 Statistical inputs, and RBD-via-FORM

The deterministic formulation also underlies RBD-via-FORM. However, RBD-via-FORM also requires standard deviations and parametric correlations (if any), which are not used in deterministic analysis.

6.2.2.1 Quantifying the scatter/uncertainty in c_u

The undrained shear strength profile is modelled as a random variable by (random field will be considered later):

$$c_u = \xi \mu_{cu} \tag{6.5}$$

where

$$\mu_{cu} = c_0 + by = 63 + 8.41y \tag{6.5a}$$

The average trendline μ_{cu} is the solid line in Figure 6.2a, obtained from regression analysis of the 19 c_u data points in the stiff clay layer (depth $y >$ 4.5 m). Nineteen values of $\xi = c_u/\mu_{cu}$ were then computed, obtaining $\mu_\xi = 1.0$, and $\sigma_\xi = 0.13$. This means $\xi = 1.0$ stands for the average trendline, and $\xi = 1 \pm 0.13$ yields the one-standard-deviation dashed lines left and right of the solid trendline in Figure 6.2a. The scatter in c_u can therefore be represented by an average trendline, Eq. 6.5a, and a coefficient of variation (c.o.v., or σ/μ) of 0.13, shown in the first row of Figure 6.2b. This c_u is treated as a single random variable at the midpoint of the pile embedment length (L_p) in the stiff clay (which will change when different pile lengths are tried). A random field modelling of c_u will be compared shortly.

6.2.2.2 Quantifying the scatter/uncertainty in adhesion factor α

Charts were given in Tomlinson (2001) for determining adhesion factor α. This study uses the Kolk and van der Velde (1996) empirical equation for the mean value of α:

$$\alpha = 0.55 \left(\frac{40}{L_p/D_0} \right)^{0.2} \left(\frac{c_u}{\sigma'_{v0}} \right)^{-0.3} \tag{6.6}$$

where
L_p is that portion of pile length in the stiff clay layer,
D_0 is the pile diameter, and
$\dfrac{c_u}{\sigma'_{v0}}$ is based on the average value at the mid-point of L_p

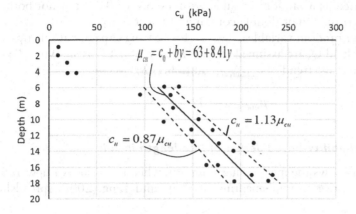

(a) Mean trend of c_u, and $\pm 1\sigma$ of c_u in dashed lines

	Mean value (μ)		Coefficient of variation
c_u	$\mu_{cu} = 63 + 8.41y$ (kPa)		0.13
α	$\mu_\alpha = 0.55\left(\dfrac{40}{L_p/D_0}\right)^{0.2}\left(\dfrac{\mu_{cu}}{\sigma'_{v0}}\right)^{-0.3}$		0.25
$c_{u,\text{fissured}}$	$\mu_{cu,\,\text{fissured}} = 34.5 + 8.5y$ (kPa)		0.15
Load Q	400 kN		0.15

(b) Uncertainties in c_u, adhesion factor α, $c_{u,\text{fissured}}$, and applied load Q

Figure 6.2 Statistics of shear strengths, adhesion factor α, applied load Q and correlations, for subsequent RBD in the next figure.

Other empirical equations (e.g., Randolph and Murphy 1985; Semple and Rigden 1984) can be used without invalidating the discussions and inferences in this study.

The scatter of actual data points (Semple and Rigden 1984; Karlsud et al.1993) around the fitted equation of Kolk and van der Velde (1996) is substantial. A coefficient of variation (c.o.v.) of 0.25 on α is used in this study. This has the same effect as multiplying α by a model uncertainty factor ξ_α, that is $\xi_\alpha\alpha$, where ξ_α has mean of 1.0 and c.o.v. of 0.25. Either α or ξ_α

can be treated as a random variable with a c.o.v. of 0.25, but not both, to avoid overcorrection or double-counting of uncertainty.

Two other random variables, $c_{u,fissured}$ (for bearing capacity at pile tip) and the applied load Q, are assumed to have a c.o.v. of 0.15, as shown in Figure 6.2b. The average trend of $c_{u,fissured}$ (Figure 6.1) is:

$$\mu_{cu,fissured} = 34.5 + 8.5y \tag{6.7}$$

6.2.2.3 Reliability-based design of pile length in soil

Figure 6.3a shows the mean-value situation, when the values of n_i were initially zeros, prior to implementing the Low and Tang (2007) spreadsheet

	L	D_0	γ_1'	γ_2'	$\mu_{\sigma v0}$
	11.78	0.5	8	10	64.4

		x*	μ	σ	n	Correlation matrix R			
Normal	c_u	131.46	131.46	17.09	0.000	1	0	0	0
Normal	α	0.543	0.543	0.136	0.000	0	1	0	0
Normal	$c_{u,fissured}$	134.63	134.63	20.19	0.000	0	0	1	0
Normal	Q	400.00	400	60	0.000	0	0	0	1

	R_s	R_b	g(x)	β
	816.98	237.91	654.89	0.00

(a) Mean resistance of Rs = 817 kN and R_b = 238 kN compared with mean loading 400 kN, prior to RBD in (b) below.

	L	D_0	γ_1'	γ_2'	$\mu_{\sigma v0}$
	11.78	0.5	8	10	64.4

		x*	μ	σ	n	Correlation matrix R			
Normal	c_u	123.46	131.46	17.09	-0.468	1	0	0	0
Normal	α	0.165	0.543	0.136	-2.786	0	1	0	0
Normal	$c_{u,fissured}$	124.16	134.63	20.19	-0.518	0	0	1	0
Normal	Q	452.29	400	60	0.871	0	0	0	1

	R_s	R_b	g(x)	β
	232.87	219.41	0.00	3.00

(b) RBD outcome: Design resistance R_s = 233 kN and R_b = 219 kN. Pile length L = 11.78m achieves the target reliability index β of 3.0.

Figure 6.3 Extending Tomlinson's deterministic case into RBD-via-FORM. The uncertainties in shear strengths, adhesion factor and loading (Figure 6.2) have been incorporated.

FORM computational procedure. The pile length L of 11.78 m was obtained by trial to achieve a target β of 3.0 in Figure 6.3b.

For a trial length L, the mean value of c_u under the column labelled μ in Figure 6.3a is calculated at the midpoint of the pile embedment length in the stiff clay layer below a depth of 4.5 m, using Eq. 6.5a. For a pile length L of 11.78 m, the mean $c_u = 63 + 8.41*(4.5 + L)/2 = 131.46$ kPa, as shown in Figure 6.3a. The standard deviation (σ) of c_u is equal to its c.o.v. of 0.13 (from Figure 6.2) multiplied by its mean value.

Other formulas have been entered in the boxed cells of Figure 6.3a, for calculating the mean value of effective vertical stress (μ_{ovo}') and the mean adhesion factor α (Eq. 6.6) at the midpoint of the pile embedment length in the stiff clay layer, and mean value of $c_{u,fissured}$ at pile toe (Eq. 6.7). For normal variates, the x^* column contains the equations $\mu_i + n_i\sigma_i$. (The Low and Tang VBA code to compute x_i from n_i is necessary only when the variates are nonnormal.)

The cells R_s and R_b in Figure 6.3a contain formulas to calculate the shaft and base resistance based on the x^* values. The performance function g(x) cell contains the formula $R_s + R_b - Q^*$, where Q^* is the x^* value of Q.

The reliability index β cell contains the array formula for the quadratic form $\sqrt{n^T R^{-1} n}$ of Eq. 2.9. In this case where the variables are uncorrelated, R reduces to an identity matrix and the quadratic form reduces to $\sqrt{\sum n_i^2}$. Hence the simple formula '=sqrt(sumsq(n))' can also be used in the β cell, in which 'n' is entered by selecting the column of 4 values at one go.

With the n column values initially equal to zeros in Figure 6.3a, the x^* values of the four random variables result in a mean shaft resistance R_s of 817 kN, and a mean end bearing resistance R_b of 238 kN, with a total mean resistance of $817 + 238 = 1055$ kN, which is 655 kN (shown in the g(x) cell) higher than the mean applied load Q under the x^* column. This is a safe situation. Hence, the β value to be obtained shortly is treated as positive.

Figure 6.3b shows that, after trying different L values (and each time invoking the Solver tool to obtain the β solution), a pile length of 11.78 m in soil (excluding the length above seabed) is required to achieve the target reliability index of 3.0. The values under the column labelled x^* now represent the most probable point (MPP) of failure, also referred to as the design point. This design point is reached when the values of the resistance parameters (c_u, α, $c_{u,fissured}$) decrease, and the Q value increases, from their mean values of Figure 6.3a, to the values shown under the x^* column in Figure 6.3b. This is the point where an expanding 4D dispersion ellipsoid, defined by Eq. 6.8, just touches the limit state surface (LSS), defined by Eq. 6.9:

$$\beta = \min_{x \in F} \sqrt{n^T R^{-1} n} \tag{6.8}$$

Table 6.1 Back-calculated partial factors with respect to mean values μ

γ_{cu}	γ_α	$\gamma_{cu,fissured}$	γ_Q	LF	RF	RF_{shaft}	RF_{base}
1.07	3.3	1.08	1.13	1.13	0.43	0.29	0.92

in which R is the correlation matrix in Figure 6.3.

$$g(\mathbf{x}) = R_s + R_b - Q = 0 \qquad (6.9)$$

in which R_s, R_b and Q are based on the x* values.

The n_i values in Figure 6.3b are sensitivity indicators: they represent the difference between the MPP of failure values and the mean values, normalized by the respective standard deviations. A negative value of n_i means that the design value (x_i*) is smaller than its mean value, and vice versa. The adhesion factor α is the most sensitive parameter for the case in hand, with an n_α value of –2.786, which means that the design point value of x_α* is $2.786 \times \sigma_\alpha$ *below* the μ_α value. (The above intuitive picture of β and MPP of failure remains valid for nonnormals, when viewed from the perspective of equivalent normals, as explained in Low and Tang 2007.)

Partial factors, and load and resistance factors, can be back-calculated from the outcome of RBD-via-FORM, as shown in Table 6.1. However, such back-calculated partial factors reflect case-specific sensitivities and target β and are also affected by the level of conservatism in the characteristic and nominal values in EC7 and LRFD, The back-calculated partial factors γ_{cu}, γ_α, $\gamma_{cu,fissured}$ and γ_Q in Table 6.1 would be lower if based on conservative characteristic values instead of mean values, and the back-calculated load factor (LF) would be lower and resistance factor (RF) higher if based on conservative nominal values.

It is more illuminating to conduct RBD-via-FORM in parallel with EC7 or LRFD design, instead of attempting to calibrate partial factors and LF and RF from RBD-via-FORM.

6.2.2.4 Discretized random field

Spatial autocorrelation arises in geological material by virtue of its formation by natural processes acting over unimaginably long time. This endows geomaterial with some unique statistical features (e.g., spatial autocorrelation) not commonly found in structural material manufactured under strict quality control.

In the RBD-via-FORM of Figure 6.3, a single representative value of c_u at the mid-point of the embedment length in the stiff clay is treated as a random variable. The question may be asked: how will the RBD be affected if a series of spatially autocorrelated c_u values along the linear trend of Figure 6.2 are modelled as random variables each with a c.o.v. of 0.13?

	L	D_0	γ_1'	γ_2'	$\mu_{\sigma vo}$	δ (m)
	11.78	0.5	8	10	64.4	1.5

		x*	μ	σ	n	Correlation matrix R							α	$c_{u,f}$	Q
Normal	c_{u1}	103.74	105.22	13.68	-0.108	1.00 0.50 0.25 0.12 0.06 0.03 0.02						0	0	0
Normal	c_{u2}	111.88	113.96	14.82	-0.141	0.50 1.00 0.50 0.25 0.12 0.06 0.03						0	0	0
Normal	c_{u3}	120.16	122.71	15.95	-0.160	0.25 0.50 1.00 0.50 0.25 0.12 0.06						0	0	0
Normal	c_{u4}	128.52	131.46	17.09	-0.172	0.12 0.25 0.50 1.00 0.50 0.25 0.12						0	0	0
Normal	c_{u5}	136.99	140.20	18.23	-0.176	0.06 0.12 0.25 0.50 1.00 0.50 0.25						0	0	0
Normal	c_{u6}	145.70	148.95	19.36	-0.168	0.03 0.06 0.12 0.25 0.50 1.00 0.50						0	0	0
Normal	c_{u7}	154.83	157.70	20.50	-0.140	0.02 0.03 0.06 0.12 0.25 0.50 1.00						0	0	0
Normal	α	0.157	0.543	0.14	-2.844	0 0 0 0 0 0 0						1	0	0
Normal	$c_{u,fiss.}$	124.39	134.63	20.19	-0.507	0 0 0 0 0 0 0						0	1	0
Normal	Q	451.13	400	60.00	0.852	0 0 0 0 0 0 0						0	0	1

R_s	R_b	g(x)	β	$\sqrt{n^T R^{-1} n}$
231.3	219.8	0.00	3.02	

$$R_s + R_b - Q*$$

Figure 6.4 Random field modelling with autocorrelation distance δ = 1.5 m, with 7 segments in the stiff clay of segmental length about 1 m.

Figure 6.4 shows the RBD for the case in hand, accounting for spatial autocorrelation of the seven c_u values by the established exponential autocorrelation function, between c_{ui} and c_{uj} at depths y_i and y_j, respectively:

$$\rho_{ij} = e^{-\frac{|y_i - y_j|}{\delta}} \qquad (6.10)$$

where the autocorrelation distance δ in the vertical direction is assumed to be 1.5 m.

The computed β value of 3.02 is practically the same as the 3.0 in Figure 6.3b which did not model random field. This may be attributed to the low sensitivity of the c_u values relative to α for the case in hand, where the c.o.v. of c_u is 0.13 while that of α is 0.25. In general, discretized random field (like Figure 6.4, with spacing between c_u values smaller than 2 × δ) can be conducted if necessary. Table 6.2 shows that failure probabilities from FORM β agree well with those from Monte Carlo simulations.

The single random variable of c_u in Figure 6.3 represents the entire pile length of 7.28 m (= 11.78 m – 4.5 m) in the stiff clay layer, in contrast to the length of 1.04 m for each of the seven spatially autocorrelated c_u values of Figure 6.4. In principle, one can use the following approximate equation of the reduction factor Γ_R (Vanmarcke 1980) to reduce the standard deviation σ_{cu} of Figure 6.3 to a smaller local average value when the length is

Table 6.2 Comparing failure probabilities from FORM β with Monte Carlo simulations (MCS)

Figure 6.3: $P_f = \Phi(-\beta)$	*MCS, 800,000*	*Figure 6.4: $P_f = \Phi(-\beta)$*	*MCS, 800,000*
0.135%	0.144%	0.125%	0.127%

greater than the scale of fluctuation δ_R (equal to twice the autocorrelation distance δ):

$$\Gamma_R(l) = \sqrt{(\delta_R/l)} \quad \text{for } l \geq \delta_R \tag{6.11}$$

For the case in Figure 6.4 with $\delta_R = 2 \times 1.5$ m, the reduction factor Γ_R for the standard deviation of c_u in Figure 6.3 is $\sqrt{3/7.28} = 0.642$. This will reduce the σ_{cu} of Figure 6.3 from 17.09 to 10.97 kPa, obtaining $\beta = 3.02$ as in Figure 6.4. Alternatively, with this reduced value of 10.97 kPa for the standard deviation c_u in Figure 6.3, a target β of 3.00 (instead of 3.02) is achieved if L is 11.70 m (instead of 11.78 m), the same $\beta = 3.0$ in Figure 6.4 if L = 11.70 m.

The above RBD-via-FORM example is for a driven pile in clay. The next example involves a driven pile in sand, with somewhat different formulations and random variables. The EC7 Design Approach 1 Combination 2 (DA1b) is summarized, before extending it to RBD-via-FORM.

6.3 A DRIVEN PILE IN SAND

6.3.1 Maximum allowable design load based on EC7-DA1b

The EC7-DA1b calculations below are from Knappett and Craig (2019, Example 10.1). A single close-ended steel tubular pile of outer diameter 0.3 m and length 10 m is driven into a site at Canada. The soil profile consists of 15 m of silica sand overlying soft clay, Figure 6.5. Based on an assumed coefficient of earth pressure K = 1.0, a characteristic value of $\phi' = 36°$ (from CPTU data) and an interface friction angle $\delta'_{design} = \lambda\varphi'_{design}$ where $\lambda = 0.75$, the allowable permanent design load was determined sequentially, based on EC7 DA1b, as shown below.

$$\phi'_{des} = \tan^{-1}\left(\tan\phi' / \gamma_{\tan\phi}\right) = \tan^{-1}\left(36° / 1.25\right) = 30.2° \tag{6.12}$$

from which $\delta'_{des} = 0.75\varphi'_{des} = 22.7°$.

The $\gamma_{\tan\phi}$ above is the EC7 partial factor for $\tan\phi'$. The design shaft capacity $R_{s,des}$ is:

$$R_{s,des} = R_s / \gamma_{Rs} = \left[\pi D_0 \int_0^{10} K\left(\gamma' / \gamma_\gamma\right) z \tan\delta'_{des} dz\right] / \gamma_{Rs} \tag{6.13}$$

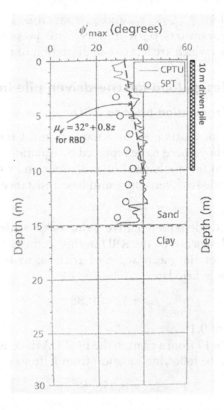

Figure 6.5 The mean angle of friction based on CPTU is modelled by a linear trend for reliability-based design, with a c.o.v. of 0.1. (CPTU curve and SPT data points from Figure 7.29b of Craig and Knappett 2020.)

$$= \left[\pi \times 0.3 \times 1 \times \frac{(20 - 9.8)}{1.0} \times \tan 22.7° \times \left[\frac{z^2}{2} \right]_0^{10} \right] \bigg/ 1.3 = 155 \, \text{kN}$$

The design base capacity $R_{b,des}$ is:

$$R_{b,des} = \frac{R_b}{\gamma_{Rb}} = \frac{A_b N_q (\gamma'/\gamma_\gamma) L_p}{\gamma_{Rb}} = \frac{0.0707 \times 17 \times (10.2/1.0) \times 10}{1.30} = 94 \, \text{kN} \quad (6.14)$$

where γ_{Rs}, γ_{Rb} and γ_γ are partial factors form EC7, and $N_q \approx 17$ is read from a chart. The maximum allowable permanent design load Q is:

$$Q_d = R_d = R_{s,des} + R_{b,des} = 155 + 94 = 249 \, \text{kN} \quad (6.15)$$

Apart from the four $\gamma_{\tan\phi}$, γ_{Rs}, γ_{Rb} and γ_γ from tables in EC7, the values of other important parameters (λ, K and N_q) are presumably conservative characteristic values which are left to the discretion of the designer.

6.3.2 RBD of design load for the driven pile in sand

6.3.2.1 Refined formulation prior to RBD

For RBD-via-FORM, charts used by Knappett and Craig (2019, Example 10.1), mentioned above, need to be replaced by equations. The following are refinements made to the mechanical formulations and treatment of parameters affecting the pile shaft resistance and base resistance for the purpose of conducting RBD:

(a) Instead of a single representative value of ϕ' at depth of 5 m in the EC7-DA1b design above, the RBD in this section adopts a linear trend (Figure 6.5) for the mean angle of friction, to be treated as a discretized random field later:

$$\mu_{\phi'} = 32° + 0.8z \tag{6.16}$$

with a c.o.v. of 0.1

(b) Instead of $N_q = 17$ from a chart in the EC7-DA1b design of Knappett and Craig (2019), the following equation from Jeffrey et al. (2016) is used:

$$N_q = 0.55e^{0.11\phi'} \tag{6.17}$$

Equation 6.17 is only one of several possible N_q models suggested in the literature. Hence, a model uncertainty factor M_{Nq} with a mean of 1.0 and a c.o.v. of 0.2 is assumed for N_q.

(c) Instead of K = 1 in the EC7-DA1b design, the coefficient of earth pressure K is a random variable with a mean value of 1.0 and a c.o.v. of 0.15.

(d) Instead of $\lambda = 0.75$ for computing interface friction angle $\delta = \lambda\phi'$ in the EC7-DA1b design, in RBD, the parameter λ is a random variable with a mean value of 0.75 and a standard deviation of 0.10.

(e) The buoyant unit weight γ' is treated as random variable with a mean value of 10.2 kN/m³ and a standard deviation of 0.5 kN/m³.

(f) An autocorrelation distance of $\xi = 2$ m is assumed for the case in hand. The symbol ξ is used, because δ in this case stands for interface friction angle as mentioned in (d) above.

The performance function is:

$$g(x) = R_{shaft} + R_{base} - Q \tag{6.18}$$

where

$$R_{shaft} = \pi D_0 \int_0^{L_p} K\sigma_v' \tan\lambda\phi' dz \tag{6.18a}$$

and

$$R_{base} = M_{Nq}N_q\sigma'_{v,base}A_{base} \tag{6.18b}$$

6.3.2.2 RBD involving discretized random field

Figure 6.6a shows that, for the 10 m long-driven pile, the design load (Q_{design}), top row, should not exceed 247 kN (obtained by trial) in order to achieve the target β of 3.0. Ten spatially autocorrelated ϕ' random variables were modelled, each for a vertical length of 1 m. The sensitivity indicators (n_i in Figure 6.6a) of ϕ' are plotted in Figure 6.7a and show the sensitivity of ϕ' increasing nonlinearly with depth. This means that ϕ' values at greater depth affect shaft resistance more than those at shallower depth, due to the dual effects of increasing $\mu_{\phi'}$ with depth (Figure 6.5) and increasing σ_v' with depth, both affecting shaft resistance according to Eq. 6.18a. The RBD outcome in Figure 6.6b, where four autocorrelated ϕ' values are used, each representing a length of 2.5 m, is practically the same as Figure 6.6a with respect to Q_{design} and the design values of R_{shaft}, $M_{nQ}N_q$ and R_{base} (to the left of the Q_{design} cell in the top row), because in both Figure 6.6a and 6.6b the discretized distances (1 m and 2.5 m) are both smaller than the scale of fluctuation (δ_R in Eq. 6.11, equal to twice the autocorrelation distance ξ of 2 m, i.e., 4 m).

The values of the sensitivity indicators (n_i) in Figures 6.6 and 6.7 suggest that, for the case in hand, reliability is most affected by the ϕ' values near pile base, K, λ and M_{Nq}, where the symbols are defined in connection with Eqs. 6.12–6.14 and 6.16–6.18.

In Figure 6.7a, the two sensitivity indicators of soil friction angles along pile shaft (from Figure 6.6c, at depth 2.5 m and 7.5 m) are plotted in triangular markers, without linking them with a line, because a minimum of three points are needed to capture nonlinearity in sensitivities. The shaft length represented by each of these two friction-angle random variables is 5 m, slightly greater than the scale of fluctuation of 4 m. Nevertheless, the reliability index is practically the same as Figure 6.6a and 6.6b, without variance reduction.

The R_{shaft} and R_{base} values of 150.5 kN and 96.47 kN in Figure 6.6a (and similar values in Figure 6.6b and 6.6c) are design resistance values based on the design point values under the x* column. Their sum is the total design resistance of 247 kN, equal to the design load. The mean values of R_{shaft}, R_{base} and other random variables are displayed (in the top row and under the x_i^* column) when the n_i values were initially zeros. The ratios of mean values to design values (or their reciprocals) are akin to the partial factors in EC7 and load and resistance factors in LRFD. RBD-via-FORM needs mean values and standard deviations, but not nominal/characteristic values and partial factors. Back-calculated partial factors, LF and RF, vary with target β, assumed characteristic/nominal values, and are also case dependent. It is more

L_p	Δz	D_0	ξ	R_{shaft}	$M_{nq}N_q$	R_{base}	Q_{design}	$g(x)$	β	$\Phi(-\beta)$
10	1	0.3	2	150.5	13.75	96.47	247.0	0.00	3.00	0.135%

$$\xi = 2.0 \text{ m}$$
$$\rho_{\phi'_i \cdot \phi'_{i+1}} = e^{-|z_{i+1}-z_i|/\xi}$$

ξ is autocorrelation distance

		μ	σ	x_i^*	n_i		Depth	0.5	1.5	2.5	3.5	4.5	5.5	6.5	7.5	8.5	9.5	10
Normal	M_{Nq}	1	0.20	0.78	-1.084		1	0	0	0	0	0	0	0	0	0	0	0
Normal	γ	10.2	0.50	9.93	-0.548		0	1	0	0	0	0	0	0	0	0	0	0
Normal	λ	0.75	0.10	0.64	-1.148	$\delta=\lambda\phi'$	0	0	1	0	0	0	0	0	0	0	0	0
Normal	K	1	0.15	0.82	-1.215	$K\gamma z \tan\delta \cdot \Delta z$	0	0	0	1	0	0	0	0	0	0	0	0
Normal	ϕ_1'	32.4	3.24	32.11	-0.090	1.51	0.50	0	0	0	0	1.00	0.61	0.37	0.22	0.14	0.08	0.05 0.03 0.02 0.01 0.01
Normal	ϕ_2'	33.2	3.32	32.73	-0.142	4.62	1.50	0	0	0	0	0.61	1.00	0.61	0.37	0.22	0.14	0.08 0.05 0.03 0.02 0.01
Normal	ϕ_3'	34	3.4	33.29	-0.208	7.85	2.50	0	0	0	0	0.37	0.61	1.00	0.61	0.37	0.22	0.14 0.08 0.05 0.03 0.02
Normal	ϕ_4'	34.8	3.48	33.79	-0.290	11.17	3.50	0	0	0	0	0.22	0.37	0.61	1.00	0.61	0.37	0.22 0.14 0.08 0.05 0.04
Normal	ϕ_5'	35.6	3.56	34.21	-0.391	14.56	4.50	0	0	0	0	0.14	0.22	0.37	0.61	1.00	0.61	0.37 0.22 0.14 0.08 0.06
Normal	ϕ_6'	36.4	3.64	34.51	-0.520	17.96	5.50	0	0	0	0	0.08	0.14	0.22	0.37	0.61	1.00	0.61 0.37 0.22 0.14 0.11
Normal	ϕ_7'	37.2	3.72	34.63	-0.690	21.32	6.50	0	0	0	0	0.05	0.08	0.14	0.22	0.37	0.61	1.00 0.61 0.37 0.22 0.17
Normal	ϕ_8'	38	3.8	34.48	-0.927	24.47	7.50	0	0	0	0	0.03	0.05	0.08	0.14	0.22	0.37	0.61 1.00 0.61 0.37 0.29
Normal	ϕ_9'	38.8	3.88	33.87	-1.270	27.20	8.50	0	0	0	0	0.02	0.03	0.05	0.08	0.14	0.22	0.37 0.61 1.00 0.61 0.47
Normal	ϕ_{10}'	39.6	3.96	32.51	-1.791	29.06	9.50	0	0	0	0	0.01	0.02	0.03	0.05	0.08	0.14	0.22 0.37 0.61 1.00 0.78
Normal	ϕ_b'	40	4	31.48	-2.129		10	0	0	0	0	0.01	0.01	0.02	0.04	0.06	0.11	0.17 0.29 0.47 0.78 1.00

(a) RBD of applied load for a target $\beta = 3.0$, with 1D random field of 10 values of ϕ' along pile shaft

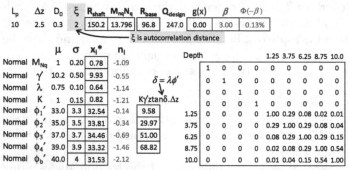

L_p	Δz	D_0	ξ	R_{shaft}	$M_{nq}N_q$	R_{base}	Q_{design}	$g(x)$	β	$\Phi(-\beta)$
10	2.5	0.3	2	150.2	13.796	96.8	247.0	0.00	3.00	0.13%

ξ is autocorrelation distance

		μ	σ	x_i^*	n_i		Depth		1.25	3.75	6.25	8.75	10.0
Normal	M_{Nq}	1	0.20	0.78	-1.09		1	0	0	0	0	0	0 0 0
Normal	γ	10.2	0.50	9.93	-0.55	$\delta=\lambda\phi'$	0	1	0	0	0	0	0 0 0
Normal	λ	0.75	0.10	0.64	-1.14		0	0	1	0	0	0	0 0 0
Normal	K	1	0.15	0.82	-1.21	$K\gamma z \tan\delta \cdot \Delta z$	0	0	0	1	0	0	0 0 0
Normal	ϕ_1'	33.0	3.3	32.54	-0.14	9.58	1.25	0	0	0	0	1.00	0.29 0.08 0.02 0.01
Normal	ϕ_2'	35.0	3.5	33.81	-0.34	29.97	3.75	0	0	0	0	0.29	1.00 0.29 0.08 0.04
Normal	ϕ_3'	37.0	3.7	34.46	-0.69	51.00	6.25	0	0	0	0	0.08	0.29 1.00 0.29 0.15
Normal	ϕ_4'	39.0	3.9	33.32	-1.46	68.82	8.75	0	0	0	0	0.02	0.08 0.29 1.00 0.54
Normal	ϕ_b'	40.0	4	31.53	-2.12		10.0	0	0	0	0	0.01	0.04 0.15 0.54 1.00

(b) Practically identical Q_{design} as (a), with 1D random field of 4 values of ϕ' along pile shaft

L_p	Δz	D_0	ξ	R_{shaft}	$M_{nq}N_q$	R_{base}	Q_{design}	$g(x)$	β	$\Phi(-\beta)$
10	5	0.3	2	149.1	13.951	97.875	247.0	0.00	3.00	0.13%

ξ is autocorrelation distance

		μ	σ	x_i^*	n_i		Depth			2.50	7.50	10.0
Normal	M_{Nq}	1	0.20	0.78	-1.11		1	0	0	0	0	0 0
Normal	γ	10.2	0.50	9.93	-0.55	$\delta=\lambda\phi'$	0	1	0	0	0	0 0
Normal	λ	0.75	0.10	0.64	-1.14		0	0	1	0	0	0 0
Normal	K	1	0.15	0.82	-1.21	$K\gamma z \tan\delta' \cdot \Delta z$	0	0	0	1	0	0 0
Normal	ϕ_1'	34.0	3.4	33.05	-0.28	39.09	2.5	0	0	0	0	1.00 0.08 0.02
Normal	ϕ_2'	38.0	3.8	33.53	-1.18	119.14	7.5	0	0	0	0	0.08 1.00 0.29
Normal	ϕ_b'	40.0	4	31.68	-2.08		10.0	0	0	0	0	0.02 0.29 1.00

Monte Carlo simulations for Fig. (a), (b), and (c), each 3x300,000, get average failure probabilities of 0.152%, 0.146% and 0.144%, respectively, compared with $P_f = \Phi(-\beta) = 0.135\%$.

(c) Practically identical Q_{design} as (a) and (b), with 1D random field of 2 values of ϕ' along pile shaft.

Figure 6.6 RBD-via-FORM of the safe applied load (Q_{design}) with different numbers of random variables ϕ' along pile shaft, for a target $\beta = 3.0$.

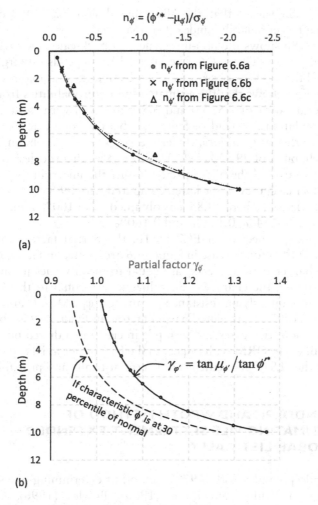

Figure 6.7 (a) The sensitivity of ϕ' increases nonlinearly with depth, (b) inferred partial factor γ_ϕ' based on $\tan\mu_{\phi'}/\tan\phi'^*$, smaller if based on conservative characteristic value instead of mean value.

illuminating to study the information and insights attainable by conducting RBD-via-FORM in parallel with LRFD or EC7 designs, than to try to calibrate partial factors for design codes.

Figure 6.7b shows that the inferred partial factor γ_ϕ', based on mean value $\mu_{\phi'}$, $\gamma_\phi' = \tan\mu_{\phi'}/\tan\phi'^*$, increases with depth, testifying to the bigger influence of the friction angles near the lower portion of the shaft. Multiple curves can be plotted for back-calculated partial factors based on characteristic value ϕ_k', corresponding to different percentiles of the probability density function of ϕ' (e.g., 5 percentile, 30 percentile). That based on ϕ_k' at 30 percentile is shown as a dashed curve in Figure 6.7b.

Figure 6.8a shows that the allowable design load Q decreases with increasing target reliability index β, as expected.

Figure 6.8b shows the non-uniqueness of back-calculated resistance factor for the case in Figure 6.6. The inferred RF depends on target reliability index and the level of conservatism in selecting the nominal resistance.

Figure 6.8c shows that the estimated failure probabilities from FORM β, for target β values of 2.5, 3.0 and 3.5, agree well with those from Monte Carlo simulations. Instead of failure probability, one can also switch perspective to that of *reliability*. Under the *Target β* = 3.0 column, FORM predicts a reliability of 99.87% (=100% – 0.135%) that the design load of 247 kN will not exceed the pile capacity despite the uncertainties of the inputs and model inaccuracy of the latter, compared with 99.85% based on Monte Carlo simulation, where 99.85% is obtained from 100% minus the average (0.152%) of 0.154%, 0.141% and 0.160%.

From the perspective of EC7 DA1b, the partial factors inferred from RBD-via-FORM for the case in Figure 6.6 are plotted in Figure 6.9a, assuming the characteristic values x_k are equal to mean values μ_x, and in Figure 6.9b, assuming the characteristic values x_k are equal to the 30 percentile values of the underlying resistance parameters ϕ_b', M_{Nq}, K and λ. The non-uniqueness of back-calculated partial factors in Figure 6.9 is obvious.

The previous two cases, a driven pile in clay and a driven pile in sand, are about RBD-via-FORM with respect to the ULS of pile axial capacity. We next consider the SLS of pile settlement, deterministically and probabilistically.

6.4 RANDOLPH AND WROTH METHOD OF ESTIMATING PILE SETTLEMENT, EXTENDED PROBABILISTICALLY

The Randolph and Wroth (1978) method of determining pile settlement is described in Poulos and Davis (1980), Bowles (1996), Geotechnical Engineering Office of Hong Kong (2006), Salgado (2008) and Knappett and Craig (2019). In the case explored below, the soil is assumed to behave elastically, and the pile to be axially rigid.

6.4.1 Deterministic estimation of pile settlement using the Randolph and Wroth method

Knappett and Craig (2019, Example 10.3) illustrated the deterministic calculation of the long-term pile settlement for a 21.5 m long under-reamed bored pile to be constructed in a clay layer. The shear modulus of the clay is modelled by G = 5 + 0.5z MPa, where z is the depth in meters below ground level, and the Poisson's ratio is ν′ = 0.2. The pile shaft has a length of 20 m above the under-ream, and a diameter of 1.5 m. The under-ream has a

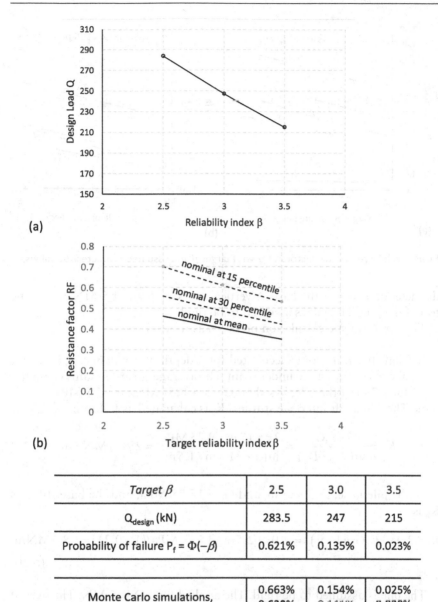

(a)

(b)

(c)

Target β	2.5	3.0	3.5
Q_{design} (kN)	283.5	247	215
Probability of failure $P_f = \Phi(-\beta)$	0.621%	0.135%	0.023%

	2.5	3.0	3.5
Monte Carlo simulations, 3 × 300,000 trials	0.663%	0.154%	0.025%
	0.638%	0.141%	0.030%
	0.619%	0.160%	0.033%

The above is based on Fig. 6.6a with 10 random variables of ϕ' along shaft. Same accuracy with 4 and 2 random variables of ϕ' along shaft (Fig. 6.6b&c).

Figure 6.8 (a) Allowable design load Q_{design} varies with target reliability index β, (b) inferred resistance factor varies with target β and assumed nominal resistance, (c) comparison of failure probabilities from FORM β and Monte Carlo simulations.

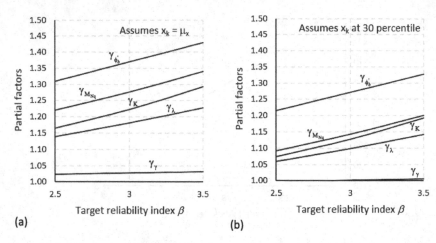

Figure 6.9 Inferred partial factors vary with target β and assumed characteristic values.

diameter of 2.5 m at the base. A vertical working load of 5 MN acts at the top of the pile. The pile is assumed to be rigid.

The deterministic calculation proceeds as follows:

(i) Shaft friction is only accounted for a depth of 2 × diameter, i.e., 3 m, above the top of the under-ream. The average shear modulus G_s on the top 17 m of the pile, based on $G = 5 + 0.5z$ MPa, is 9.25 MPa.

(ii) The soil-shaft interface stiffness K_s is calculated as follows:

$$K_s = \frac{2\pi L_s G_s}{\ln(2r_m / D_0)} = \frac{2\pi \times 17\,\mathrm{m} \times 9.25\,\mathrm{MPa}}{\ln(2 \times 21.5\,\mathrm{m} / 1.5\,\mathrm{m})} = 294.4\,\mathrm{MN/m} \quad (6.19)$$

At the pile base, $D_b = 2.5$ m, and $G_b = 15.75$ MPa, and the base stiffness K_b is:

$$K_b = 2.0 \times \left[D_b G_b / (1-v) \right] = 2.0 \times \left[2.5\,\mathrm{m} \times 15.75\,\mathrm{MPa} / (1-0.2) \right] = 98.4\,\mathrm{MN/m}$$
$$(6.20)$$

The pile is assumed to be rigid. The applied load Q is 5 MN. The overall pile settlement s at the top of the rigid pile is then:

$$s = \frac{Q}{K_s + K_b} = \frac{5\,\mathrm{MN}}{294.4\,\mathrm{MN/m} + 98.4\,\mathrm{MN/m}} = 0.0127\,\mathrm{m} = 12.7\,\mathrm{mm} \quad (6.21)$$

Knappett and Craig also illustrated (1) settlement calculation using the t-z method, obtaining 12.8 mm for this case if the pile is rigid, and 13.6 mm pile head settlement if pile is compressible with a Young's modulus E = 30 GPa, and (2) settlement of a group of eight such piles, arranged in two rows at a centre-to-centre spacing of $5D_0$, supporting a total load of 40 MN, and found the settlement of the corner and internal piles in the pile group about 2 times and 2.5 times that of the above single pile under the same nominal load.

RBD-via-FORM can consider pile group effect. This chapter confines itself to RBD-via-FORM with respect to the ULS and SLS of vertically loaded single pile and (later) laterally loaded single pile.

The above deterministic calculation of pile head settlement is extended next to reliability analysis which explicitly accounts for the input uncertainty.

6.4.2 Extending the Randolph and Wroth method to reliability analysis of pile settlement

Figure 6.10 shows the outcome of FORM reliability analysis of the axially loaded single pile. The random variables are the average soil shear modulus G_s along the upper 17 m shaft length, the shear modulus G_b at the base of the pile under-ream, the Poisson's ratio v and the applied vertical load Q. The mean values and standard deviations of these four random variables are shown under the column labelled μ and σ, with c.o.v. of 0.1, 0.1, 0.1 and 0.15, respectively. The random variables G_s and G_b are assumed to be positively correlated, with a correlation coefficient ρ of 0.5. The cells labelled K_s and K_b contain Eqs. 6.19 and 6.20, based on input values from the same row and from the x^* column, and the cells labelled s contain Eq. 6.21. Initially the column values labelled n were zeros.

The Low and Tang (2007) spreadsheet-automated constrained optimization method readily obtained the results in Figure 6.10a and 6.10b, for normal distributions and lognormal distributions, with reliability index β of 2.77 and 2.63, respectively, with respect to a limiting (i.e., maximum allowable) settlement of 20 mm. This means that although the settlement based on mean input values is about 12.7 mm, which is smaller than the allowable 20 mm, there is a small probability P_f of 0.28% and 0.43%, from $P_f = \Phi(-\beta)$, of exceeding the allowable settlement of 20 mm, given the statistical inputs in Figure 6.10a and 6.10b respectively. From the standpoint of reliability, the settlement will not exceed 20 mm with a reliability of 99.72% for the normal distributions of Figure 6.10a, and a reliability of 99.57% for the lognormal distributions of Figure 6.10b.

The point where, in the space of the four random variables (G_s, G_b, v and Q), the expanding four-dimensional dispersion ellipsoid (*equivalent* ellipsoid for lognormals) just touches the limit state surface (defined by g(x) = $s_{limiting} - s = 0$) is the most probable point (MPP) of failure (or of settlement exceedance), also called the design point if a target β is aimed at in design. This MPP of failure is represented by the values under the x^* column. The four values under column n are sensitivity indicators. In this case, the reliability is more sensitive to Q and G_s than to G_b and v.

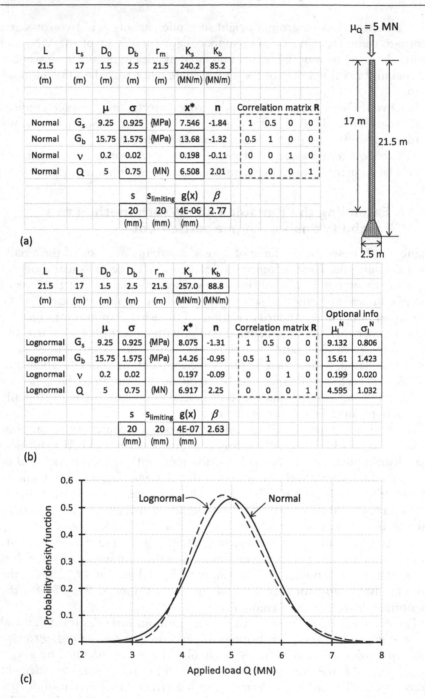

(a)

L	L_s	D_0	D_b	r_m	K_s	K_b
21.5	17	1.5	2.5	21.5	240.2	85.2
(m)	(m)	(m)	(m)	(m)	(MN/m)	(MN/m)

		μ	σ		x*	n	Correlation matrix R			
Normal	G_s	9.25	0.925	(MPa)	7.546	-1.84	1	0.5	0	0
Normal	G_b	15.75	1.575	(MPa)	13.68	-1.32	0.5	1	0	0
Normal	v	0.2	0.02		0.198	-0.11	0	0	1	0
Normal	Q	5	0.75	(MN)	6.508	2.01	0	0	0	1

s	s_limiting	g(x)	β
20	20	4E-06	2.77
(mm)	(mm)	(mm)	

(b)

L	L_s	D_0	D_b	r_m	K_s	K_b
21.5	17	1.5	2.5	21.5	257.0	88.8
(m)	(m)	(m)	(m)	(m)	(MN/m)	(MN/m)

		μ	σ		x*	n	Correlation matrix R				Optional info	
											μ_i^N	σ_i^N
Lognormal	G_s	9.25	0.925	(MPa)	8.075	-1.31	1	0.5	0	0	9.132	0.806
Lognormal	G_b	15.75	1.575	(MPa)	14.26	-0.95	0.5	1	0	0	15.61	1.423
Lognormal	v	0.2	0.02		0.197	-0.09	0	0	1	0	0.199	0.020
Lognormal	Q	5	0.75	(MN)	6.917	2.25	0	0	0	1	4.595	1.032

s	s_limiting	g(x)	β
20	20	4E-07	2.63
(mm)	(mm)	(mm)	

(c)

Figure 6.10 Reliability analysis of the settlement of an under-reamed bored pile in clay, assuming (a) normal distributions, (b) lognormal distributions, (c) comparing lognormal and normal distribution of applied load Q, at COV of 0.15.

For most geotechnical design problems, the target reliability index β for ultimate limit states is often in the range 2.5–3.5, which correspond to a probability of failure P_f of 0.6% and 0.02%, respectively. For serviceability limit state such as the case in hand, a lower target β (e.g., $\beta = 2.0$, or $P_f \approx$ 2.3%) may be acceptable. Figure 6.11 shows that, for the case in hand where settlement based on mean inputs is 12.7 mm, a target reliability index of 2.0 will be achieved when the limiting settlement is 17.5 mm. This means even if the allowable settlement ($s_{limiting}$) is the stricter requirement of 17.5 mm instead of the 20 mm in Figure 6.10, the target reliability of 97.7% (=100%–2.3%) against exceedance is satisfied.

Figure 6.10c compares the probability density function (PDF) of normally distributed and lognormally distributed Q, which has a mean value of 5 MN and a standard deviation of 0.75 MN.

The normal distribution is symmetric with respect to its mean value (i.e., zero skewness). The lognormal distribution has a positive skewness which increases with the coefficient of variation. The lognormal distribution is often used, on the ground that it excludes negative values. In fact, when the c.o.v. of a random variable is 0.25 or smaller, normal distributions can be used in most circumstances even if the random variable admits positive values only, because the probability of encroaching into negative domain is 0.003% when c.o.v. is 0.25. Of course, if physical considerations justify skewed probability density functions (PDF), the left bounded lognormal distributions or other bounded distributions can be considered. In particular, the four-parameter general beta distribution (bounded between user-defined lower and upper limits) is versatile because it can assume various PDF shapes. It can be considered in lieu of lognormals which is unbounded in the upper tail. The PERT distribution (also known as Beta-PERT distribution), defined by lower and upper bounds and most likely value (the mode), can also be considered in lieu of the lognormal distribution.

The above paragraphs and associated figures deal with the reliability analysis of the SLS of a vertically loaded pile. We next consider RBD involving both ULS and SLS of a laterally loaded pile in soil with spatially autocorrelated nonlinear strain-softening and depth-dependent lateral resistance modelled by p-y curves.

6.5 A LATERALLY LOADED PILE IN SOIL WITH STRAIN-SOFTENING AND DEPTH-DEPENDENT P-Y CURVES

The analysis of a laterally loaded pile based on the p-y curve concept is commonly done using the finite element or the finite difference method. The problem is akin to the classical problem of the beam on elastic foundation, except that in a laterally loaded pile the springs which model the resistance offered

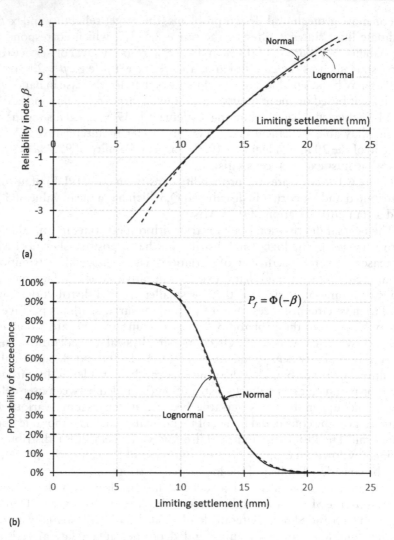

Figure 6.11 (a) Reliability with respect to SLS increases with increasing limiting settlement, and (b) probability of exceeding limiting settlement, for the axially loaded bored pile.

by the soil medium (Figure 6.12) typically follow a nonlinear resistance-deflection behaviour (nonlinear p-y curves). The nonlinear p-y curves are also likely to vary with depth. Hence, the deterministic soil–structure interaction analysis requires iterative numerical methods. Extending the deterministic analysis to a reliability analysis is a complicated task, by virtue of the implicit, numerical, and iterative nature of the performance function, and the need to account for the spatial correlation of the soil spring properties.

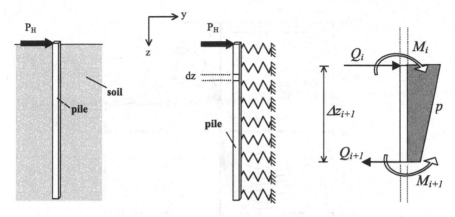

Figure 6.12 (a) Laterally loaded pile in soil, (b) soil spring idealization, (c) a segment of length dz.

A deterministic numerical analysis of a laterally loaded pile is presented next, prior to extending it into RBD-via-FORM with respect to both ULS and SLS.

6.5.1 Deterministic numerical procedure for a laterally loaded pile involving strain-softening p-y curves

This case is similar to that in Tomlinson (1994) and Tomlinson and Woodward (2015, Example 8.2). A steel tubular pile having an outside diameter d of 1.3 m and a wall thickness of 0.03 m forms part of a pile group in a breasting dolphin. A cyclic force of 421 kN is applied at 26 m above the seabed. The flexural rigidity $E_p I_p$ of the pile is 4,829,082 kNm². The pile is embedded 23 m in a stiff overconsolidated clay. Instead of Tomlinson's assumption of uniform undrained shear strength c_u of 150 kN/m², it is assumed here that $c_u = 150 + 2z$ kN/m², where z is depth below seabed. The pile protrudes 26 m above the seabed. The Matlock p-y curves for clays (Figure 6.13), as summarized in Tomlinson (1994, Eqs. 6.66–6.70), have been used to model horizontal soil resistance as a function of pile lateral deflection. These p-y curves are nonlinear, exhibit strain-softening, and vary with depth even if c_u is uniform with depth.

The rigorous iterative numerical procedure in the spreadsheet platform is based on Low et al. (2001), where its accuracy was verified. The 23 m embedded length L of the pile (Figure 6.14) is divided into 30 segments of varying length Δz. The deflection of the pile head at 26 m above the seabed can be inferred readily from statics once the deflection and rotation of the pile at seabed level are known.

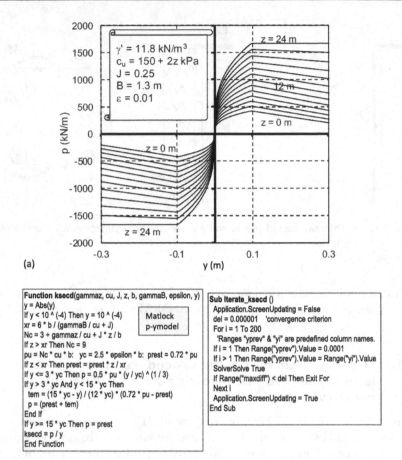

(a)

(b)

Figure 6.13 Matlock p-y curves: nonlinear, strain-softening, and depth dependent, and user-created VBA function and procedure for iterative nonlinear p-y analysis of pile.

The third column in Figure 6.14 shows the flexural rigidity $E_p I_p$ of the pile at the nodal points. The fourth column (y_i) computes the lateral pile deflection y, as follows:

$$y_i = y_{i+1} + \left(\Delta z_{i+1} / \Delta z_{i+2} \right)^2 * \left(y_{i+2} - y_{i+1} \right) - \Delta z_{i+1} * \left(1 + \frac{\Delta z_{i+1}}{\Delta z_{i+2}} \right) * y'_{i+1} \quad (6.22)$$

in which $\Delta z_{i+1} = z_{i+1} - z_i$, and $\Delta z_{i+2} = z_{i+2} - z_{i+1}$. The above equation has been derived by quadratic curve fitting over the nodes i to $i + 2$, based on the computed deflection values y_{i+1} and y_{i+2} at z_{i+1} and z_{i+2}, respectively, and the gradient y'_{i+1} at z_{i+1}.

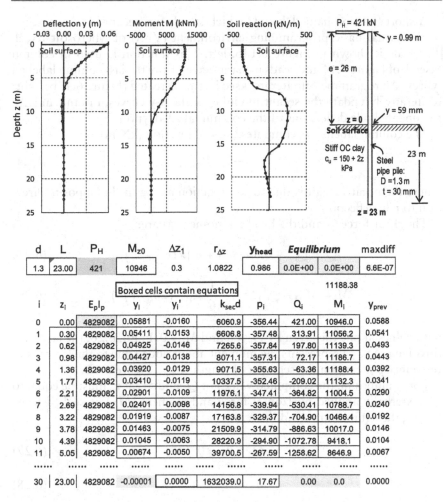

Figure 6.14 Deterministic analysis of a cantilever laterally loaded pile partially embedded in soil modelled with Matlock p-y curves.

The fifth column, labelled y_i', gives the slope dy/dz. of the pile at various z values, as follows:

$$y_i' = y_{i+1}' - 0.5 * \left[\left(\frac{M}{E_p I_p} \right)_{i+1} + \left(\frac{M}{E_p I_p} \right)_i \right] (z_{i+1} - z_i) \qquad (6.23)$$

The above equation is an expression of the first moment area theorem, which relates the change of slope between any two points on a beam to the bending moment diagram. At the pile toe, the value of y' is calculated as $(y_n - y_{n-1})/(z_n - z_{n-1})$, where n is the last nodal number, 30 in this case.

A short function named 'ksecd,' which stands for (secant k)*d, is created in the Visual Basic programming environment (VBA) of Microsoft Excel. The code is shown in Figure 6.13b, left. This function is used in column 'ksecd' of Figure 6.14 and reads its parameter y from the last column labelled 'yprev.' A programme 'Sub Iterate_ksecd' is also created (Figure 6.13b, right) to iteratively update the secant modulus of the p-y curves, in tandem with the updated pile deflections of the column labelled 'yprev.'

The column labelled p_i computes the soil reaction (kN/m) as:

$$p_i = -k_h d * y_i \tag{6.24}$$

where the negative sign is due to soil reaction acting in the opposite direction to pile deflection.

The shear force Q and the bending moment M are:

$$Q_i = Q_{i+1} - 0.5 \times (p_{i+1} + p_i) \times (z_{i+1} - z_i) \tag{6.25}$$

$$M_i = M_{i+1} - Q_{i+1}\Delta z_{i+1} + \frac{1}{6}\Delta z_{i+1}^2 (2p_{i+1} + p_i) \tag{6.26}$$

with values at pile toe given by $Q_n = 0$ and $M_n = 0$. In Eq. 6.26, the last term (involving Δz_{i+1}^2) represents the moment about node i due to the lateral soil resistance p acting on the segment of length Δz_{i+1}, Figure 6.12c.

For equilibrium, the computed values of Q_0 and M_0 at z_0 must be equal to the externally applied horizontal force and moment (M_{z0}) at seabed level. Hence, the equilibrium conditions are:

$$Q_0 - P_H = 0 \tag{6.27}$$

$$M_0 - M_{z0} = 0 \tag{6.28}$$

The above two equations have been entered in Figure 6.14 under the heading 'Equilibrium.' Note that the Q_0 and M_0 in Eqs. 6.27 and 6.28 can only be determined recursively and numerically by satisfying (at all nodal points i) the relationships among deflection y_i, slope y_i', mobilized soil resistance p_i, shear Q and bending moment M_i, according to Eqs. 6.22–6.26.

The lateral deflections y at the last two nodes of the pile are initially zeros, which rendered the pile initially straight along its entire length, to start with. The Microsoft Excel Solver tool is invoked, to preset the optimization scenario as follows: 'set' Eq. 6.27 equal to zero, 'By changing' (automatically) the values of the lateral deflections y at the bottom two nodes, 'Subject to' the constraint that Eq. 6.28 be equal to zero. Solver option 'Use automatic scaling' is also selected. Other Solver options (precision, estimates, derivatives and search method) are left at their default settings. The programme 'Sub Iterate_ksecd' (Figure 6.13b) is then executed, to invoke the Excel

Solver repeatedly based on the preset scenario. The values in the 'ksecd' column of Figure 6.14 will automatically be updated each time the 'yprev' column is iteratively changed by programme 'Iterate_ksecd.'

The converged pile deflection (y_i) value at seabed level (where $z = 0$) is 0.0588 m. The pile head deflection (at 26 m above the seabed) is, by integrating the moment-curvature equation, equal to:

$$y = \frac{P_H}{E_p I_p} \times \left(\frac{1}{2} e z^2 + \frac{1}{6} z^3 \right) + y_0' z + y_0 \quad \text{,for} \ -26 \le z \le 0 \qquad (6.29)$$

Substituting for $P_H = 421$, $E_p I_p = 4,829,082$, $e = 26$, $z = -26$, $y_0' = -0.01601$, $y_0 = 0.0588$, where the last two values are the rotation and deflection at sea bed level (Figure 6.14), the deflection at pile head (y_{head}) is calculated to be 0.986 m in Figure 6.14.

Separate analysis of a similar case (with uniform c_u of 150 kN/m^2, instead of $c_u = 150 + 2z$ kN/m^2 in this section), using a specially written Fortran programme to perform the finite element analysis using 60 equally spaced elements, yielded a pile deflection of 0.0596 m at sea bed level, compared with 0.0602 m by the spreadsheet VBA and Solver tool constrained optimization numerical procedure described in the preceding two paragraphs and Figure 6.13. The shear and moment distribution along the pile length were virtually identical.

6.5.2 From deterministic numerical procedure to probabilistic analysis of laterally loaded piles

If the parameters (applied loads and p-y curves) in the above section are average values, the computed pile head deflection and maximum pile bending moment represent, at best, only average values. These average responses alone cannot be used to judge, at the design stage, whether the pile will perform satisfactorily with respect to some specified permissible pile head deflection and tolerable bending moment. A more rational approach would need to take into account not just the mean values of the parameters but also their uncertainty. The partial factors in Eurocode 7 and the load and resistance factor design (LRFD) method are attempts in this direction.

A more rigorous and direct approach is to evaluate the reliability index as defined by Hasofer and Lind (1974) for correlated normal, or FORM for correlated nonnormals.

The FORM analysis and RBD of laterally loaded single piles below involve spatially autocorrelated undrained shear strength, implicit performance functions (incorporating nonlinear p-y model and iterative secant moduli) for ultimate and serviceability limit states. The case of a laterally loaded pile with 26 m cantilever length will be investigated first, followed by another case of smaller pile with zero cantilever length but in the same soil.

The outcome demonstrates the non-intrinsic and case-specific nature of back-calculated LF and RF which are affected not only by the assumed nominal load and resistance values and the target reliability index, but also by the different sensitivities of the performance to load and resistance as the cantilever length of the pile changes from 26 to 0 m.

The analyses and discussions below are similar to those associated with Figure 6 of Low (2017, JGGE, ASCE), but more details and information are provided in the figures here (Figures 6.12–6.17).

6.5.3 Pile with 26 m cantilever length

For reliability analysis, the random variables are the lateral load P_H at pile head and the undrained shear strength c_u at 31 nodal points along the embedded pile length of 23 m below the seabed. The P_H is assumed to be normally distributed, with mean value 421 kN and a coefficient of variation of 25%. The mean undrained shear strength c_u is $\mu_{cu} = 150 + 2z$, kPa. The standard deviations of the normally distributed 31 c_u values are assumed to be equal to 30% of their respective mean values. This c.o.v. of 30% is within the range suggested in Phoon (2017). The following established negative exponential model is adopted to model the vertical spatial variation of the c_u values:

$$\rho_{ij} = e^{-\frac{|Depth(i)-Depth(j)|}{\delta}} \tag{6.30}$$

where δ is the autocorrelation distance, equal to 2 m in this study.

A 32 × 32 correlation matrix was set up. The first entry was the diagonal value of 1.0 for P_H. The other entries are the correlation coefficients of the c_u values at the 31 nodes along the pile length, based on Eq. 6.30. A vector **n** of 32 components is also set up, with initial values of zeros, as in Figure 6.3a, together with a 32-component **x*** column containing the formulas $\mu_i + n_i\sigma_i$ (which are applicable for normal variates), where μ_i and σ_i are the mean and standard deviation of a random variable x_i. (The simple VBA code in the Appendix can be used if nonnormals are modelled; the procedure below remains the same.)

The quadratic form $\sqrt{\mathbf{n}^T\mathbf{R}^{-1}\mathbf{n}}$ in Eq. 2.9 was then entered as an array formula in a cell labelled β, using the spreadsheet matrix functions *transpose*, *mmult* and *minverse*.

The built-in optimization routine (Solver) in Microsoft Excel was invoked to predefine the optimization setting as follows: '*to minimize*' the quadratic form in the β cell (a 32-dimensional ellipsoid in original space), '*by changing*' the 32 **n** values and the last two nodal deflections (y_n and y_{n-1}, at and adjacent to the pile toe, respectively), '*subject to the constraints*' that the ellipsoid be tangent to the limit state surface (imposed by $g(\mathbf{x}) = 0$), and the equilibrium equations (Eqs. 6.27 and 6.28) be satisfied. The programme code 'Sub Iterate_ksecd' shown in Figure 6.14 was then used to invoke Solver iteratively to obtain the 32 values of the sensitivity indicators **n** and the most probable failure point **x*** (i.e., the 'design point') at where the ellipsoid touches the limit state surface. The procedure does not require orthogonal transformation of the correlation matrix.

The β index obtained was 1.514 with respect to yielding at the outer edge of the annular steel cross section. At the most probable failure point (where the hyperellipsoid touches the limit state surface), the value of the lateral load at pile head is P_H = 580.3 kN, that is, at $1.513\sigma_{PH}$ *above* the mean value of P_H, while the 31 autocorrelated c_u values deviates only very slightly from their mean values. This will be discussed further after the pile outer diameter D and wall thickness t are increased so as to reach a target reliability index of $\beta \approx 3.0$ with respect to both ultimate limit state (steel yielding due to bending moment) and serviceability limit state (defined by excessive or limiting pile head deflection, assumed to be 1.4 m for the breasting dolphin pile under consideration). The external pile diameter D and wall thickness t affect both the ultimate limit state (yield in steel due to bending) and serviceability limit state (pile head deflection), by affecting the moment of inertia I, the Matlock p-y curves (Figure 6.13, in which B = D = 1.3 m), and the yield moment $M_y = 2\sigma_y I/D$, where σ_y is the yield stress (417 MPa) of high-tensile alloy steel.

Based on the results plotted in Figure 6.15a, and with the uncertainties of P_h and spatially auto-correlated c_u profile as described in the lines preceding Eq. 6.30 and summarized at the top of Figure 6.16, the steel pile outer diameter D was increased to 1.42 m and the wall thickness increased to 0.032 m (Figure 6.16, right), for which the reliability index β was calculated to be 2.98 with respect to the serviceability limit state (that pile head deflection $\leq$ 1.4 m) and 3.00 with respect to the ultimate limit state (that bending stress in pipe pile < yield stress of steel).

The information contained in the FORM design points of this pile (D = 1.42 m, t = 32 mm, embedment 23 m, cantilever 26 m, Figure 6.16, right) and the insights for LRFD and EC 7 are as follows:

(1) The FORM design value of horizontal load P_h is 728 kN for SLS, and 736 kN for ULS. Assuming nominal load at 84 percentile, that is, $P_{h, nominal} = \mu_{Ph} + \sigma_{Ph} = 421 \times (1 + 0.25) = 526$ kN, the back-calculated load factors are LF_{SLS} = 728/526 = 1.38, and LF_{ULS} = 736/526 = 1.40, as shown in the right table of Figure 6.16. The back-calculated LF_{SLS} and LF_{ULS} will be 1.73 and 1.75 if the nominal load is taken at its mean value, also shown in the figure. These values will be compared in the subsection below with a case where pile head is at the ground surface, that is, zero cantilever length.

(2) The mean resistance and nominal resistance need to be determined if RF is to be back-calculated from the design resistance obtained in FORM. For the vertically loaded piles in the previous sections, the

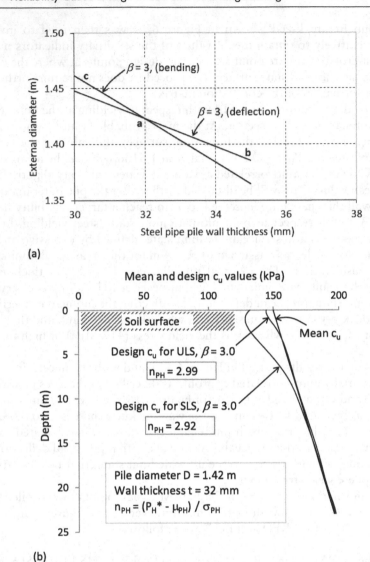

(a)

(b)

Figure 6.15 (a) RBD of pile diameter and steel wall thickness for bending ULS and deflection SLS, (b) design c_u values with depth.

mean resistance is the sum of skin and point resistance using mean shear strength parameters (e.g., mean undrained shear strength c_u). For laterally loaded pile involving nonlinear p-y curves and reversal of direction of horizontal soil pressures (as shown in Figure 6.14), a meaningful definition of mean resistance for deflection and bending limit states would be the horizontal pile head forces P_h required to reach the deflection limit state and bending limit state, respectively,

	P_h: $\mu_{Ph} = 421$ kN,	$\sigma_{Ph} = 0.25\mu_{Ph} = 105.3$ kN	
	c_u: $\mu_{cu} = 150 + 2z$ (kPa),	$\sigma_{cu} = 0.30\mu_{cu}$	Autocorrelation distance $\delta = 2$ m

<table>
<tr><td colspan="4">10 m long pile entirely embedded.
Diameter 0.88 m, wall thickness = 20 mm.
Reliability index β = 3.0 for SLS</td><td colspan="4">Pile with 26 m cantilever and 23 m
Diameter 1.42 m, wall thickness = 32 mm
Reliability index β = 3.0 for SLS & ULS</td></tr>
<tr><td></td><td>LF</td><td>mLF</td><td>mRF</td><td></td><td>LF</td><td>mLF</td><td>mRF</td></tr>
<tr><td>SLS</td><td>1.12</td><td>1.41</td><td>0.58</td><td>SLS</td><td>1.38</td><td>1.73</td><td>0.97</td></tr>
<tr><td>ULS</td><td colspan="3">Bending not an issue, $\beta \gg 3.0$.</td><td>ULS</td><td>1.40</td><td>1.75</td><td>1.00</td></tr>
</table>

SLS = Serviceability limit state, pile head deflection y_{limit}
ULS = Ultimate limit state, defined by bending-induced yield in steel pipe.
LF = Back-calculated load factor = Design load / Nominal load, where nominal load = $\mu_{load} + \sigma_{load}$
mLF, mRF = Back-calculated resistance factor w.r.t. mean load and mean resistance

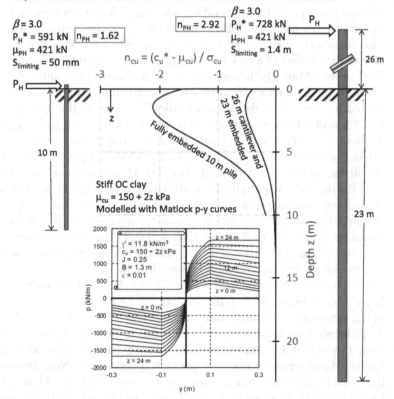

$\beta = 3.0$
$P_H^* = 591$ kN $n_{PH} = 1.62$
$\mu_{PH} = 421$ kN
$S_{limiting} = 50$ mm

$\beta = 3.0$
$n_{PH} = 2.92$ $P_H^* = 728$ kN P_H
$\mu_{PH} = 421$ kN
$S_{limiting} = 1.4$ m

$n_{cu} = (c_u^* - \mu_{cu})/\sigma_{cu}$

Figure 6.16 Context-dependent nature of back-calculated RF and LF, and variation of sensitivity indicator n_{cu} with depth.

based on mean shear strength values of soil. For the case in hand, these values are 748 kN and 737 kN, which are only slightly higher than the design loads of 728 kN and 736 kN. This suggests that, for the case in hand with a big cantilever length of 26 m, more critical design scenarios are reached when loads are factored up (via nominal value and LF) much more than when the resistance is factored down. In fact, for the case in hand, one is inclined to suggest that nominal resistance values be equal to mean resistance values (for SLS and ULS), for which the back-calculated mRF, defined as design resistance/mean resistance (in contrast with RF = design resistance/nominal resistance), will be 728/748 or 0.97 for SLS and 736/737 or 0.999 for ULS, both close to 1.0. A moment's thought would render these RF values consistent with the physical behaviour of a laterally loaded pile with large cantilever length: pile head deflection (with 26 m cantilever length) and pile bending (critical within a few meters below seabed) are more sensitive to pile head horizontal load than to soil resistance.

(3) Figure 6.15b shows that the difference between the mean values of c_u and the design c_u values depend on whether SLS or ULS is being considered for the partially embedded pile with 26 m of cantilever length. This means that the sensitivities of c_u not only vary with depth, but also with whether SLS or ULS is being considered, and on the type of p-y curves (Matlock, hyperbolic, etc.). Adding to the difficulty of applying partial factors from EC7 to c_u rationally is that, for laterally loaded piles in soil modelled with Matlock p-y curves, the impact of c_u on SLS and ULS is nonlinear: even if c_u is uniform with depth, the Matlock p-y curves will vary with depth in a pattern similar to Figure 6.13a. Regardless of the type of p-y curves (Matlock, hyperbolic, etc), the design point found by RBD-via-FORM, comprising the lateral load P_H and the 31 spatially autocorrelated c_u values, reflects context-dependent parametric sensitivities and autocorrelations in a way design point via EC7 cannot. In this case, the values of the sensitivity indicator n_{pH}, equal to 2.99 for ULS and 2.92 for SLS, indicate that both the ULS and SLS of this long cantilever pile are much more sensitive to uncertainty in P_H than to uncertainty in c_u.

(4) The LF for the horizontal load here (LF_{Ph} of 1.38 and 1.40 in the right-hand table of Figure 6.16) is higher than the LF for the horizontal load of the retaining wall foundation case of Figure 5.15b. That is, LF can be different for different target β (2.5, 3.0 or 3.5), and even for the same target β of 3.0. Back-calculated LF (and RF) also depend on the values of conservative nominal loads and resistance. Note however that LRFD does not aim at any target β (or target P_f). Hence, for LRFD to apply the same LF for the P_h here and the Q_h of the retaining wall foundation case will result in designs of different reliability levels (and different P_f) that will still be acceptable if the different reliability levels are both sufficiently high (or the two different P_f are both sufficiently low). When the statistical information of the inputs is available, an LRFD design can be subjected to FORM analysis in order to determine the reliability index and probability of failure.

Next, the information contained in the design point of a smaller pile (with no cantilever length, in the same soil, subjected to the same lateral load, and with same reliability index of 3.0) and the insights for LRFD and EC7 are considered.

6.5.4 Fully embedded pile with no cantilever length

The long pile with a 26 m cantilever length analyzed in the preceding subsection forms part of a pile group in a breasting dolphin. This section studies a shorter and thinner pile subjected to the same mean horizontal load P_h of 421 kN acting at the pile head but with zero cantilever length, in the same soil with mean undrained shear strength $c_u = 150 + 2z$ kPa, and the same uncertainties in the P_h and c_u profile as summarized at the top of Figure 6.16. For the engineering function of the pile, suppose that the limiting pile head deflection is 50 mm. The target reliability index is 3.0, as before. A 10 m long tubular pile of diameter 0.88 m and wall thickness 0.020 m (Figure 6.16, left) will have a reliability index of 3.0 against the assumed serviceability limit state of 0.05 m pile head deflection. The design value of P_h is 591 kN. The P_h load that will cause pile head deflection of 0.05 m (the serviceability limit state), based on mean soil shear strength values, is 1025 kN, which is then the mean resistance with respect to SLS. The LF_{SLS} is 591/$P_{h,\text{nominal}}$ = 591/526 = 1.12, whereas the mRF_{SLS} = 591/1025 = 0.58, compared with the mRF_{SLS} of 0.97 of the case with 26 m cantilever length. In other words, for the case in the previous section with large cantilever length, the SLS and ULS are much more sensitive to the pile head load than to the p-y resistance of soil, whereas for the case with zero cantilever length, the SLS is sensitive to both the pile head load and p-y resistance of soil. This is shown by the two sensitivity indicator curves of c_u for SLS in Figure 6.16, in which the n_{cu} curve of the zero-cantilever pile is greater in absolute value than that of the 26-m cantilever pile.

The back-calculated mRF_{SLS} of 0.97 for the case with large cantilever length and 0.58 for the case with zero cantilever length are based on nominal resistance equal to mean resistance, bypassing the need to estimate the standard deviation of resistance via Monte Carlo simulation. The insights and conclusions on different sensitivities with and without cantilever length remain valid even if RF_{SLS} values are back-calculated based on nominal resistance being smaller than mean resistance.

Unlike the previous case of large cantilever length, bending limit state is a much less critical mode for the present case in which the entire pile length is embedded in the soil. In other words, the reliability index of the ULS against bending is much higher than 3.0 (the target reliability index of the SLS). Figure 6.16 summarizes the RBD insights that sensitivities vary with the pile cantilever length.

Figure 6.17 shows the variation of partial factors of c_u back-calculated from the FORM design point with respect to the mean values of c_u. Not only

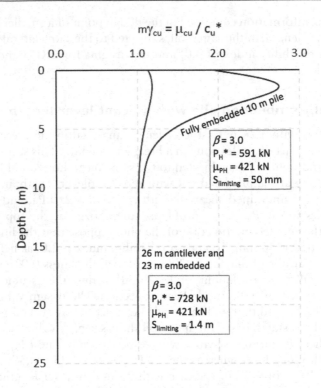

$$m\gamma_{cu} = \mu_{cu} / c_u{}^*$$

Fully embedded 10 m pile

$\beta = 3.0$
$P_H{}^* = 591$ kN
$\mu_{PH} = 421$ kN
$S_{limiting} = 50$ mm

26 m cantilever and
23 m embedded

$\beta = 3.0$
$P_H{}^* = 728$ kN
$\mu_{PH} = 421$ kN
$S_{limiting} = 1.4$ m

Depth z (m)

Figure 6.17 Back-calculated partial factors for c_u (with respect to mean values) are not uniform but vary with depth. They are also context dependent.

do these partial factors vary with depth, they are also very different for the fully embedded 10 m long pile and the 49 m long pile with lower 23 m embedded.

To sum up this section: a numerical method based on constrained optimization has been presented for nonlinear strain-softening p-y analysis of laterally loaded single piles. Only a minimal amount of VBA programming in the spreadsheet environment is required. The deterministic numerical procedure was then extended into reliability analysis, in which the soil springs were modelled stochastically to reflect spatial autocorrelation of soil resistance. An intuitive expanding ellipsoid perspective in the original space of the random variables (Low and Tang 2007) greatly simplifies the computation of the FORM reliability index (which includes the Hasofer–Lind index as a special case). The transition from the numerical soil–pile interaction deterministic analysis to a stochastic nonlinear and strain-softening p-y analysis requires little additional effort, despite the highly implicit nature of the performance functions for the ULS of pile bending and the SLS of pile head deflection. Multicriteria reliability-based design was illustrated for a steel tubular pile that forms part of a pile group in a breasting dolphin. Both ULS and SLS were considered. Context-dependent parametric sensitivities were demon-strated for a pile with long cantilever length and another with zero cantile-ver length. Advantages of reliability-based design over design based on partial factors were discussed, including how RBD-via-FORM can comple-ment EC7 and LRFD method. The procedure for cross-correlated and auto-correlated variables does not involve orthogonal transformation of the correlation matrix. If desired, RBD-via-FORM of the laterally loaded pile with 32 random variables (including autocorrelated undrained shear strength c_u values) can be modelled by correlated nonnormal distributions, using the same spreadsheet RBD-via-FORM procedure.

6.6 LATERALLY LOADED PILE: VERIFICATION OF SPREADSHEET NUMERICAL PROCEDURE WITH HETENYI SOLUTION FOR LINEAR P-Y MODEL

The numerical procedure for analysis of laterally loaded single pile, based on constrained optimization in spreadsheet, was presented in the preceding section in connection with Figures 6.12, 6.13 and 6.14 of the strain-soften-ing Matlock p-y curves. Its accuracy can be verified by comparison with the finite element method.

For the case of a free-head pile of length L subjected to a horizontal load P_H applied at ground level, the following solution for horizontal displace-ment y at a depth z below the surface was given by Hetenyi (1946), for constant k_h with depth, as cited in Poulos and Davis (1980, pp. 166–167):

$$y = \frac{2P_H\lambda}{k_h d} \times \left[\frac{\sinh \lambda L \cos \lambda z \cosh \lambda(L-z) - \sin \lambda L \cosh \lambda z \cos \lambda(L-z)}{\sinh^2 \lambda L - \sin^2 \lambda L} \right] \quad (6.31)$$

where

$$\lambda = \left[k_h d / \left(4E_p I_p \right) \right]^{1/4} \quad (6.32)$$

The values computed using Eq. 6.31 are shown in Figure 6.18 at the last column labelled '$y_{Hetenyi}$.' Hetenyi's solutions for moment M and shear Q at depth z are of similar form as (6.31). They are given in Poulos and Davis (1980), Eqs. 8.14 and 8.15.

Hetenyi's solutions for y, M and Q are plotted as solid curves in Figure 6.18, for comparison with values (plotted as open dots) from the numerical procedure presented in the previous section. The two sets of plots of y, M and Q are practically identical. This verifies the accuracy of the spreadsheet numerical procedure based on constrained optimization.

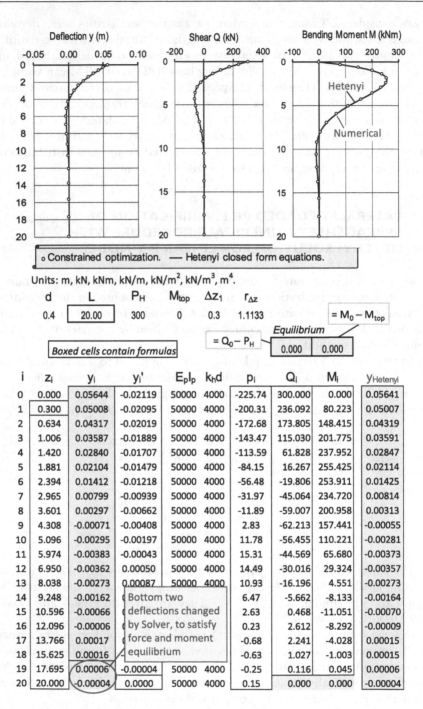

Units: m, kN, kNm, kN/m, kN/m², kN/m³, m⁴.

d	L	P_H	M_{top}	Δz_1	$r_{\Delta z}$	
0.4	20.00	300	0	0.3	1.1133	$= M_0 - M_{top}$

Boxed cells contain formulas $= Q_0 - P_H$ Equilibrium

| | | | | 0.000 | 0.000 |

i	z_i	y_i	y_i'	$E_p I_p$	$k_h d$	p_i	Q_i	M_i	$y_{Hetenyi}$
0	0.000	0.05644	-0.02119	50000	4000	-225.74	300.000	0.000	0.05641
1	0.300	0.05008	-0.02095	50000	4000	-200.31	236.092	80.223	0.05007
2	0.634	0.04317	-0.02019	50000	4000	-172.68	173.805	148.415	0.04319
3	1.006	0.03587	-0.01889	50000	4000	-143.47	115.030	201.775	0.03591
4	1.420	0.02840	-0.01707	50000	4000	-113.59	61.828	237.952	0.02847
5	1.881	0.02104	-0.01479	50000	4000	-84.15	16.267	255.425	0.02114
6	2.394	0.01412	-0.01218	50000	4000	-56.48	-19.806	253.911	0.01425
7	2.965	0.00799	-0.00939	50000	4000	-31.97	-45.064	234.720	0.00814
8	3.601	0.00297	-0.00662	50000	4000	-11.89	-59.007	200.958	0.00313
9	4.308	-0.00071	-0.00408	50000	4000	2.83	-62.213	157.441	-0.00055
10	5.096	-0.00295	-0.00197	50000	4000	11.78	-56.455	110.221	-0.00281
11	5.974	-0.00383	-0.00043	50000	4000	15.31	-44.569	65.680	-0.00373
12	6.950	-0.00362	0.00050	50000	4000	14.49	-30.016	29.324	-0.00357
13	8.038	-0.00273	0.00087	50000	4000	10.93	-16.196	4.551	-0.00273
14	9.248	-0.00162				6.47	-5.662	-8.133	-0.00164
15	10.596	-0.00066				2.63	0.468	-11.051	-0.00070
16	12.096	-0.00006				0.23	2.612	-8.292	-0.00009
17	13.766	0.00017				-0.68	2.241	-4.028	0.00015
18	15.625	0.00016				-0.63	1.027	-1.003	0.00015
19	17.695	0.00006	-0.00004	50000	4000	-0.25	0.116	0.045	0.00006
20	20.000	-0.00004	0.0000	50000	4000	0.15	0.000	0.000	-0.00004

Bottom two deflections changed by Solver, to satisfy force and moment equilibrium

Figure 6.18 Laterally loaded pile, comparison of spreadsheet-automated numerical method with the Hetenyi solution.

Chapter 7

Earth retaining structures

7.1 INTRODUCTION

This chapter illustrates practical reliability-based design procedures for retaining walls based on the Hasofer–Lind index, for correlated normal variates, and the first-order reliability method (FORM), for correlated non-normal variates. The differences between a reliability-based design (RBD) and one based on partial factors will be discussed. Sensitivity information as conveyed in a reliability analysis will be studied. The probabilities of failure inferred from reliability indices will be compared with Monte Carlo simulations. Connections are made with LRFD and EC7 design methods. System reliability analysis involving multiple failure modes is illustrated. The cases presented include a solution procedure for a problem after Terzaghi et al. (1996) that requires searching for the critical quadrilateral soil wedge in both deterministic analysis and probability-based design, and deterministic examples of seepage in retained fills from Clayton et al. (2013) and Lambe and Whitman (1979), with extensions to probability-based designs. Insights for consistent application of load and resistance factors in LRFD and EC7-DA2 are discussed to resolve a paradox in a semi-gravity wall design that arises by segregating the active earth thrust into a destabilizing horizontal component and a stabilizing vertical component. Possible multiple outcomes in partial factor design of anchored sheet pile involving stabilizing–destabilizing duality of soil unit weight are elaborated.

The RBD-via-FORM examples here are implemented using the Low and Tang (2007) spreadsheet FORM procedure, which obtains the same solution as the traditional FORM procedure, but more efficient and succinct. The emphasis here is on the information and insights contained in the design point of RBD-via-FORM, useful to readers regardless of the platform (e.g. MatLab, or standalone software) and procedure (e.g. classical FORM computational approach) which readers may already know about the implementation of FORM.

DOI: 10.1201/9781003112297-9

7.2 ROTATIONAL ULTIMATE LIMIT STATE OF A SEMI-GRAVITY WALL

Consider the semi-gravity retaining wall shown in Figure 7.1. A long-established deterministic approach evaluates the lumped factor of safety (F_s) against rotational failure as:

$$F_s = \frac{W_1 \times Arm_1 + W_2 \times Arm_2}{P_{ah} \times Arm_{ah} - P_{av} \times Arm_{av}} = f\left(\phi', \delta, \dots\right) \qquad (7.1)$$

where W_1 and W_2 are the component weights of the semi-gravity wall, with horizontal lever distances Arm_1 and Arm_2, respectively, measured from the

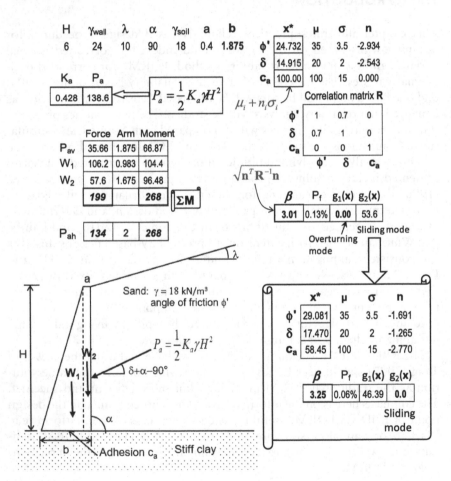

Figure 7.1 Reliability analysis of the rotational ultimate limit state and sliding ultimate limit state of a semi-gravity retaining wall, for correlated normal random variables, using Microsoft Excel spreadsheet.

toe of the wall; P_{ah} and P_{av} are the horizontal and vertical components of the active earth thrust P_a, with levers Arm_{ah} and Arm_{av}, respectively; and ϕ' and δ are the internal friction angle of the retained soil and the soil/wall interface friction angle, respectively. With the notations as defined in Figure 7.1, if $H = 6$ m, $\gamma_{wall} = 24$ kN/m^3, $\lambda = 10°$, $\alpha = 90°$, $\gamma_{soil} = 18$ kN/m^3, $a = 0.4$ m, $b = 1.88$ m, $\phi' = 35°$ and $\delta = 20°$, then the lumped (or overall) factor of safety against overturning is $F_s \approx 1.82$, by Eq. 7.1. In the two-dimensional space of ϕ' and δ, one can plot the F_s contours for different combinations of ϕ' and δ, as shown in Figure 7.2, where the average point ($\phi' = 35°$ and $\delta = 20°$) is situated on the contour (not plotted) of 1.82. Design is considered satisfactory with respect to overturning if the factor of safety by Eq. 7.1 is not smaller than a certain value (e.g., when $F_s \geq 1.5$).

A more recent and logical approach (e.g., Eurocode 7, LRFD) applies partial factors to the parameters in the evaluation of resisting and overturning moments. Design is acceptable if:

$$\Sigma(\text{Resisting Moments, factored}) \geq \Sigma(\text{Overturning moments, factored}) \quad (7.2)$$

A third approach is reliability-based design, where the uncertainties and correlation structure of the parameters are represented by a one-standard-deviation dispersion ellipsoid (Figure 7.2a) centred at the mean-value point, and safety is gauged by a reliability index which is the shortest distance (measured in units of directional standard deviations, R/r) from the safe mean-value point to the most probable failure combination of parametric values ('the design point') on the limit state surface (defined by $F_s = 1.0$). Other aspects of Figure 7.2 will be discussed shortly.

The probability of failure (P_f) can be estimated from the reliability index β using the following established equation:

$$P_f = 1 - \Phi(\beta) = \Phi(-\beta) \quad (7.3)$$

where Φ is the cumulative distribution function (CDF) of the standard normal variate.

The relationship is exact when the limit state surface is planar and the parameters follow normal distributions, and approximate otherwise.

The merits of a reliability-based approach over the lumped factor-of-safety approach are illustrated in Figure 7.3a, in which case A and case B (with different average values of soil shear strength parameters c' and ϕ') show the same values of lumped factor of safety, yet case A is clearly safer than case B. The higher reliability of case A over case B will correctly be revealed when the reliability indices are computed. On the other hand, a slope may have a computed lumped factor of safety of 1.5, and a particular foundation (with certain geometry and loadings) in the same soil may have a computed lumped factor of safety of 2.5, as in case C of Figure 7.3b. Yet a reliability analysis will reveal that they both have similar levels of reliability.

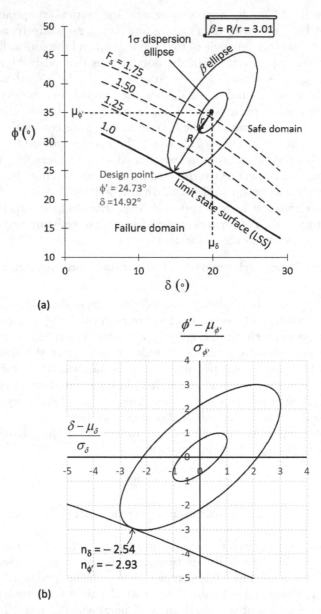

Figure 7.2 The expanding dispersion ellipse (or ellipsoid in higher dimensions) and the design point, for the rotational (overturning) limit state: (a) in original space of ϕ' and δ, and (b) in the dimensionless unrotated space of sensitivity indicators $n_{\phi'}$ and n_{δ}.

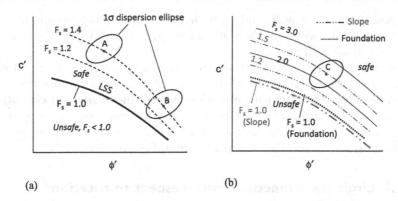

Figure 7.3 Schematic scenarios showing possible limitations of lumped factor of safety: (a) cases A and B have the same lumped F_s = 1.4, but case A is clearly more reliable than Case B; (b) case C may have F_s = 1.5 for a slope and F_s = 2.5 for a foundation, and yet have a similar level of reliability.

As explained in Chapter 2, the design point (Figure 7.2a) is the most probable failure combination of parametric values. The ratios of the respective parametric values at the centre of the dispersion ellipsoid (corresponding to the mean values) to those at the design point are similar to the partial factors in limit state design, except that these factored values at the FORM design point are arrived at automatically (i.e., as by-products, or corollaries) in RBD-via-FORM. The reliability-based approach is thus able to reflect varying parametric sensitivities from case to case in the same design problem (e.g., Figure 7.3a) and across different design realms (e.g., Figure 7.3b).

In the next section, the Low and Tang (2007) spreadsheet-automated FORM computational procedure is first illustrated and discussed for a simple retaining wall with only two random variables. The intuitive expanding dispersion ellipsoid perspective and the meaning of reliability index are explained. This is followed by the RBD of other walls, including an anchored sheet pile wall (in Section 7.6) involving six random variables, which are treated as correlated nonnormal variates. Comparisons are made with Monte Carlo simulations, and similarities and differences between RBD and limit state design using partial factors are discussed. Reasonable statistical properties are assumed for the hypothetical cases presented. The focus is on obtaining useful information and insights from RBD-via-FORM for enhancing partial factor design methods.

7.3 RELIABILITY-BASED DESIGN OF A SEMI-GRAVITY RETAINING WALL

At least three geotechnical failure modes need to be considered in the design of a semi-gravity retaining wall, namely: rotation about the toe of the wall, horizontal sliding along the base of the wall, and bearing capacity failure of the soil beneath the wall, under the inclined and eccentric resultant load derived from the weight of the wall and the active earth thrust P_a acting on the back of the wall.

The reliability analysis of the semi-gravity retaining wall in Figure 7.1 was performed using Microsoft Excel software and its built-in optimization programme Solver. For simplicity, it is assumed that the *in situ* stiff clay offers ample reliability against bearing capacity failure; hence, only the rotational and sliding modes will be considered. (Reliability analysis with respect to bearing capacity limit state was dealt with in Chapter 5.)

7.3.1 Limit state functions with respect to rotation and sliding

The Coulomb active earth pressure coefficient is used, which is based on the assumption of a plane slip surface in the soil. The value is practically the same as the more rigorous Caquot and Kerisel active earth pressure coefficient which assumes a logarithmic spiral slip surface, as given in British Standards BS 8002 (1994), for example. The Coulomb active coefficient K_a is (e.g., Lambe and Whitman 1979):

$$K_a = \left(\frac{\sin(\alpha - \phi')/\sin\alpha}{\sqrt{\sin(\alpha + \delta)} + \sqrt{\sin(\phi' + \delta)\sin(\phi' - \lambda)/\sin(\alpha - \lambda)}} \right)^2 \quad (7.4)$$

where α and λ are the inclinations (Figure 7.1) of the back of the wall and the retained fill surface with respect to the horizontal, ϕ' is the angle of internal friction of the soil and δ is the interface friction angle between the concrete wall and the soil. The water table is below the base of the retaining wall. The active earth thrust P_a (kN/m) is taken to act at a height of H/3 above the base of the wall and at an angle δ with the normal to the back of the wall; that is, at an angle $(\delta + \alpha - 90°)$ with the horizontal.

Although Excel solution files are downloadable from the book's website for the reliability cases of this book, readers are strongly encouraged to start from scratch on a blank worksheet to set up Figure 7.1 and conduct reliability analysis hands-on following the steps below, in order to have a deeper personal feel of the Low and Tang (2007) spreadsheet FORM procedure.

In Figure 7.1, the following equations have been set up, for a wall with a vertical back (i.e., $\alpha = 90°$):

$$P_{av} = P_a \sin\delta, \quad \text{Arm}_{av} = b, \quad W_1 = 0.5\gamma_{wall}(b-a)H, \quad \text{Arm}_1 = \frac{2}{3}(b-a)$$

$$W_2 = \gamma_{wall}aH, \quad \text{Arm}_2 = b - \frac{a}{2}, \quad P_{ah} = P_a \cos\delta, \quad \text{Arm}_{ah} = \frac{H}{3}$$

where ϕ', δ and c_a are read from the column labelled x^*.

Note that trigonometric functions in Excel require inputs in radians, not degrees. The spreadsheet function *radians(degree)* will convert degree to radian.

The performance functions $g_1(\mathbf{x})$ and $g_2(\mathbf{x})$ with respect to rotational ULS and sliding ULS are, respectively:

$$\text{Rotational ULS:} \quad g_1(\mathbf{x}) = W_1 Arm_1 + W_2 Arm_2 + P_{av} Arm_{av} - P_{ah} Arm_{ah} \quad (7.5)$$

$$\text{Sliding ULS:} \quad g_2(\mathbf{x}) = b \times c_a - P_{ah} \quad (7.6)$$

If the base resistance to sliding has a frictional component $(W_1 + W_2 + P_{av})$ $\tan\phi_a$, it can be added to the adhesion component $b \times c_a$ without affecting the solution procedure described below.

7.3.2 Hasofer–Lind reliability index for correlated normal variates

The Hasofer–Lind index β for correlated normal random variables is:

$$\beta = \min_{\mathbf{x} \in F} \sqrt{(\mathbf{x} - \mathbf{\mu})^T \mathbf{C}^{-1} (\mathbf{x} - \mathbf{\mu})} \quad (7.7)$$

or, equivalently:

$$\beta = \min_{\mathbf{x} \in F} \sqrt{\left[\frac{x_i - \mu_i}{\sigma_i}\right]^T \mathbf{R}^{-1} \left[\frac{x_i - \mu_i}{\sigma_i}\right]} = \min_{\mathbf{x} \in F} \sqrt{\mathbf{n}^T \mathbf{R}^{-1} \mathbf{n}} \quad (7.8a)$$

where

$$n_i = \frac{x_i - \mu_i}{\sigma_i}, \quad (7.8b)$$

from which

$$x_i = \mu_i + n_i \sigma_i \quad (\text{for normal variates only}) \quad (7.8c)$$

where $\mathbf{x}$ is a vector representing the set of random variables x_i, $\mathbf{\mu}$ the vector of mean values μ_i, $\mathbf{C}$ the covariance matrix, $\mathbf{R}$ the correlation matrix, σ_i the standard deviation, F the failure domain.

For the reliability example in Figure 7.1, the random variables are soil friction angle ϕ', the interface friction angle δ and the base adhesion c_a. It is expected that ϕ' and δ are positively correlated, and a correlation coefficient of 0.7 is adopted, as shown by the correlation matrix in Figure 7.1. The parameters ϕ', δ and c_a in the above equations read their values from the column labeled $\mathbf{x}^*$. The $\mathbf{x}^*$ values, and the functions dependent on them, change during the optimization search for the most probable failure point.

The column labeled $\mathbf{x}^*$ in Figure 7.1 contains Eq. 7.8c. (If nonnormal variates are modelled, one simply enter the VBA function $@x\_i(\ldots \ ni)$ which obtains x_i from n_i via $x_i = F^{-1}[\Phi(n_i)]$, as explained in Chapter 2 and the Appendix.)

The quadratic form $\sqrt{\mathbf{n}^T \mathbf{R}^{-1} \mathbf{n}}$ in Eq. 7.8a for the reliability index is entered in the cell labelled β, as an *array formula*:

$$= \mathrm{sqrt}\Big(\mathrm{mmult}\big(\mathrm{transpose}(\mathbf{n}), \mathrm{mmult}\big(\mathrm{minverse}(\mathbf{R}),\mathbf{n}\big)\big)\Big) \qquad (7.9)$$

followed by '*Enter*' while holding down the '*Ctrl*' and '*Shift*' keys. In the above formula, *mmult*, *transpose* and *minverse* are Microsoft Excel's built-in functions, each being a container of programme codes for matrix operations. The quadratic form depicts an ellipse when there are two variates, an ellipsoid when there are three variates, and an hyperellipsoid when the number of variates is more than three. This ellipsoid perspective is valid even for nonnormal variates when viewed as equivalent normal ellipsoids, if one recalls that the Rackwitz-Fiessler equivalent normal transformation is used in the classical FORM computational approach involving nonnormals.

The column values labelled '**n**' were initially zeros. The Excel spreadsheet's built-in *Solver* optimization routine was then invoked, to '*Minimize*' the β cell, '*By Changing*' the three values under the '**n**' column, '*Subject To*' the constraints $g_1(\mathbf{x}) \le 0$ and the $\mathbf{x}^*$ value of $\phi' \ge$ the $\mathbf{x}^*$ of δ. The Solver option 'Use automatic scaling' can also be activated. (The constraint $g_2(\mathbf{x}) \le 0$ would be specified instead when performing reliability analysis with respect to the sliding mode.) This setting of constrained optimization literally instructs the *Solver* tool to find the smallest expanding dispersion ellipsoid that just touches the LSS, defined by $g(\mathbf{x}) = 0$. The nearest point of contact between the ellipsoid and the LSS is the most probable point (MPP) of failure, also called the design point when there is a target reliability index.

A retaining wall of base width $b = 1.875$ m was found, after a few trials, to achieve a target reliability index of $\beta = 3.0$ against the overturning mode, Figure 7.1a. For this width, the solution for the sliding mode ($\beta = 3.25$) is also shown in the insert at the bottom right of the figure. The spreadsheet approach is simple and intuitive because it works in the original space of the variables. It does not involve the orthogonal transformation of the correlation matrix, and iterative numerical partial derivatives are done automatically by *Excel Solver* on spreadsheet objects which may be implicit or contain VBA codes.

The '$\mathbf{x}^*$' values obtained in the top right of Figure 7.1 represent the most probable point (MPP) of failure on the rotational LSS. It is the point of tangency (Figure 7.2a) of the expanding dispersion ellipse (or ellipsoid when more random variables are involved) with the rotational LSS and, in the bottom inset of Figure 7.1, with the sliding LSS. In this case where the MPP of failure corresponds to a target value (3.0) of reliability index, it is also referred to as the design point. Figure 7.2b shows an alternative way of plotting the dispersion ellipse and the design point in the dimensionless n_1 and n_2 space, where sensitivities are portrayed more objectively, similar in procedure and perspective to the initial figures of Chapter 5 when the bearing capacity limit state was investigated. In the bearing capacity case of Figures 5.1 and 5.2, the shear strength parameters c' and ϕ' were negatively correlated, and in Figure 7.2 the internal friction angle ϕ' and soil-wall interface friction angle δ are positively correlated based on physical considerations. Hence, the ellipses in the Figures 5.1 and 5.2 and those in Figure 7.2 are rotated in different directions.

The following may be noted:

(a) For the rotational limit state, the $\mathbf{x}^*$ values shown in the top right of Figure 7.1 render the performance function for overturning mode (Eq. 7.5) equal to zero. Hence, the point represented by these $\mathbf{x}^*$ values lies on the overturning LSS. That the dimensionless sensitivity indicator of c_a under the column labelled $\mathbf{n}$ turns out to be zero (resulting in the $\mathbf{x}^*$ value of c_a not deviating at all from the mean c_a) implies that the overturning mode is insensitive to the base adhesion c_a, as one would expect when the overturning mode does not depend on c_a at all. This means that the overturning ULS and its associated β index can be plotted in the two-dimensional space of ϕ' and δ, as shown in Figure 7.2, to scale. The limit state surface separating the safe domain from the unsafe domain is described by Eq. 7.5 when it is equal to zero, that is $g_1(\mathbf{x}) = 0$. The one-standard-deviation ellipse and the β-ellipse are rotated because the correlation coefficient between ϕ' and δ is 0.7. The *design point* is where the expanding dispersion ellipse touches the LSS, as shown in Figure 7.2a and 7.2b, at the point represented by the $\mathbf{x}^*$ values and the $n_{\phi'}$ and n_δ values of Figure 7.1.

(b) As a multivariate normal dispersion ellipsoid expands, its expanding surfaces are contours of decreasing probability values, according to the established probability density function of the multivariate normal distribution (Eq. 2.4 of Chapter 2). Hence, to minimize β according to Eq. 7.7 or 7.8 is tantamount to finding the smallest ellipsoid tangent to the LSS at the MPP of failure (the *design point*).

(c) Higher values of ϕ' and δ reduce the Coulomb active earth pressure coefficient K_a, thereby enhancing reliability against both the overturning and sliding ULS of Eqs. 7.5 and 7.6, respectively. One may investigate the significance of assuming uncorrelated ϕ' and δ, by replacing the two 0.7 values in the correlation matrix (Figure 7.1) with 0, reinitializing $\mathbf{n}$ values to zeros, and re-invoking Solver. A higher reliability index of $\beta = 3.7$ is obtained, as shown in Figure 7.4, to scale. The two ellipses in Figure 7.4a are canonical (i.e., not rotated) because the correlation coefficient is zero. Thus, in this case when ϕ' and δ are both resistance parameters, ignoring their positive correlation leads to an

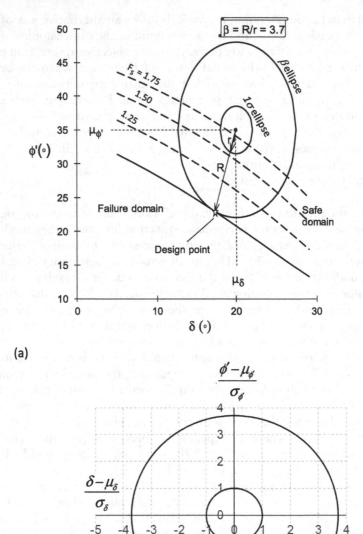

(a)

(b)

Figure 7.4 The consequence of ignoring positive correlation between angle of internal friction φ' of the retained fill and the wall-soil interface friction angle δ, for the rotational ULS. This figure is to be compared with Figure 7.2, where positive correlation between φ' and δ was modelled.

unconservative estimate of reliability against overturning mode: β = 3.0 in Figure 7.2 versus β = 3.7 in Figure 7.4, consistent with Table 5.1 in page 125. On the other hand, had the two random variables been ϕ' (a resistance parameter) and soil unit weight γ (a load parameter for earth retaining wall), which logically should also be positively correlated, ignoring the correlation would result in an underestimation of the reliability index β against the overturning limit state.

(d) A graphical appreciation of positive and negative correlations for ϕ' and δ, or any other two parameters x_1 and x_2, in the dimensionless space of n_1 and n_2, is shown in Figure 5.4 of Chapter 5. For uncorrelated parameters, the correlation coefficient ρ is equal to zero, and the canonical ellipse in the n_1 and n_2 space reduces to a sphere, as shown in Figure 7.4b.

(e) The results of reliability analysis for the sliding limit state given by Eq. 7.6 are shown in the bottom-right inset in Figure 7.1. A reliability index of 3.25 was obtained. The x^* values and the three nonzero values of n imply that the sliding limit state is sensitive to all three parameters, ϕ', δ and c_a, with c_a being the highest on the sensitivity scale. The mean value point, at (35, 20, 100), is safe against sliding, manifested by the $g_2(x)$ cell displaying a positive value when the n values were initially zeros; but sliding failure occurs when the ϕ', δ and c_a values were decreased (automatically by the *Solver* tool, via the n values) to the values shown: (29.08, 17.47, 58.45). The distance from the safe mean-value point to this MPP of failure, in units of directional standard deviations, is the reliability index β, equal to 3.25 in this case. One may further note that the ratios of μ_i/x_i^* (akin to the partial factors of EC7) are not pre-defined. This ability of the reliability analysis to seek the most probable x^* values (corresponding to a target reliability level) without presuming any partial factors is an important merit of conducting FORM and RBD-via-FORM.

(f) The probability of failure (P_f) can be estimated from the reliability index β, using Eq. 7.3. Microsoft Excel's built-in function NormSDist(.) can be used to compute $\Phi(.)$ and hence P_f. Thus, for the overturning mode of Figure 7.1, P_f = NormSDist(–3.01) = 0.13%, and for the sliding mode P_f = NormSDist(–3.25) = 0.06%. These values compare remarkably well with the values 0.132% and 0.062% in Figure 7.5a, obtained from Monte Carlo simulation with 800,000 trials using the commercial simulation software @RISK version 8.0 (www.palisade. com). The correlation matrix was accounted for in the simulation. The excellent agreement between 0.13% from reliability index and 0.132% from Monte Carlo simulation is hardly surprising given the almost linear limit state surface and normal variates shown in Figure 7.2. However, for the sliding ULS involving three random variables, and especially the anchored wall (Section 7.6) involving six random variables and nonnormal distributions, the six-dimensional equivalent hyperellispoid and the limit state hypersurface can only be perceived in the mind's eye. Nevertheless, as shown later, the probabilities of failure inferred from reliability indices for the higher dimensional cases are again in close agreement with Monte Carlo simulations. Computing the reliability index and $P_f = \Phi(-\beta)$ by the present approach takes only a few seconds. In contrast, the time needed to obtain the probability of failure by Monte Carlo simulation is several orders of magnitude longer, particularly when the probability of failure is small and many trials are needed. It is also a simple matter to investigate sensitivities by re-computing the reliability index β (and P_f) for different mean values and standard deviations in numerous what-if scenarios. (Note that the probability of failure as used here means the probability that, in the presence of parametric uncertainties in ϕ' and δ, the factor of safety, Eq. 7.1, will be ≤ 1.0, or, equivalently, the probability that the performance function, Eq. 7.5, will be ≤ 0.)

(g) Monte Carlo simulations provide additional information on the number of simultaneous overturning-and-sliding failures, from which the probability of simultaneous failure and the total failure probability (Columns 4 and 5, Figure 7.5a) can be inferred. The value of approximately 0.188% for 800,000 trials is deemed to be more accurate than the corresponding value (0.179%) for 200,000 trials. The two-mode total failure probability of approximately 0.188% is near the upper end of the range indicated by first-order series bounds for different failure modes F_i (Cornell 1967):

$$\max\left[P(F_i)\right] \leq P(\text{failure}) \leq 1 - \Pi\left[1 - P(F_i)\right] \qquad (7.10a)$$

which, for the reliability indices of 3.01 and 3.25 shown in Figure 7.1, translates to:

$$0.13\% \leq P(\text{failure}) \leq 0.19\% \qquad (7.10b)$$

A convenient alternative for estimating the system failure probability more accurately than Eq. 7.10 is afforded by the system reliability procedure of the Kounias-Ditlevsen bimodal bounds for systems with multiple failure modes, implemented in the Excel spreadsheet platform in Low et al. (2011). The estimated system failure probability for the case in hand, with the overturning ULS and the sliding ULS, is in the range $0.181\% \leq P_f \leq 0.184\%$, or about 0.18%, as shown in Figure 7.5b, which compares very well with the Monte Carlo P_f of about 0.19% at 800,000 trials.

Another system reliability analysis example involving eight failure modes (page 257) is presented in the chapter on soil slope reliability.

(h) Figure 7.2a defines the reliability index β as the dimensionless ratio R/r, in the direction from the mean-value point to the design point.

Monte Carlo simulations	Overturning failure (1)	Sliding failure (2)	Simultaneous overturning & sliding (3)	Total Failure probability (1)+(2)–(3)
50,000 trials	0.146%	0.058%	0.017%	0.187% ±
200,000 trials	0.119%	0.065%	0.006%	0.179% ±
800,000 trials	0.132%	0.062%	0.007%	0.188% ±

Note: "±" means approximate, as outcomes of repeated MC simulations will vary slightly

(a)

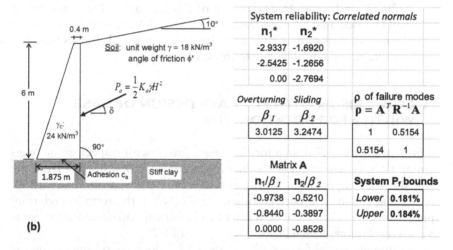

(b)

Figure 7.5 (a) Probabilities of failure of Figure 7.1's semi-gravity retaining wall, using Monte Carlo simulations via @RISK 8.0 software; (b) Bimodal bounds of system failure probability based on component FORM results.

This is the axis ratio of the β-ellipsoid (tangential to the limit state surface) to the one-standard-deviation dispersion ellipsoid. This axis ratio is dimensionless and independent of orientation, when R and r are co-directional. For instance, a vertical line through the mean-value point (centre of the ellipses in Figure 7.2a) will intersect the upper part of the ellipses at $\phi' = 37.50$ and 42.53, respectively, and the dimensionless ratio $(42.53–35)/(37.50–35)$ is equal to 3.01, identical to the R/r ratio from the mean-value point to the design point, and in fact identical to the co-directional R/r ratios in any other direction. This definition of FORM β in terms of the invariant axis-ratio R/r also applies to the normalized dimensionless n_1 and n_2 space of Figure 7.2b. This axis-ratio interpretation in the original space of the variables overcomes a drawback in Shinozuka's (1983) standardized variable space that 'the interpretation of β as the shortest distance between the origin (of the standardized space) and the (transformed) limit state surface is no longer valid' if the random variables are correlated. A further advantage of the original space, apart from its intuitive transparency, is that it renders feasible and efficient computational approach involving nonnormals as presented in Low and Tang (2004, 2007).

(i) The soil/wall interface friction angle δ is often empirically given as a fraction ζ of the internal friction angle ϕ' of the retained soil, $\delta = \zeta\phi'$, where ζ may depend on several factors and cannot be quantified precisely. Instead of treating ϕ' and δ as two highly correlated random variables as in Figure 7.1, one may instead choose to model the uncertainties of ζ and ϕ', for example, with a mean of ζ equal to 0.57 (=20/35) and a standard deviation of ζ equal to 0.1. If this is done, δ will be calculated by using the formula $\delta = (\zeta\phi')^*$, where '*' means the values of ζ and ϕ' under the x^* column in Figure 7.1. In this way, the uncertainty of δ will be due to those of ζ and ϕ'.

7.4 COMPARISON WITH EC7 DA I B DESIGN OF BASE WIDTH b FOR ROTATION ULS

Figure 7.6 shows EC7 design for the base width b with respect to the overturning ULS, via characteristic values and partial factors, starting from the same statistical inputs of mean values and standard deviations, but without considering parametric correlations in EC7. (When the term 'correlation' appears in the EC7 document, it refers to empirical correlations, not parametric correlations, which is not dealt with in EC7).

Even though partial factors are specified, EC7 does not produce a unique design but depends on how conservative the characteristic values are determined. This is not objectionable, for it allows flexibility in design to match the consequence of failure, in the same way that target reliability index can be lower or higher (2.5 or 3.0) depending on the consequence of failure. Analogous non-unique design outcome situation exists for LRFD's nominal values and load and resistance factors.

Figure 7.6 shows that using ϕ' and δ (instead of $\tan\phi'$ and $\tan\delta$) as random variables does not impede comparisons with EC7 in which partial factors are applied to the characteristic values of $\tan\phi_k'$ instead of ϕ_k'. After deciding on the characteristic values of ϕ_k' and δ_k, one computes the design values in a way consistent with EC7 procedure:

$$\phi_d' = \tan^{-1}\left[\left(\tan\phi_k'\right)/1.25\right] \tag{7.11a}$$

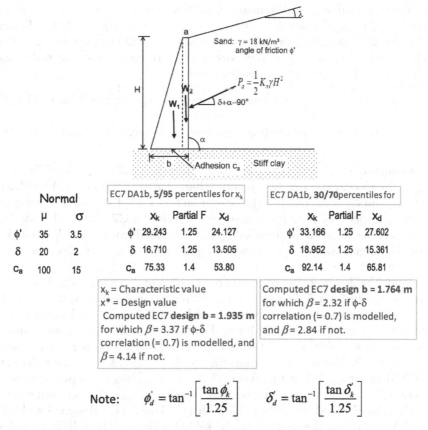

Note: $\phi'_d = \tan^{-1}\left[\dfrac{\tan\phi'_k}{1.25}\right]$ $\delta'_d = \tan^{-1}\left[\dfrac{\tan\delta'_k}{1.25}\right]$

Ambiguities and non-uniqueness of back-calculated partial factors from RBD-via-FORM:

- Affected by assumed characteristic values
- Affected by target reliability index (whether 2.5 or 3.0)
- Affecetd by context-dependent sensitivities even for the same parameter
- Affected by statistical inputs including parametric correlations, probability distribution types

Figure 7.6 Eurocode 7 DA1b design of the base width b for rotation ULS of a gravity retaining wall, and non-uniqueness of partial factor back-calculated from RBD.

and

$$\delta'_d = \tan^{-1}\left[\left(\tan\delta'_k\right)/1.25\right] \tag{7.11b}$$

(The partial factor for the interface friction angle δ_k is assumed the same as that for ϕ_k')
which can then be compared with the design value of ϕ_d' in this example.

The mean values μ and standard deviations σ of ϕ', δ and c_a shown at the top left of Figure 7.6 are the same as those in the RBD-via-FORM of Figure 7.1. For EC7 Design Approach Combination 2 (or DA1b), the conservative characteristic values x_k of the normally distributed favorable resistance parameters ϕ', δ and c_a are calculated from:

$$x_k = \mu - 1.645\sigma \qquad \text{(for the more conservative 5 percentile value)} \qquad (7.12a)$$

$$x_k = \mu - 0.524\sigma \qquad \text{(for the less conservative 30 percentile value)} \qquad (7.12b)$$

yielding the 5 percentile characteristic values $\phi_k' = 29.24°$, $\delta = 16.71°$ and $c_{a,k} = 75.33$ kPa, and the 30 percentile characteristic values $\phi_k' = 33.17°$, $\delta = 18.95°$ and $c_{a,k} = 92.14$ kPa. The partial factors 1.25, 1.25 and 1.4, for ϕ', δ and c_a, respectively, are those given for DA1b of EC7 (2004). The design values under the columns labelled x_d were obtained from $x_d = x_k$/partial factor. Although adhesion c_a is shown in Figure 7.6, it does not affect the over-turning ULS.

Using the design values of ϕ' and δ, one obtains a design width b = 1.935 m that satisfies (Eq. 7.5) $g_1(x) = 0$ for the more conservative 5/95 percentile criterion for characteristic values x_k, and b = 1.764 m for the less conservative 30/70 percentile criterion for x_k. For the case in hand, the more conservative design width of 1.935 m is bigger, and the less conservative design width of 1.764 m is smaller, than the width b = 1.875 m in the RBD-via-FORM of Figure 7.1. If the two EC7-designed widths are input in turn in Figure 7.1, and accounting for positive correlation $\rho_{\phi\delta} = 0.7$ between ϕ' and δ, the reliability index will be $\beta = 3.37$ for the larger EC7-designed width b, and $\beta = 2.32$ for the smaller EC7-designed width b. The reliability index values of 4.14 and 2.84 for uncorrelated ϕ' and δ are also shown but should be ignored because the β values of 3.37 and 2.32, which reflect positive correlation between ϕ' and δ, are more realistic.

We have seen from the above that the design outcome from EC7 varies with the level of conservatism in selecting the characteristic values and that parametric correlation is not considered in EC7. It is useful to conduct FORM analysis on the design outcome of EC7, to gain information on the reliability index and failure probability of the EC7 design, and to compare the EC7 design point values (x_d) with the RBD-via-FORM design point values. Such comparisons will be elaborated in the anchored sheet pile wall example of Section 7.6. Parametric correlations must be incorporated in the FORM analysis of β on EC7 design outcome, if the correlations are justified by physical considerations, notwithstanding that EC7 does not consider parametric considerations.

EC7 DA1a requires characteristic values of resistance and actions, on which partial factors are applied. If characteristic values are based on percentiles, one needs to know the probability distributions of actions and resistance in order to estimate the upper tail (e.g., 70 percentile) characteristic value of actions and lower tail characteristic values of resistance (e.g., 30 percentile). Whether based on 5%/95% or 30%/70%, DA1a is satisfied for the case in hand; DA1b governs.

In RBD-via-FORM, one can account for higher consequence of failure by aiming at a higher reliability index. For the case in Figure 7.1, a design width b = 1.875 m is required to achieve a target reliability index of 3.0. On the other hand, a lower target β = 2.5 is achieved by a smaller design width B = 1.79 m, and a higher target β = 3.5 is achieved by a larger design width B = 1.96 m, as plotted in Figure 7.7a for $\rho_{\phi\delta}$ = 0.7. The results for $\rho_{\phi\delta}$ = 0.0 are plotted in dashed curve, which should be ignored if one deems it logical to assume positively correlated ϕ' and δ based on physical considerations.

RBD-via-FORM needs statistical inputs (mean values, standard deviations, parametric correlations if any, and probability distribution type) but does not require characteristic values and partial factors. In principle, one can back-calculate the implied partial factors $\gamma_{\phi'}$ and γ_δ from the design point of RBD-via-FORM:

$$\gamma_{\phi'} = \tan\left(\phi_k'\right) \text{from EC7} \,/\, \tan\left(\phi'^*\right) \text{from RBD-via-FORM} \quad (7.13a)$$

$$= \frac{\tan\phi_k'}{\tan\mu_{\phi'}} \times \frac{\tan\mu_{\phi'}}{\tan\phi'^*} \quad (7.13b)$$

and similar equations for interface friction angle δ.

Although the second term in Eq. 7.13b is an unambiguous entity in RBD-via-FORM, as plotted in Figure 7.7b, the back-calculated partial factors $\gamma_{\phi'}$ and γ_δ are ambiguous, non-unique because the first term in Eq. 7.13b varies with the level of conservatism in selecting characteristic values (e.g., 5/95 percentiles or 30/70 percentiles). Hence, it is more enlightening to compare the design points of RBD-via-FORM and to consider the RBD sensitivity information revealed in $n_{\phi'}$ and n_δ (Figure 7.7c). When partial factors are back-calculated from RBD-via-FORM, its context-dependency must be borne in mind.

7.5 VECTOR COMPONENTS OF ACTIVE EARTH THRUST ARE ALLIED, NOT ADVERSARIAL

The paragraphs below describe an enlightening apparent paradox which arises when the results of RBD-via-FORM of Figure 7.1 are interpreted from the perspectives of the Load and Resistance Factor Design (LRFD) and the Eurocode 7 Design Approach 2 (EC7-DA2). The apparent paradox is resolved when the horizontal and vertical earth thrusts, P_{ah} and P_{av} (and their moments), are recognized as constituents of the single entity P_a (and the overturning moment caused by P_a). The resulting insights for consistent application of load factors in LRFD and EC7-DA2 are discussed.

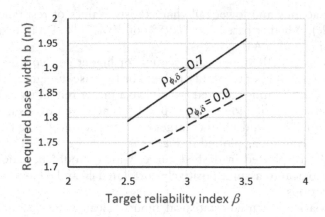

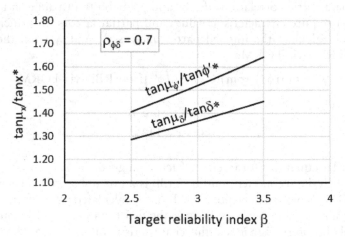

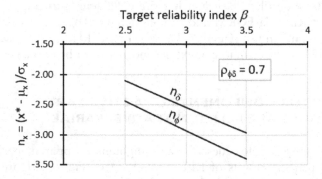

Figure 7.7 The target level of reliability in RBD affects the required width b, the back-calculated mean partial factors, and the sensitivity indicators n.

The load and resistance factor design approach (LRFD) of North America and the EC7-DA2 (and to some extent also EC7-DA1a) are alike in applying partial factors to the nominal/characteristic loads (e.g., overturning moment) and nominal/characteristic resistance (e.g., resisting moment) instead of to the underlying parameters (e.g., ϕ', δ ...) affecting load and resistance. Again, because the nominal loads/resistance in LRFD (or characteristic loads/resistance in EC7) depend on the level of conservatism adopted by designers, which renders back-calculated partial factors non-unique, comparisons below are made in terms of the ratios *Design value/Mean value* of loads and resistance, from RBD-via-FORM and from LRFD or EC7-DA2, without attempting to back-calculate the load and resistance factors from RBD-via-FORM.

Misinterpretation will also arise if one does not realize that P_{ah} and P_{av} are components of the active earth thrust P_a, equal to $P_a \cos\delta$ and $P_a \sin\delta$, respectively. In the lumped factor of safety approach, Eq. 7.14a below (the same as Eq. 7.1) is more consistent than Eq. 7.14b, because its denominator represents the overturning moment of the active earth thrust, computed more conveniently via vertical lever arm (Arm_{ah}) of P_{ah} and horizontal lever arm (Arm_{av}) of P_{av}:

$$F_s = \frac{W_1 \times Arm_1 + W_2 \times Arm_2}{P_{ah} \times Arm_{ah} - P_{av} \times Arm_{av}} \qquad (7.14a)$$

is more consistent than

$$F_s = \frac{W_1 \times Arm_1 + W_2 \times Arm_2 + P_{av} \times Arm_{av}}{P_{ah} \times Arm_{ah}} \qquad (7.14b)$$

Likewise, treating P_{ah} and P_{av} as components of the same entity P_a is more consistent in the interpretation of load and resistance factors from RBD-via-FORM. To illustrate, Figure 7.8a, with the n values initially zeros, shows the mean values of resisting and overturning moments of the case in Figure 7.1, and Figure 7.8b shows the design values of the resisting and overturning moments obtained by RBD-via-FORM, where a design width of 1.875 m achieves a target $\beta = 3.0$, as in Figure 7.1.

The design resisting moment M_R of 200.9 kNm/m is the same as the mean value of M_R for the case in hand where the underlying parameters (geometry and concrete unit weight γ_{wall}) are treated as deterministic with no uncertainty. (If widths a and b and unit weight γ_{wall} are treated as random variables, the design M_R will be smaller than the mean-value M_R.)

The design overturning moment is $P_{ah} \times Arm_{ah} - P_{av} \times Arm_{av} = 200.9$ kNm/m. This means that although the mean values of overturning and resisting moments, at 110.3 kNm/m and 200.9 kNm/m, respectively, do not result in the overturning ULS, the overturning failure occurs when the

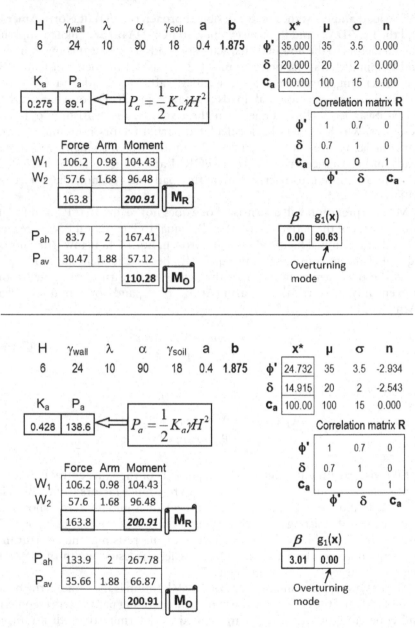

Figure 7.8 (a) Information at mean-value point when the values under the column labelled **n** are initially zeros; (b) RBD-via-FORM solution for target β = 3.0, achieved by width b = 1.875 m as in Figure 7.1. P_{ah} and P_{av} are here treated logically as components of the same entity P_a, a required condition which may be missed in LRFD or EC7-DA2.

overturning moment increases from the mean value of 110.3 kNm/m to its design value of 200.9 kNm/m, equal to that of the design resisting moment. The load factor for overturning moment, in terms of mean value, is 200.9/100.3 = 2.0. The load factor in terms of conservative nominal overturning moment will be smaller than 2.0, depending on the ratio (<1.0) of Mean M_o/Nominal M_o in the first term on the right below:

$$\text{Implied load factor:} \quad \gamma_{MO} = \frac{\text{Mean } M_o}{\text{Nominal } M_o} \times \frac{\text{Design } M_o}{\text{Mean } M_o} \quad (7.15)$$

In the RBD-via-FORM of Figure 7.1 and Figure 7.8, the limit state surface is when the performance function is equal to zero, that is when Eq. 7.5 for $g_1(\mathbf{x})$ = Resisting moment M_R − Overturning moment M_o, is equal to zero. This leads to the same mathematical requirement whether $P_{av} \times Arm_{av}$ appears as increasing M_R (Figure 7.1) or reducing M_o (Figure 7.8). Hence, the RBD-via-FORM solutions in Figures 7.1 and 7.8 are identical. However, in drawing information from either figure from the perspective of LRFD or EC7-DA2, an apparent paradox will arise if one did not recognize $P_{ah} \times Arm_{ah} - P_{av} \times Arm_{av}$ as constituents of overturning moment M_o due to active thrust P_a and instead treat them as separate adversarial entities. That is, if $P_{av} \times Arm_{av}$ is treated as an additional resisting moment, the mean value of M_R is 200.91 + 57.12 = 258 kNm/m, and the design value of M_R is 200.91 + 66.87 = 267.78 kNm/m, while the mean and design values of M_o are 167.4 and 267.78 kNm/m, respectively. The design value of M_R (267.78) being greater than the mean value M_R (258) is an apparent paradox, due to concomitant increase in both P_{ah} and P_{av} when P_a increases. Such counter-intuitive phenomenon does not appear when $P_{av} \times Arm_{av}$ is treated as reducing M_o instead of increasing M_R.

The above discussion also means that it is inconsistent in LRFD and EC7-DA2 to regard P_{av} and $P_{av} \times Arm_{av}$ as favorable resistance entities behaving independently of P_{ah} and $P_{ah} \times Arm_{ah}$, which are unfavorable load and unfavorable overturning moment. In other words, in LRFD or EC7-DA2, it is a questionable procedure to factor down a favorable component (P_{av} or $P_{av} \times Arm_{av}$) and factor up an unfavorable component (P_{ah} or $P_{ah} \times Arm_{ah}$) when both are components of the same vector (P_a and $P_a \times Arm_a$).

It is one of the merits of RBD-via-FORM that its design point values (e.g., ϕ' and δ, or M_o and M_R) change logically with the target reliability index and reflect context-dependent sensitivities.

The design point values in EC7 and LRFD change with level of conservatism in the characteristic values and convey no information on reliability, failure probability or sensitivities.

The next example applies RBD-via-FORM to an anchored sheet pile wall involving six correlated lognormal random variables, some of which affect both the active earth pressure (loadings) and passive earth pressure (resistance). The resulting load-resistance duality is resolved automatically in RBD-via-FORM without the need for trial-and-error favourable and unfavourable assumptions as required in a partial factor design approach.

7.6 RBD-VIA-FORM FOR AN ANCHORED SHEET PILE WALL

The embedded sheet pile wall example in this section illustrates RBD involving six logically correlated random variables reflecting uncertainties in soil shear strength parameter, soil unit weights, wall friction, surcharge and dredge level. The need to observe consistency in partial factor design method when soil of the same source contributes to both load and resistance is explained. This example was discussed in Low (2017) with respect to LRFD. In this section, it is discussed also with respect to EC7. Eurocode 7 is evolving, the latest version being Eurocode 7 (EC7) (2020). However, Clause 1.1 of the EC7 (2020) document 'EN 1997-3' states that it is intended to be used in conjunction with EN 1997-1 (EC7, 2004).

In the anchored sheet pile wall of Figure 7.9, the six random variables are soil unit weight γ (kN/m³) above water, soil saturated unit weight γ_{sat} (kN/m³) below water table, surcharge q_s (kPa) acting on top of the retained soil, angle of internal friction ϕ' (degree) of the sandy soil, wall friction angle δ (degree), and distance z between water table and the dredge level. Their mean values and standard deviations are as shown in the labelled columns of Figure 7.9. These six random variables are assumed to be lognormally distributed. Given the uncertainties and correlation structure as shown in Figure 7.9, one wishes to find the required total wall height H so as to achieve a reliability index of 3.0 against rotational failure about the anchor point 'A.' The stability formulation was based on the free-earth support method, with K_a based on Coulomb formula, and K_p based on Kerisel–Absi chart coded as an Excel VBA function, Figure 7.10. Coulomb equation for K_p is not used because its plane slip surface assumption is known to overestimate the K_p value.

The performance function in the g(**x**) cell is equal to the sum of the five moments M_1 to M_5.

The RBD in Figure 7.9a shows that a total height H (= 6.4 + z + d) of 12.19 m would give a reliability index β of 3.0 against rotational failure. The corresponding probability of failure is estimated as $P_f = \Phi(-\beta) = 0.14\%$.

For comparison, three Monte Carlo simulations using the @RISK software (www.palisade.com) with 800,000 realizations each gives P_f of 0.153%, 0.159%, 0.151%.

The **x*** values denote the *design point* on the limit state surface and represent the most likely combination of parametric values that will cause failure. The expected embedment depth is d = 12.19–6.4 – μ_z = 3.39 m. At the

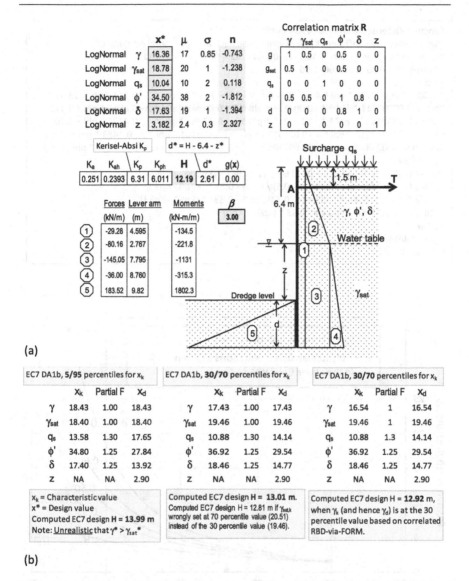

		x*	μ	σ	n
LogNormal	γ	16.36	17	0.85	-0.743
LogNormal	γ_{sat}	18.78	20	1	-1.238
LogNormal	q_s	10.04	10	2	0.118
LogNormal	ϕ'	34.50	38	2	-1.812
LogNormal	δ	17.63	19	1	-1.394
LogNormal	z	3.182	2.4	0.3	2.327

Correlation matrix R

	γ	γ_{sat}	q_s	ϕ'	δ	z
g	1	0.5	0	0.5	0	0
g_{sat}	0.5	1	0	0.5	0	0
q_s	0	0	1	0	0	0
f	0.5	0.5	0	1	0.8	0
d	0	0	0	0.8	1	0
z	0	0	0	0	0	1

Kerisel-Absi K_p $d^* = H - 6.4 - z^*$

K_a	K_{ah}	K_p	K_{ph}	H	d^*	g(x)
0.251	0.2393	6.31	6.011	12.19	2.61	0.00

	Forces (kN/m)	Lever arm (m)	Moments (kN-m/m)	β
①	-29.28	4.595	-134.5	3.00
②	-80.16	2.767	-221.8	
③	-145.05	7.795	-1131	
④	-36.00	8.760	-315.3	
⑤	183.52	9.82	1802.3	

(a)

	EC7 DA1b, **5/95** percentiles for x_k			EC7 DA1b, **30/70** percentiles for x_k			EC7 DA1b, **30/70** percentiles for x_k		
	x_k	Partial F	x_d	x_k	Partial F	x_d	x_k	Partial F	x_d
γ	18.43	1.00	18.43	17.43	1.00	17.43	16.54	1	16.54
γ_{sat}	18.40	1.00	18.40	19.46	1.00	19.46	19.46	1	19.46
q_s	13.58	1.30	17.65	10.88	1.30	14.14	10.88	1.3	14.14
ϕ'	34.80	1.25	27.84	36.92	1.25	29.54	36.92	1.25	29.54
δ	17.40	1.25	13.92	18.46	1.25	14.77	18.46	1.25	14.77
z	NA	NA	2.90	NA	NA	2.90	NA	NA	2.90

x_k = Characteristic value x* = Design value Computed EC7 design H = **13.99 m** Note: <u>Unrealistic</u> that $\gamma^* > \gamma_{sat}^*$	Computed **EC7 design H = 13.01 m**. Computed EC7 design H = 12.81 m if γ_{satk} wrongly set at 70 percentile value (20.51) instead of the 30 percentile value (19.46).	Computed EC7 design H = **12.92 m**, when γ_k (and hence γ_d) is at the 30 percentile value based on correlated RBD-via-FORM.

(b)

Figure 7.9 (a) RBD-via-FORM of anchored wall; (b) EC7-DA1b design, with different outcomes depending on the level of conservatism in characteristics values.

failure combination of parametric values, the design value of z is $z^* = 3.182$, and $d^* = 12.19 - 6.4 - z^* = 2.61$ m. This corresponds to an 'overdig' allowance of 0.78 m. This 'overdig' is determined automatically and reflects uncertainties and sensitivities from case to case in a way that specified 'overdig' cannot.

The values of **n** in Figure 7.9 underlie the computed reliability index β and are indicators of parametric sensitivities. For the case in hand, the values of

```
Function KpKeriselAbsi(phi, del) As Double
'Passive pressure coefficient Kp, for vertical wall back and horizontal retained fill
'Based on Tables in Kerisel & Absi (1990), for beta = 0, lamda = 0, and
' del/phi = 0, 0.33, 0.5, 0.66, 1.00
x = del / phi
Kp100 = 0.00007776 * phi ^ 4 - 0.006608 * phi ^ 3 + 0.2107 * phi ^ 2 - 2.714 * phi + 13.63
Kp66 = 0.00002611 * phi ^ 4 - 0.002113 * phi ^ 3 + 0.06843 * phi ^ 2 - 0.8512 * phi + 5.142
Kp50 = 0.00001559 * phi ^ 4 - 0.001215 * phi ^ 3 + 0.03886 * phi ^ 2 - 0.4473 * phi + 3.208
Kp33 = 0.000007318 * phi ^ 4 - 0.0005195 * phi ^ 3 + 0.0164 * phi ^ 2 - 0.1483 * phi + 1.798
Kp0 = 0.000002636 * phi ^ 4 - 0.0002201 * phi ^ 3 + 0.008267 * phi ^ 2 - 0.0714 * phi + 1.507
Select Case x
    Case 0.66 To 1: Kp = Kp66 + (x - 0.66) / (1 - 0.66) * (Kp100 - Kp66)
    Case 0.5 To 0.66: Kp = Kp50 + (x - 0.5) / (0.66 - 0.5) * (Kp66 - Kp50)
    Case 0.33 To 0.5: Kp = Kp33 + (x - 0.33) / (0.5 - 0.33) * (Kp50 - Kp33)
    Case 0 To 0.33: Kp = Kp0 + x / 0.33 * (Kp33 - Kp0)
End Select
KpKeriselAbsi = Kp
End Function
```

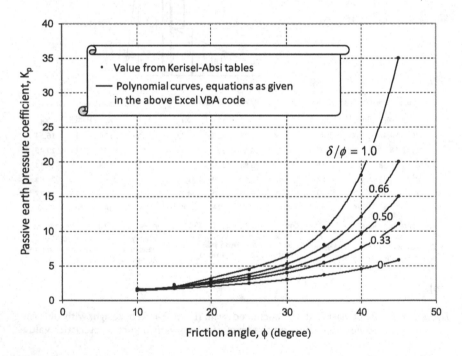

Figure 7.10 User-created Excel VBA function for K_p.

$n_z = 2.327$ and $n_{\phi'} = -1.812$ indicate that, for the given mean values and uncertainties, rotational stability is, not surprisingly, most sensitive to the dredge level (which affects z and d and hence the passive resistance) and ϕ'. It is interesting to note that at the design point where the six-dimensional dispersion ellipsoid touches the limit state surface, both unit weights γ and

γ_{sat} (16.36 and 18.78 kN/m³) are lower than their corresponding mean values, contrary to the expectation that higher unit weights will increase active pressure and hence greater instability. This apparent paradox is resolved if one notes that smaller γ_{sat} will (via smaller γ') reduce passive resistance, smaller ϕ' will cause greater active pressure and smaller passive pressure, and that γ, γ_{sat} and ϕ' are logically positively correlated and thus modelled with positive correlation coefficients in the correlation matrix. It is also logical to assume a positive correlation between the soil angle of friction ϕ' and the interface friction angle δ between the soil and the sheet pile, and this also has been modelled in the correlation matrix.

In this case where some of the six parameters are (logically) correlated, as shown in the correlation matrix of Figure 7.9, and where uncertainties in unit weights of the positively correlated γ and γ_{sat} are modeled and the sand source is the same in the active and passive zone (i.e., single-source or same-source), RBD can play a complementary role to partial factor design methods in automatically deciding whether the design values of γ and γ_{sat} should be lower or higher than their respective mean values, yielding the corresponding design (factored) active thrust and design (factored) passive resistance. In contrast, applying code-prescribed partial factors may produce factored active thrust and factored passive resistance which connote different soil unit weights left and right of the sheet pile, violating the same-source principle. The implications for EC7 design will be considered shortly.

Prior to RBD, the overturning and resisting moments evaluated at the equivalent mean-value point (corresponding to initial values of 0.0 in the column labelled n) are M_R = 4324 kNm/m, and M_o = 1629 kNm/m. After RBD design, the diminished resisting moment is equal to the amplified overturning moment, both equal to 1802 kNm/m. This means that the resistance factor (with respect to moments at mean-value point) is 1802/4324 = 0.42, and the load factor (in this case overturning moment factor) is 1802/1629 = 1.11. (The corresponding LRFD resistance factor is higher than 0.42, and the LRFD load factor is smaller than 1.11, because the RF and LF in LRFD are defined in terms of nominal resistance (< 4324 kNm/m) and nominal loads (>1629 kNm/m).) Note that in this example the overturning and resisting moments are affected probabilistically in a complexly intertwined manner due to uncertainty in dredge level z, parametric correlations, and same-source principle on the unit weights. The back-calculated RF and LF will differ from the above values of 0.42 and 1.11 if the uncertainty model (number and type of random variables, parametric correlations, probability distributions) is different. RBD can play a complementary role to LRFD by directly basing the design on the uncertainty model for a target reliability index or probability of failure, and, if desired, providing back-calculated RF and LF.

EC7 allows statistical methods to be used in the selection of characteristic values for ground properties. The characteristic value of a geotechnical parameter is a cautious estimate of the value affecting the occurrence of the limit state. Characteristic values can be lower values, which are less than the most probable values, or upper values, which are greater. EC7 requires that, for each calculation, the most unfavorable combination of lower and upper values of independent parameters shall be used. Parametric correlations (e.g., those in the correlation matrix of Figure 7.9) cannot be considered in EC7.

We next consider the design of the anchored sheet pile wall of Figure 7.9 based on EC7 procedure which applies partial factors to conservative characteristic values.

7.6.1 Possible multiple outcomes in EC7 (2004) DA1b design of anchored sheet pile involving stabilizing–destabilizing unit weight

Figure 7.9b illustrates possible design outcomes for the anchored sheet pile wall, based on the EC7 (2004) Design Approach 1 Combination 2 (referred to as EC7-DA1b), which ignores any parametric correlations. EC7 has an 'unforeseen overdig' allowance for z, to account for the uncertainty of the dredge level. The design value of z is obtained from $\mu_z + 0.5$ m $= 2.4 + 0.5 = 2.9$ m, where 0.5 m is the 'overdig.' This design value $z^* = 2.90$ m is shown in the last row of Figure 7.9b.

In EC7, the partial factor of soil unit weight is specified to be 1.0, but conservative characteristic values of γ and γ_{sat} still need to be estimated. If soils on both side of the sheet pile originate from the same source, it is not logical to increase the unit weight on the active side and decrease the unit weight on the passive side. Also, assuming 5/95 percentiles for characteristic values leads to the design values $\gamma_d > \gamma_{sat,d}$ (18.43 > 18.40 in the third column of Figure 7.9b left panel), which violates soil physics. This may arise when a practitioner judges that the soil unit weight above the water table (γ) is unambiguously a load parameter (causing active pressure in the upper 6.4 m only of the sheet pile), and, given its μ of 17 kN/m^3 and σ of 0.85 kN/m^3 (based on $\sigma/\mu = 0.5$), decides on the characteristic value γ_k as 18.43 kN/m^3, being the 95 percentile value of its lognormal distribution. Suppose the same practitioner judges that γ' ($= \gamma_{sat} - \gamma_w$) affects the passive resistance more than the active pressure and decides to take the characteristic value of γ_{sat} ($\mu = 20$ kN/m^3 and $\sigma = 1$ kN/m^3) at its 5-percentile value of 18.40 kN/m^3. This leads to the physically impossible design value of γ_{sat} being smaller than the design value of γ, 18.40 < 18.43. The higher-than-average value of γ_d (18.43 > 17) is also incompatible with the lower-than-average value of ϕ_d' (27.84°, which is logical because smaller ϕ_d' leads to larger active pressure and smaller passive resistance). In contrast, RBD-via-FORM automatically determines that the design values of γ and γ_{sat}, 16.36 and 18.78, are at 0.743 and 1.238 (Figure 7.9a, values under the column labelled n) times the respective equivalent standard deviations *below* their respective equivalent mean values, thereby reflecting the logical positive correlations between γ, γ_{sat} and ϕ'.

This means that smaller unit weights, γ and $(\gamma_{sat} - \gamma_w)$, tend to occur with smaller internal angle of friction ϕ'. Parametric correlations are not part of the EC procedure and may be difficult to be incorporated as EC7 evolves. It is in this and other aspects that RBD-via-FORM can provide guidance and insights to EC7 practitioners.

The central panel of Figure 7.9b shows the EC7 design outcome based on the less conservative 30/70 percentiles for characteristic values, resulting in a smaller total design height H than the 5/95 percentiles assumption. The characteristic values of γ and γ_{sat} are now chosen by the designer to be at their 70 and 30 percentile values (17.43, 19.46), respectively, with the characteristic values of the other parameters q_s, ϕ' and δ chosen at their 70, 30 and 30 percentiles, respectively. The resultant EC7-DA1b design H is 13.01 m, more critical than the design H of 12.81 m had the designer chosen the characteristic value of γ_{sat} to be at the 70 percentile level. Such calculations involving trial combinations of lower and upper values of parameters (assumed independent in EC7) are necessary in designs based on EC7. The RBD-via-FORM n_γ and $n_{\gamma sat}$ values of -0.743 and -1.238 in Figure 7.9a, which reflects parametric correlations between γ, γ_{sat} and ϕ', not only suggest that a design value of γ_{sat} smaller than its average value (given that $n_{\gamma sat} = -1.238$) is more critical than a design value of γ_{sat} greater than its average value, but also suggest that even the unfavorable sand unit weight γ above water table should have a characteristic value (same as design value when partial factor is 1.0) smaller than its average value of 17 kN/m^3, due to the realistic positive correlation of γ with γ_{sat} and ϕ', both having design values smaller than their average values in RBD-via-FORM. These considerations are shown in the right panel of Figure 7.9b, where γ_d ($= \gamma_k$) is at the 30-percentile value (16.54) instead of the 70-percentile value (17.43) for γ_d in the central panel. The resulting design H of 12.9 m is slightly less critical than the 13.0 m of the central panel of Figure 7.9b but is physically more realistic in using smaller-than-average design value of γ to go with smaller-than-average design values of γ_{sat} and ϕ'.

In summary, the design outcome based on EC7 is not unique but depends on:

(a) The level of conservatism in selecting the characteristic values, e.g., left and central panels of Figure 7.9b.
(b) Physically incompatible design values of properties may arise and need to be resolved, e.g., design value of γ and γ_{sat} in the left panel of Figure 7.9b.
(c) Design values of properties which contradict logical parametric correlations will need to be resolved, for example, it is a contradiction to have significantly higher-than-average design values of sand unit weights coexisting with significantly lower-than-average design value of internal friction angle. RBD-via-FORM can model parametric correlations to provide insights to EC7 and LRFD, which cannot deal with parametric correlations and load-resistance correlations easily.

The design point values (the x^* values in Figure 7.9a) in RBD-via-FORM reflect case-specific statistical inputs, correlation matrix, and target reliability index value ($\beta = 3.0$ for the case in hand) and hence will not be identical to the uncorrelated design point values in the EC7 design based on code-recommended partial factors applied to subjective characteristic values. The design outcomes from RBD-via-FORM and from EC7 are also likely to be different, e.g., 12.19 m in Figure 7.9a versus 12.92 m in Figure 7.9b right panel. It is suggested that RBD-via-FORM be conducted in tandem with EC7 and LRFD in the following two ways:

(1) Conduct RBD-via-FORM, with parametric correlations if justified by physical considerations, for a target reliability index of β equal to 2.5 or 3.0, for example, as done in Figure 7.9a. The outcome of the RBD-via-FORM design (e.g., H = 12.19 m in Figure 7.9a) can be compared with that from partial factor design (e.g., H = 12.92 m in Figure 7.9b right panel). Insights can be obtained from the MPP of failure (the column labelled x^* in Figure 7.9a) and from the sensitivity indicators (the column labelled n in Figure 7.9a) to guide EC7 design.
(2) Conduct FORM analysis on the outcome of EC7 design, for example, H = 12.92 m from Figure 7.9b, to obtain an estimate of the failure probability, and to compare the MPP of FORM with the design point of EC7. For this purpose, parametric correlations (if justified by physical considerations) should be modelled in FORM analysis of EC7 design outcome, even though EC7 cannot consider parametric correlations.

7.7 POSITIVE RELIABILITY INDEX ONLY IF THE MEAN-VALUE POINT IS IN THE SAFE DOMAIN

In reliability analysis and reliability-based design, one needs to distinguish negative from positive reliability index. The computed β index can be regarded as positive only if the performance function value is positive at the mean value point. Although the discussions in the next paragraph assume normally distributed random variables, they are equally valid for the equivalent normals of nonnormal random variables in FORM. The same principle applies to RBD involving multidimensional space.

In the two-dimensional schematic illustration of Figure 7.11a, the LSS is defined by performance function $g(x) = 0$. The safe domain is where $g(x) > 0$, and the unsafe domain where $g(x) < 0$. The mean-value point of case I is in the safe domain and at the centre of a one-standard-deviation dispersion ellipse, or ellipsoid in higher dimensions. As the dispersion ellipsoid expands, the probability density on its surface diminishes. The first point of contact with the LSS is the most-probable failure point, also called the design point. The reliability index β of case I is therefore positive and represents the distance (in units of directional standard deviations) *from the safe mean-value*

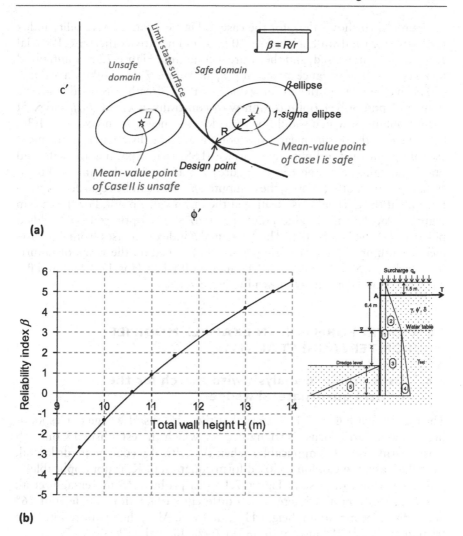

Figure 7.11 (a) Distinguishing negative from positive reliability index. Reliability index of case I is positive, that of case II is negative, (b) For the anchored sheet pile wall example, β is positive for H >10.6 m, and negative for smaller H.

point to the unsafe boundary (the LSS) in the space of the random variables. In contrast, case II's mean-value point is already in the unsafe domain, and the computed β must be given a negative sign because it is the distance *from the unsafe mean-value point to the safe boundary* (the LSS). In RBD, one can easily determine whether the computed reliability index should be positive or negative by checking whether the performance function is positive or negative when the random variables are at their mean values or equivalent mean values (when the n column values in Figures 7.1 and 7.9a are initially zeros prior to FORM implementation).

Figure 7.11b shows that, for the case in Figure 7.9a, the reliability index is positive for total wall height H > 10.6 m, and negative otherwise. If a trial *H* value of 10 m is used, and the entire **n** column in Figure 7.9a is initialized to zeros, the performance function g(**x**) in Figure 7.9a exhibits a value – 437.3, meaning that the mean value point is already inside the unsafe domain. Upon Solver optimization with constraint g(**x**) = 0, a β index of 1.36 is obtained, which should be regarded as a negative index, i.e., –1.36, meaning that the *unsafe* mean value point is at some distance from the nearest safe point on the limit state surface (LSS) that separates the safe and unsafe domains (e.g., the LSS in Figures 7.2 and 7.11a, but in six dimensional space). In other words, the computed β index can be regarded as positive only if the g(**x**) value is positive at the mean value point. For the case in Figure 7.9a, the mean value point (prior to Solver optimization) yields a positive g(**x**) for *H* > 10.6 m. The computed β index increases from 0 (equivalent to a lumped factor of safety equal to 1.0, i.e., on the verge of failure, for which P_f = 50%) when *H* is 10.6 m, to 3.0 when *H* is 12.19 m, and β = 4.0 when H is 12.85 m, as shown in Figure 7.11b.

7.8 RBD-VIA-FORM FOR AN EXAMPLE MODIFIED FROM TERZAGHI ET AL. (1996)

7.8.1 Deterministic analysis with search for the critical quadrilateral wedge

The example in Figure 7.1 assumes that the critical soil wedge corresponding to active earth thrust is triangular, for which the established Coulomb closed form Eq. 7.4 is applicable. When the critical wedge is quadrilateral, as in this case, the Coulomb closed form equation of K_a is not applicable.

The deterministic case in Figure 7.12 is a problem from Terzaghi et al. (1996, p. 253), modified here with a different surface slope angle ψ of 26° and a different maximum height H_{crest} of 14 m. Also, the wall-ore interface friction angle δ is 28° instead of 36° in Terzaghi et al. (1996).

The critical ore wedge may be quadrilateral instead of triangular. Although Eq. 7.4 can be used to obtain an approximate solution, one can obtain a more correct solution for the active earth thrust P_a by trying different slip surface inclination angle θ. The wedge weight W is computed for each trial wedge, to obtain the thrust P from the geometry of the force polygon, Figure 7.12a. The active earth thrust P_a is the largest P of all the trial wedges. Automatic search for the critical wedge and the active earth thrust P_a is possible using the *Solver* tool in spreadsheet, as shown in Figure 7.12b.

The unboxed cells in Figure 7.12b contain numerical inputs of H, γ, H_{crest}, and, in degrees and radians, δ, α, ϕ, θ and ψ, where the notations and symbols are as defined in the diagram and text of Figure 7.12a. The boxed cells

Deterministic analysis with search for the critical quadrilateral wedge: The stem of a cantilever retaining wall is 11 m high. It retains a storage pile of cohesionless iron ore. The wall has a cross-section symmetrical about its vertical center line. At the top its width is 1.83 m and at the base of the stem is 3.66 m. From a point on the back of the wall 1.2 m below the crest, the surface of the ore pile rises at an angle of $\psi = 26°$ with the horizontal to a maximum height of 14 m above the base of the stem. The remainder of the pile is level. If ϕ' and δ are 36° and 28°, respectively, and γ of the iron ore is 26 kN/m³, what is the total active thrust of the ore against the wall, based on the Coulomb trial wedge method? (after Terzaghi et al., 1996).

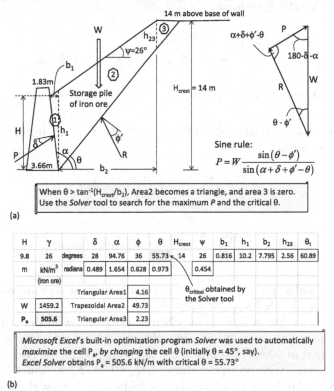

(a)

When $\theta > \tan^{-1}(H_{crest}/b_2)$, Area2 becomes a triangle, and area 3 is zero. Use the Solver tool to search for the maximum P and the critical θ.

H	γ		δ	α	ϕ	θ	H_{crest}	ψ	b_1	h_1	b_2	h_{23}	θ_t
9.8	26	degrees	28	94.76	36	55.73	14	26	0.816	10.2	7.795	2.56	60.89
m	kN/m³	radians	0.489	1.654	0.628	0.973		0.454					
	(iron ore)												

		Triangular Area1	4.16	$\theta_{critical}$ obtained by	
W	1459.2	Trapezoidal Area2	49.73	the Solver tool	
P_a	505.6	Triangular Area3	2.23		

Microsoft Excel's built-in optimization program Solver was used to automatically maximize the cell P_a, by changing the cell θ (initially $\theta = 45°$, say). Excel Solver obtains $P_a = 505.6$ kN/m with critical $\theta = 55.73°$

(b)

Figure 7.12 Deterministic analysis by trial wedge method is necessary to obtain the active earth thrust P_a. The closed form Coulomb equation for K_a cannot be used for this case because the critical wedge could be a quadrilateral.

of b_1, h_1, b_2, h_{23} and θ_t contain the following equations based on geometrical considerations:

$$b_1 = H \tan(\alpha - 90°) \tag{7.16a}$$

$$h_1 = H + b_1 \tan\psi \tag{7.16b}$$

$$b_2 = (h_{crest} - h_1)/\tan(\psi) \tag{7.16c}$$

$$h_{23} = H_{crest} - b_2 \tan\theta \tag{7.16d}$$

$\theta_t = \tan^{-1}(H_{crest}/b_2)$ (threshold θ when triangular transits to quadrilateral)
$$\text{(7.16e)}$$

The triangular areas 1 and 3 and trapezoidal area 2 are automatically calculated in the cells so labelled, by the following formulas:

$$\text{Area1} = 0.5 \times h_1 \times b_1 \qquad\qquad\qquad\qquad (7.16f)$$

$$\begin{aligned}
\text{Area2} &= 0.5 \times [h_1 + h_{23}] \times b_2, && \text{if } \theta \le \theta_t \\
&= 0.5h_1^2 \cos\psi \cos\theta/\sin(\theta - \psi), && \text{if } \theta > \theta_t
\end{aligned} \qquad (7.16g)$$

$$\text{Area3} = 0.5 \times h_{23}^2/\tan\theta, \qquad\qquad \text{zero if } \theta > \theta_t \quad (7.16h)$$

The weight of the trial wedge W and the thrust P are then calculated by the following formulas in cells:

$$W = (\text{Area1} + \text{Area2} + \text{Area3}) \times \gamma_{IronOre} \qquad (7.17)$$

By sine rule:

$$P = W \frac{\sin(\theta - \phi')}{\sin(\alpha + \delta + \phi' - \theta)}, \text{ need to try different } \theta. \qquad (7.18)$$

The trial wedge procedure is automatized using the optimization programme Solver inside Excel, *to maximize* the cell labelled P_a, *by changing* the cell of θ. A solution $P_a = 505.6$ kN/m is obtained, with the critical ore wedge having a θ angle of 55.73°. Since this critical θ angle is less steep than the threshold θ_t of 60.89° (rightmost cell in Figure 7.12b), the critical wedge is a quadrilateral, with nonzero values (4.16, 49.73, 2.23, Figure 7.12b) of the two triangular Areas 1 and 3 and trapezoidal area 2. Now that the deterministic P_a is known, one can proceed to consider the required horizontal resistance P_{resist}, from the LRFD and EC7-DA2 perspective, and from the RBD-via-FORM perspective, as discussed next.

7.8.2 Design resistance from the perspective of LRFD and EC7-DA2

The load and resistance factors in LRFD and EC7-DA2 are applied to conservative nominal (characteristic) values to obtain the design values. The procedure is as follows:

1. Determine the conservative nominal (characteristic) value of the unfavourable load P_a, such that $P_{a, nominal} > P_a$ obtained in Figure 7.12.
2. Apply load factor to $P_{a, nominal}$, to obtain the design value of $P_{a, design}$, which is greater than $P_{a, nominal}$.

3. Design retaining wall with horizontal resisting force P_{resist} equal to the horizontal component of the design value of P_a:

$$P_{resist,design} = \text{Horizontal component of } P_{a,design},$$

such that $P_{resist, design} < P_{resist, nominal} < \text{mean resistance } P_{resist}$.

that is, $P_{a,average} < P_{a,nominal} < \mathbf{P_{a,design}} = P_{resist,design} < P_{resist,nominal} < \text{mean } P_{resist}$

$$(7.19)$$

where P_{resist} is used instead of R, to avoid confusion with the R which denotes (i) correlation matrix and (ii) the resultant force on the slip surface in Figure 7.12a.

Even if code-recommended load and resistance factors are used, the design outcome from LRFD or EC7-DA2 is not unique but varies with the level of conservatism in selecting $P_{a, nominal}$ and $P_{resist,nominal}$.

An alternative is the reliability-based design (RBD), which does not need nominal values or partial factors but requires statistical inputs reflecting parametric uncertainty and correlation, and a target reliability index (e.g., 2.5 or 3.0), higher for greater consequence of failure. RBD also provides information on the probability of failure, and on parametric sensitivities. This is presented next.

7.8.3 Extending the modified Terzaghi et al. (1996) deterministic example to RBD-via-FORM

The P_a value of 505.6 kN/m obtained in Figure 7.12 is a deterministic outcome, because it is based on mean values of the parameters ϕ', δ and γ_{ore}. The design resistance P_{resist} provided by a retaining wall should be greater than this average P_a value of 505.6 kN/m, to account for parametric uncertainties which can increase P_a to a value higher than its average value.

In the RBD for the resistance P_{resist} in Figure 7.13, the mean values μ of ϕ', δ and γ_{ore} are $36°$, $28°$ and 26 kN/m^3, respectively, the same as those in the deterministic evaluation of Figure 7.12. A model correction factor λ_{Pa} has been introduced to account for the idealizations underlying the assumption of plane slip surface. The standard deviations under the column labelled σ correspond to coefficients of variation (σ/μ) of 0.1, 0.1, 0.1 and 0.05, for λ_{Pa}, ϕ', δ and γ_{ore}, respectively. The angle of friction ϕ' of the iron ore and the wall-ore interface friction angle δ are assumed to be positively correlated, with $\rho = 0.7$, as shown in the correlation matrix **R**.

Initially the column values labelled **n** were zeros. The Solver tool was used, *to minimize* the β cell, *by changing* the four cells under the **n** column *and* the θ cell, *subject to the constraint* that the g(**x**) cell be ≤ 0. After a few trials of the resistance value of P_{resist}, the spreadsheet procedure obtains a design resistance value $P_{resist} = 700$ kN/m that achieves the target β value of

2.5. This P_{resist} value is equal to the design value of $\lambda_{Pa}P_{ah}$. The design values of the four random variables are $\lambda_{Pa}^* = 1.116$, $\phi'^* = 28.40°$, $\delta^* = 23.47°$ and $\gamma_{ore}^* = 26.82$ kN/m³. The sensitivity indicators ($\mathbf{n}$) suggest that the RBD is most sensitive to ϕ' for the case in hand.

Note that the critical slip surface inclination (θ) will change with the design point values of ϕ', δ and γ_{ore}. This means that the critical ore wedge in the RBD of Figure 7.13, with $\theta = 50.56°$, is different from the critical wedge of the deterministic case of Figure 7.15, where $\theta = 55.73°$. Instead of simultaneously maximizing P (by changing θ) and minimizing the β cell (by changing the values under the column labelled $\mathbf{n}$), the same objective and correct solution is achieved by using the Solver tool to minimize the β cell, by changing the four $\mathbf{n}$ values *and* also the θ value. This is because the performance function g($\mathbf{x}$) is equal to $P_{resist} - \lambda_{Pa}P_{ah}$. Therefore, minimizing β subject to the constraint g($\mathbf{x}$) = 0 by changing $\mathbf{n}$ values AND θ will also maximize P_a, which can be verified by deterministic maximization of the cell P_a by varying θ using the design values ($\mathbf{x}^*$) of the random variables from Figure 7.13.

In summary, for the case in hand, with parametric uncertainties and correlations as shown in Figure 7.13, the retaining wall should be designed with a horizontal resistance of 700 kN (from steel tie rods, for example, as suggested in Terzaghi et al. 1996, p. 253), which is greater than the horizontal component of the deterministic average inclined P_a of 505.6 kN/m of Figure 7.12, in order to achieve a target β of 2.5.

The RBD case in Figure 7.13 has not accounted for the uncertainty of P_{resist}. This is done next. Another failure mode of surface sliding related to angle of repose is also considered.

7.8.4 RBD-via-FORM of tie rod force to resist lateral thrust

The required P_{resist} force of 700 kN in Figure 7.13 has not accounted for uncertainty in P_{resist}. If the resisting force is to be provided by steel tie rods, one should model the mean value and standard deviation of their strength. In Figure 7.14, the available total force of the tie rods (P_{resist}) is considered to have a c.o.v. of 0.15. The designer wants to determine the mean value of P_{resist} for a target reliability index of 2.5 that P_{resist} will not be smaller than the horizontal component of the active thrust. The RBD-via-FORM analysis in Figure 7.14a shows that a mean P_{resist} of 827 kN/m is required, bigger than the P_{resist} of 700 kN/m in Figure 7.13 where the uncertainty of P_{resist} is ignored.

7.8.5 Another failure mode involving surface rolling along the slope of the stockpiled iron ore

In the Terzaghi et al. (1996, p. 253) example problem, the surface of the iron ore behind the wall was 1.22 m below the top of the 11 m high retaining wall. This is likely a safety precaution against surface rolling failure mode.

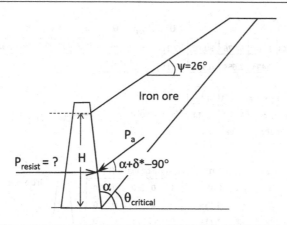

H	γ_{ore}		δ	α	ϕ'	θ	H_{crest}	ψ	b_1	h_1	b_2	h_{23}	θ_t
9.8	26.82	degrees	23.47	94.76	28.40	50.56	14	26	0.816	10.2	7.795	4.523	60.89
m	kN/m^3	radians	0.410	1.654	0.496	0.882		0.454					
	(iron ore)												

		Triangular Area1	4.16
W	1876.0	Trapezoidal Area2	57.38
P_a	711.6	Triangular Area3	8.41

		x*	μ	σ	n	Correlation matrix R			
Normal	λ_{Pa}	1.116	1	0.1	1.165	1	0	0	0
Normal	ϕ'	28.40	36	3.6	-2.111	0	1	0.7	0
Normal	δ	23.47	28	2.8	-1.617	0	0.7	1	0
Normal	γ_{ore}	26.82	26	1.3	0.630	0	0	0	1

$\lambda_{Pa}P_{ah}$	P_{resist}	g(x)	β
700	700	0.00	2.50

Figure 7.13 RBD of the resistance P_{resist} for a target reliability index β, accounting for the model uncertainty λ_{Pa} and random variables, ϕ', δ and γ_{ore}.

Assuming that the angle of internal friction of the iron ore is equal to its angle of repose, the performance function for the rolling of iron ore along its inclined surface is:

Surface sliding mode: $g(\mathbf{x}) = \phi' - \psi$

The above performance function is similar to that of a dry infinite slope (Lambe and Whitman 1979, p. 192). If the desired angle of slope inclination (ψ) of the stockpiled iron ore cannot be controlled precisely, one should account for its uncertainty. Suppose ψ has a mean value of 26° and a c.o.v. of

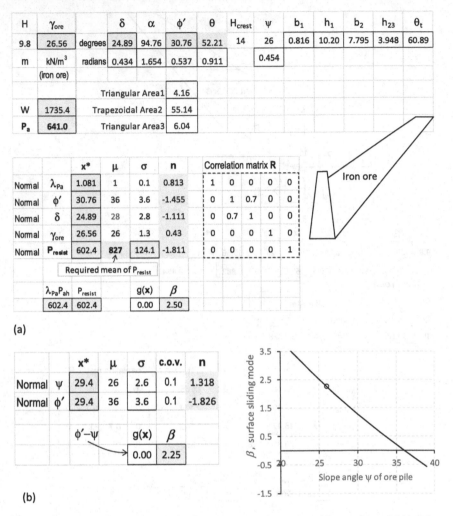

H	γ_{ore}		δ	α	ϕ'	θ	H_{crest}	ψ	b_1	h_1	b_2	h_{23}	θ_t
9.8	26.56	degrees	24.89	94.76	30.76	52.21	14	26	0.816	10.20	7.795	3.948	60.89
m	kN/m³	radians	0.434	1.654	0.537	0.911		0.454					
	(iron ore)												

		Triangular Area1	4.16		
W	1735.4	Trapezoidal Area2	55.14		
P_a	641.0	Triangular Area3	6.04		

		x*	μ	σ	n	Correlation matrix R				
Normal	λ_{Pa}	1.081	1	0.1	0.813	1	0	0	0	0
Normal	ϕ'	30.76	36	3.6	-1.455	0	1	0.7	0	0
Normal	δ	24.89	28	2.8	-1.111	0	0.7	1	0	0
Normal	γ_{ore}	26.56	26	1.3	0.43	0	0	0	1	0
Normal	P_{resist}	602.4	827	124.1	-1.811	0	0	0	0	1

Required mean of P_{resist}

$\lambda_{Pa}P_{ah}$	P_{resist}		g(x)	β
602.4	602.4		0.00	2.50

(a)

		x*	μ	σ	c.o.v.	n
Normal	ψ	29.4	26	2.6	0.1	1.318
Normal	ϕ'	29.4	36	3.6	0.1	-1.826

$\phi'-\psi$	g(x)	β
	0.00	2.25

(b)

Figure 7.14 (a) RBD of the required *mean* resisting tie-rod force P_{resist} (kN/m) for a target β of 2.5, (b) reliability against surface sliding of ore slope.

0.1. Figure 7.14b shows that the reliability against surface rolling mode of failure is 2.25. Although this is lower than 2.5, it can be regarded as acceptable in view of the wall being 1.22 higher than the retained ore at the wall-ore interface. Should tumbling of the ore occur, it will gradually fill up the top of the wall, and result in a slightly flatter ψ, and greater reliability index.

It is good practice to reduce the water pressure in the soil retained by a wall. One way is to put a granular vertical drainage layer behind the wall, as illustrated in two deterministic examples below, from Clayton et al. (2013) and Lambe and Whitman (1979), before extending them into probability-based design accounting for the uncertainties of input values and the approximation in determining the active earth thrust.

7.9 RETAINING WALLS WITH STEADY-STATE SEEPAGE TOWARDS A VERTICAL DRAINAGE LAYER BEHIND THE WALL

7.9.1 Deterministic example from Clayton et al. (2013), extended into probability-based design

This deterministic example is from Clayton et al. (2013, pp. 146–148). The wall is 6.3 m high, as shown Figure 7.15. The effective angle of friction of the retained fill is $\phi' = 30°$. The interface friction angle δ between the granular drain layer and the wall is assumed to be zero. The horizontal active earth thrust is evaluated for a soil wedge with a failure surface inclined at 60° to the horizontal. There is steady-state seepage from the surface of the retained fill to the vertical drainage layer behind the wall, with the flownet as shown. The saturated unit weight of the retained fill is 20.2 kN/m³. Based on the difference in elevations between intersection points of same equipotential lines on the vertical drain layer and the 60° inclination slip surface, the integrated pore water pressure along the slip surface results in a water force U of 49 kN/m perpendicular to the failure surface. A force polygon can be drawn for estimating the horizontal active earth thrust E_a. The equation in Figure 7.15, from Lambe and Whitman (1979, p. 337), based on the geometry of the force polygon, can also be used.

Based on geometry, the weight of the triangular soil wedge is 231.4 kN/m. The horizontal earth thrust is then calculated to be 162 kN/m using the Lambe and Whitman equation.

7.9.2 Extending the Clayton et al. example to probability-based design

It is required to find the mean value of the horizontal resisting force E_{resist} for a target reliability index of 2.5 against the destabilizing active thrust E_a. In Figure 7.16a, the mean values μ of ϕ', U and γ_t are 30°, 49 kN/m and 20.2 kN/m³, respectively, the same as those in the deterministic analysis of Figure 7.15. A model correction factor λ_{Ea} has been introduced to account for the idealizations underlying the assumption of plane slip surface and water pressure on the surface. The standard deviations under the column labelled σ correspond to coefficients of variation (σ/μ) of 0.1, 0.1, 1/7 and 0.07 for λ_{Pa}, ϕ', U and γ_t, respectively. The angle of friction ϕ' of the fill and the saturated unit weight γ_t are assumed to be positively correlated, with $\rho = 0.5$, as shown in the correlation matrix R. The performance function is $g(x) = E_{resist} - \lambda_{Ea}E_a$, based on the values in the x^* column.

Deterministic example from Clayton et al. (2013)

Calculate the total force acting on the retaining structure shown in the figure below, for the failure surface shown (inclined at 60° to the horizontal). There is a vertical drainage layer at the back of the wall. The flownet due to heavy rainfall is as shown. The estimated soil parameters are $\phi' = 30°$, $\delta' = 0°$, and $\gamma_t = 20.2$ kN/m³.

Ans. 160 kN/m run

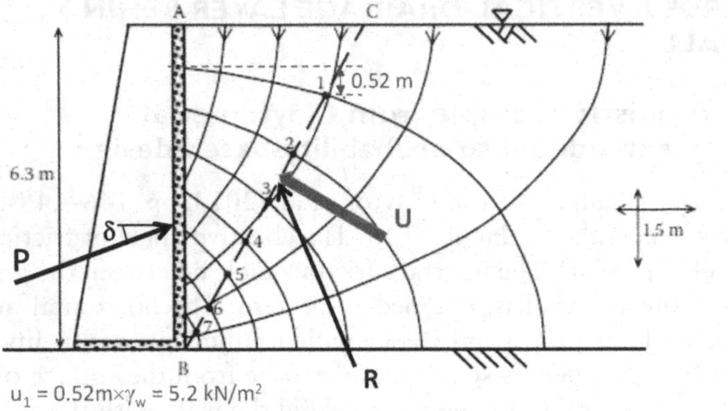

$u_1 = 0.52$m$\times\gamma_w = 5.2$ kN/m²
$u_2 = 9.0$ kN/m²
$u_3 = 10.1$ kN/m²
$u_4 = 9.9$ kN/m²
$u_5 = 8.9$ kN/m²
$u_6 = 7.0$ kN/m²
$u_7 = 4.1$ kN/m²

$U = \frac{1}{2}(0 + 5.2)1.5 + \frac{1}{2}(5.2 + 9.0)1.31 + \frac{1}{2}(9.0 + 10.1)1.06$

$\quad + \frac{1}{2}(10.1 + 9.9)0.92 + \frac{1}{2}(9.9 + 8.9)0.74$

$\quad + \frac{1}{2}(8.9 + 0.7)0.70 + \frac{1}{2}(7.0 + 4.1)0.58$

$\quad + \frac{1}{2} \times 4.4 \times 0.43 = 49$ kN/m run

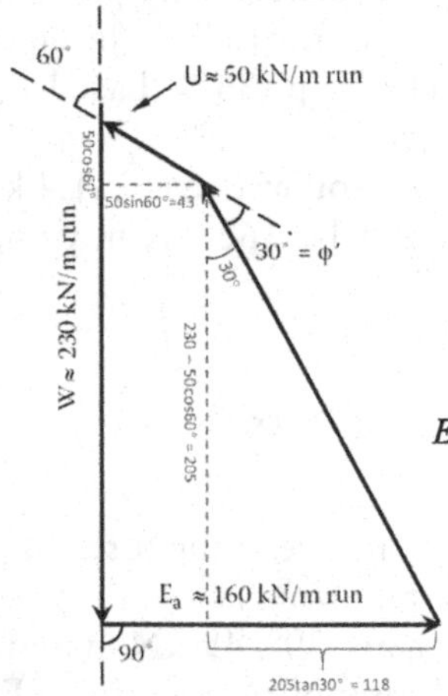

Lambe and Whitman (1979, p337):

$$P = \frac{(W - U\cos\theta)\tan(\theta - \phi') + U\sin\theta}{\sin\delta\tan(\theta - \phi') + \cos\delta}$$

$$E_a = P\cos\delta$$

In this case, $\delta = 0$, $\theta = 60°$, $W = 231.4$ kN/m

$$E_a = \frac{(231.4 - 49\cos 60°)\tan(60° - 30°) + 49\sin 60°}{0 + 1}$$

$$= 162 \text{ kN/m}$$

Figure 7.15 Deterministic analysis from Clayton et al. (2013) of a retaining wall with a vertical drainage layer at the back of the wall.

Initially the column values labelled **n** were zeros, for which the mean values are displayed in the **x*** column. The Solver tool was used, *to minimize* the β cell, *by changing* the five cells under the **n** column, *subject to the constraint* that the g(x) cell be ≤ 0. After a few trials of the mean resistance value of

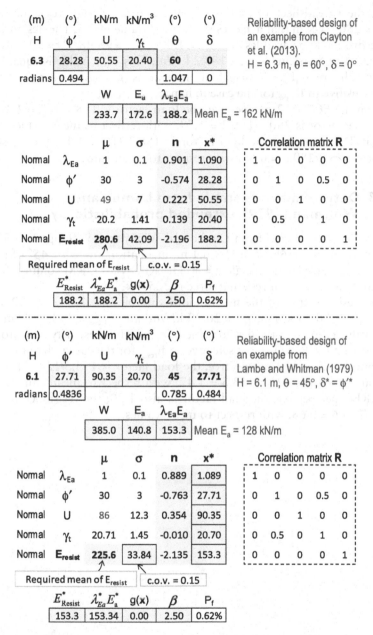

(m)	(°)	kN/m	kN/m³	(°)	(°)
H	ϕ'	U	γ_t	θ	δ
6.3	28.28	50.55	20.40	**60**	**0**
radians	0.494			1.047	0

Reliability-based design of an example from Clayton et al. (2013). H = 6.3 m, θ = 60°, δ = 0°

W	E_a	$\lambda_{Ea}E_a$
233.7	172.6	188.2

Mean E_a = 162 kN/m

		μ	σ	n	x*
Normal	λ_{Ea}	1	0.1	0.901	1.090
Normal	ϕ'	30	3	-0.574	28.28
Normal	U	49	7	0.222	50.55
Normal	γ_t	20.2	1.41	0.139	20.40
Normal	E_{resist}	280.6	42.09	-2.196	188.2

Correlation matrix **R**

1	0	0	0	0
0	1	0	0.5	0
0	0	1	0	0
0	0.5	0	1	0
0	0	0	0	1

Required mean of E_{resist} | c.o.v. = 0.15

E^*_{Resist}	$\lambda^*_{Ea}E^*_a$	g(x)	β	P_f
188.2	188.2	0.00	2.50	0.62%

(m)	(°)	kN/m	kN/m³	(°)	(°)
H	ϕ'	U	γ_t	θ	δ
6.1	27.71	90.35	20.70	**45**	**27.71**
radians	0.4836			0.785	0.484

Reliability-based design of an example from Lambe and Whitman (1979) H = 6.1 m, θ = 45°, δ* = φ'*

W	E_a	$\lambda_{Ea}E_a$
385.0	140.8	153.3

Mean E_a = 128 kN/m

		μ	σ	n	x*
Normal	λ_{Ea}	1	0.1	0.889	1.089
Normal	ϕ'	30	3	-0.763	27.71
Normal	U	86	12.3	0.354	90.35
Normal	γ_t	20.71	1.45	-0.010	20.70
Normal	E_{resist}	225.6	33.84	-2.135	153.3

Correlation matrix **R**

1	0	0	0	0
0	1	0	0.5	0
0	0	1	0	0
0	0.5	0	1	0
0	0	0	0	1

Required mean of E_{resist} | c.o.v. = 0.15

E^*_{Resist}	$\lambda^*_{Ea}E^*_a$	g(x)	β	P_f
153.3	153.34	0.00	2.50	0.62%

Figure 7.16 RBD of the required mean resisting horizontal force E_{resist} (kN/m) for a target β of 2.5, (a) for the retaining wall of Figure 7.15, with H = 6.3 m, θ= 60°, δ= 0°, (b) for a similar wall in Lambe and Whitman (1979), with H = 6.1 m, θ= 45°, δ* = φ*.

E_{resist}, the spreadsheet procedure obtains a mean design resistance value E_{resist} = 280.6 kN/m that achieves the target β value of 2.5. The design values of the five random variables are $\lambda_{Ea}{}^*$ = 1.090, ϕ'^* = 28.28°, U = 50.55 kN/m, $\gamma_t{}^*$ = 20.40 kN/m³ and E_{Resist}^*= 188.2 kN/m. This E_{Resist}^* value is equal to the design value of $\lambda_{Ea}E_a$. The sensitivity indicators (n) suggest that the RBD is most sensitive to E_{resist} for the case in hand.

From the EC7-DA2 perspective, the load factor is 188.2/162 = 1.16; the resistance factor is 280.6/188.2 = 1.49, with respect to mean values. From the LRFD perspective, the load factor is 188.2/162 = 1.16; the resistance factor is 188.2/280.6 = 0.67, with respect to mean values.

7.9.3 Deterministic example from Lambe and Whitman (1979), extended probabilistically

A slightly different case from Lambe and Whitman (1979, pp. 336–337) is shown in Figure 7.16b, with H = 6.1 m, inclination angle θ = 45° of failure surface, interface friction angle δ = ϕ. The RBD-via-FORM solution for the required mean value of resisting force is 225.6 kN/m.

The design values of the five random variables are $\lambda_{Ea}{}^*$ = 1.089, ϕ'^* = 27.71°, U = 90.35 kN/m, $\gamma_t{}^*$ = 20.70 kN/m³ and E_{Resist}^*= 153.3 kN/m. This E_{Resist}^* value is equal to the design value of $\lambda_{Ea}E_a$. The sensitivity indicators (n) suggest that the RBD is most sensitive to E_{resist} for the case in hand.

From the EC7-DA2 perspective, the load factor is 153.3/128 = 1.20; the resistance factor is 225.6/153.3 = 1.47, with respect to mean values. From the LRFD perspective, the load factor is also 1.20; the resistance factor is 153.3/225.6 = 0.68, with respect to mean values.

Chapter 8

Soil slope stability

8.1 INTRODUCTION

We will begin this chapter with a slope stability problem from Terzaghi et al. (1996). Other originally deterministic slope stability examples from the literature, including some renowned cases, will be extended to probabilistic slope stability analysis. Focus is on insights and information from FORM analysis, for enhancing partial factor design approach. There are some novelties in the author's deterministic solution procedures (which underlies the performance function in FORM), which could be of broader appeal beyond the probabilistic focus of this book.

Noncircular slip surface will result in lower factor of safety than circular slip surface, sometimes significantly when it is deep seated, as shown by Duncan et al. (2014) for one of the planned dikes of the James Bay hydroelectric project. For the first example of this chapter, we will assume circular slip surface, to demonstrate the extension of deterministic analysis to reliability-based design via FORM. The spreadsheet-based Spencer method for noncircular slip surfaces, and extension to reliability analysis, will be presented later in the chapter.

The cases in this chapter include an excavated slope failure in stiff fissured London clay, an underwater slope failure in San Francisco Bay Mud, a clay slope in Norway with anisotropic and spatially autocorrelated undrained shear strength, a sloping core dam example from Lambe and Whitman (1979) which is refined deterministically and extended probabilistically, and reliability analysis of a hypothetical embankment on soft ground with discretized 1D random field and with search for critical noncircular slip surface based on a reformulated Spencer method in spreadsheet.

8.2 A DETERMINISTIC EXAMPLE FROM TERZAGHI ET AL. (1996), EXTENDED PROBABILISTICALLY

8.2.1 Analytical $\phi_u = 0$ procedure

The slope stability problem shown in Figure 8.1 is from Terzaghi et al. (1996, p. 277, Problem 5). Consider a circular slip surface traversing the

DOI: 10.1201/9781003112297-10

A bed of clay consists of three horizontal strata, each 4.5 m thick. The values for c_u for the upper, middle, and lower strata are, respectively, 30, 20, and 140 kPa. The unit weight is 18.4 kN/m³. A cut is excavated with side slopes of 1 (vertical) to 3 (horizontal) to a depth of 6 m. What is the factor of safety of the slope against failure?

(Question from *Terzaghi, Peck and Mesri, 1996, p277*)

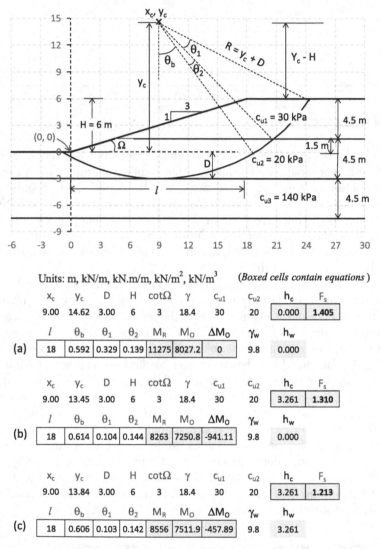

Units: m, kN/m, kN.m/m, kN/m², kN/m³   (*Boxed cells contain equations*)

	x_c	y_c	D	H	$\cot\Omega$	γ	c_{u1}	c_{u2}	h_c	F_s
	9.00	14.62	3.00	6	3	18.4	30	20	0.000	1.405

	l	θ_b	θ_1	θ_2	M_R	M_O	ΔM_O	γ_w	h_w
(a)	18	0.592	0.329	0.139	11275	8027.2	0	9.8	0.000

	x_c	y_c	D	H	$\cot\Omega$	γ	c_{u1}	c_{u2}	h_c	F_s
	9.00	13.45	3.00	6	3	18.4	30	20	3.261	1.310

	l	θ_b	θ_1	θ_2	M_R	M_O	ΔM_O	γ_w	h_w
(b)	18	0.614	0.104	0.144	8263	7250.8	-941.11	9.8	0.000

	x_c	y_c	D	H	$\cot\Omega$	γ	c_{u1}	c_{u2}	h_c	F_s
	9.00	13.84	3.00	6	3	18.4	30	20	3.261	1.213

	l	θ_b	θ_1	θ_2	M_R	M_O	ΔM_O	γ_w	h_w
(c)	18	0.606	0.103	0.142	8556	7511.9	-457.89	9.8	3.261

Figure 8.1 Factor of safety of a cut slope, closed form solution, with search for critical slip circle: (a) ignore tension crack, (b) dry tension crack equal to $2c_u/\gamma$, (c) water-filled tension crack.

upper two clay layers of weaker undrained shear strength than the underlying stiff third layer. By nature of the undrained analysis of circular slip surface with ϕ_u equal to zero, the computed factor of safety (F_s) will be the same regardless of whether an engineer uses the ordinary method of slices (OMS), the Bishop simplified method (BSM) or the circular slip surface version of the Spencer method. It is shown below that a simple analytical formulation is possible for the case in hand, which offers lucidity of concepts and affords a simple way to search for the critical slip circle.

For the excavated slope in Figure 8.1, the origin of the coordinate system is at the toe of the slope. The arbitrary deep-seated slip circle is centred at x_c and y_c, of radius R, and with a tangent depth D (measured from the toe level of the slope). The overturning moment M_o and resisting moment M_R can be expressed by *exact* closed form equations, to facilitate subsequent search for the critical slip circle, as described below.

For circular slip surface, the exact overturning moment M_o for a slope *without tension crack* is, from Low and Tang (1997, CGJ):

$$M_o = \left[\frac{x_c}{2} \times (l - x_c) - \frac{l^2}{6} + (y_c + D) \times \left(D + \frac{H}{2} \right) - \frac{1}{2} \left(D + \frac{H}{2} \right)^2 - \frac{H^2}{24} \right] \gamma H \quad (8.1)$$

in which l is the width between the crest and the toe of the slope, equal to:

$$l = H / \tan \Omega \quad (8.2)$$

The above equation for M_o is based on mathematical integration of thin horizontal strips, from $y = 0$ (toe level of slope) to $y = H$ (crest level of slope). The circular sector below the base, extending a vertical distance D below the toe level, is in self-equilibrium, because its centre of gravity is directly below the centre of the slip circle. Hence, this sector contributes zero overturning moment.

The resisting moment M_R for the case in hand is given by:

$$M_R = \left[c_{u1} R \theta_1 + c_{u2} \left(R \theta_2 + 2 R \theta_b \right) \right] \times R, \text{with} \, \theta_i \text{ in radians} \quad (8.3)$$

in which

$$\theta_b = \cos^{-1} \left(y_c / R \right) \quad (8.3a)$$

$$\theta_2 = \cos^{-1} \left((y_c - H + h_1) / R \right) - \theta_b \quad (8.3b)$$

$$\theta_1 = \cos^{-1} \left((y_c - H + h_c) / R \right) - \left(\theta_b + \theta_2 \right) \quad (8.3c)$$

where h_1 is the thickness of layer 1 (equal to 4.5 m in this case), and h_c is the height of a vertical tension crack shown in Figure 8.2.

The equations for the subtended angles θ_b, θ_1, θ_2, ... are case dependent. For another case with different layer thickness or different excavation depth H, the number of subtended angles and the corresponding equations of θ_i can be rewritten based on simple geometry. For example, for the same three-layer soil each of thickness 4.5 m, if the excavation depth is smaller than the top layer thickness, then θ_2 does not exist, and θ_b will consist of two sub-angles, one for c_{u1} and another for c_{u2}. This must be borne in mind when the excavation depth H is changed, either in parametric studies, or in changing H to find a design value that meets the EC7, LRFD or RBD criterion. These different conditions requiring different equations of θ_i can be dealt with conveniently by creating an Excel VBA function for the resisting moment M_R.

With H equal to 6 m, the factor of safety based on circular slip surface is:

$$F_s = \frac{M_R}{M_o} \quad \text{using Eqs. 8.3 and 8.1} \tag{8.4}$$

The above Eqs. 8.1–8.4 were entered in the respective spreadsheet cells as shown in Figure 8.1a. The tension crack (h_c) and water height (h_w) in the tension crack are considered in Figures 8.1b and 8.3c. For now, both h_c and h_w are equal to zero. Initially, the slip circle geometry is arbitrary, say $x_c = 6$ m, $y_c = 20$ m and D = 2 m. The Solver tool was then used *to minimize* the F_s cell, *by changing* the x_c, y_c and D cells, *subject to the constraints* that $y_c \geq H$ and $0 \leq D \leq 3$. A solution of $F_s = 1.405$ was obtained, with the critical slip circle located, by the Excel Solver tool, at $(x_c, y_c, D) = (9, 14.62, 3)$. Not surprisingly, this is a mid-point circle tangent to the top surface of the stiff clay layer.

Figure 8.1a was for the slope with no tension crack, using Eqs. 8.1–8.4.

Figure 8.1b shows the deterministic analysis for the same slope with a dry vertical tension crack of height $h_c = 2c_u/\gamma$, and Figure 8.1c when the tension crack is filled with water. With search for the critical slip circle, the computed F_s values are 1.31 and 1.21, respectively, based on:

$$F_s = \frac{M_R}{M_o + \Delta M_o} \quad \left(M_o \text{ and } M_R \text{ from Eqs. 8.1 and 8.3} \right) \tag{8.5}$$

in which

$$\Delta M_o = \frac{\gamma_w h_w^2}{2} \times \left(y_c - H + h_c - \frac{h_w}{3} \right) - \frac{\gamma h_c^2}{2} \times \left(y_c - H + \frac{2}{3} h_c \right) \tag{8.6}$$

The above equation for ΔM_o was derived mathematically by Low and Tang (1997, CGJ), to account for an additional overturning moment due to water pressure in the tension crack and a reduced overturning moment due to the exclusion of the little block to the right of the tension crack, Figure 8.2.

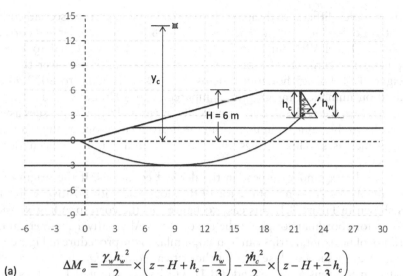

$$\Delta M_o = \frac{\gamma_w h_w^2}{2} \times \left(z - H + h_c - \frac{h_w}{3} \right) - \frac{\gamma h_c^2}{2} \times \left(z - H + \frac{2}{3} h_c \right)$$

(a)

Verification of the derived equations of M_o and ΔM_o:
For the case in Fig.8.1c, $M_o + \Delta M_o = 8030.2899...$
Using Excel VBA code for numerical integration of 25 horizontal strips, get:
TotalMoment = 8032.9395...
When nstrip = 8000, get   TotalMoment = 8030.2899..., approaching $M_o + \Delta M_o$

```
'VBA code for obtaining the overturning moment by numerical integration:
Function TotalMoment(xc, yc, H, hc, R, cotomega, gammaSoil, gammaW, hw)
  'xc and yc are the centre of slip circle,
  'H the embankment height, hc the depth of tension crack
  'R the radius, cotomega the cotangent of the slope angle
  'gw the unit weight of water, and hw the height of water in tension crack
  nstrip = 300
  dz = H / nstrip
  m = 0
  For i = 1 To nstrip
   z = (i - 0.5) * dz
   xLeft = z * cotomega
   If z <= (H - hc) Then xCircle = xc + Sqr(R ^ 2 - (yc - z) ^ 2)
   If z > (H - hc) Then xCircle = xc + Sqr(R ^ 2 - (yc - H + hc) ^ 2)
   Area = (xCircle - xLeft) * dz
   LeverArm = 0.5 * (xCircle + xLeft) - xc
   m = m + Area * LeverArm
  Next i
  Mw = 0.5 * gammaW * hw ^ 2 * (yc - H + hc - hw / 3)
  TotalMoment = m * gammaSoil + Mw
End Function
```

(b)

Figure 8.2 Verifying the accuracy of the derived equations (M_o and ΔM_o) for the overturning moment of a slope with a water-filled tension crack.

It is however difficult to decipher the physical meaning of the second component in Eq. 8.6. To confirm that Eqs. 8.1 and 8.6 have been derived correctly, a simple Excel VBA function code (Figure 8.2b) was created to perform numerical integration of moments of thin horizontal strips over the slope height of H. For 25 horizontal strips, $M_o + \Delta M_o$ is equal to 8032.9395… based on numerical integration, compared with 8030.2899…based on the derived Eqs. 8.1 and 8.6. The numerically integrated value converges asymptotically to the value computed from the derived equations as the number of strips increases. This verifies that Eqs. 8.1 and 8.6 have been derived correctly.

Note that x_c and y_c appear in the derived equations. If the origin of the coordinate system is not at the toe of the slope and with positive x direction as shown in Figure 8.1, it is easy to figure out the corresponding x_c and y_c values for coordinate origin at the toe of slope. Alternatively, numerical integration of horizontal strips can be used, similar to the procedure in Figure 8.2b.

Spencer method with search for the critical *noncircular* slip surface (illustrated in Sections 8.4, 8.5 and 8.8.1) will have computed F_s values somewhat smaller than the three critical slip circles in Figure 8.1a, 8.1b and 8.1c.

Although the F_s values of the three scenarios in Figure 8.1 are all bigger than 1.0, and hence 'safe' based on the deterministic inputs, there is still a risk of slope failure. The failure probability can be estimated by accounting for uncertainties in the input values, as illustrated next.

8.2.2 From deterministic analysis to reliability-based design of the Terzaghi et al. problem

All the three deterministic cases of 8.1a can be easily extended into reliability analysis and RBD by adding a few rows beneath it, as shown in Figure 8.3a, rows 12–14, for the Figure 8.1a case of no tension crack. The numerical inputs are in the unboxed cells, namely the mean values μ and standard deviation σ of c_{u1} and c_{u2}, and their correlations in the R matrix. It is assumed that, for geological reasons, c_{u1} and c_{u2} are positively correlated, with a correlation coefficient (ρ) equal to 0.5, and that they obey the lognormal distributions. The n column values were initially zeros. The performance function g(x) is:

$$g(\mathbf{x}) = F_s - 1 \quad (F_s \text{ as in Eq. 8.5}) \tag{8.7}$$

Further steps are:

(1) The VBA function @x_i(… ni) is entered in the column labelled x*. It obtains x_i from n_i via $x_i = F^{-1}[\Phi(n_i)]$, as explained in Chapter 2 and the Appendix.
(2) The quadratic form $\sqrt{\mathbf{n}^T \mathbf{R}^{-1} \mathbf{n}}$ for reliability index is entered in the cell labelled β, as a spreadsheet *array formula*, '=sqrt(mmult(transpose(n),

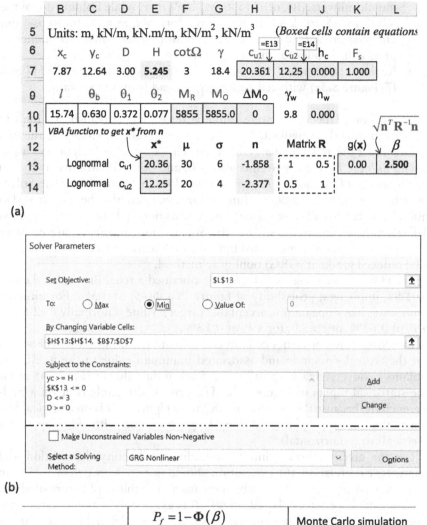

(a)

(b)

(c)

	$P_f = 1 - \Phi(\beta)$ $= \text{NormSDist}(-\beta)$	Monte Carlo simulation 800,000 realizations
Failure probability	0.62%	0.58%

Figure 8.3 (a) RBD of excavation depth for a target β of 2.5, (b) setting spreadsheet Solver parameters, (c) comparison of failure probabilities estimated from β and Monte Carlo simulation.

mmult(minverse(R), n))),' followed by '*Enter*' while holding down the '*Ctrl*' and '*Shift*' keys. In the above formula, *mmult*, *transpose* and *minverse* are spreadsheet built-in functions, each being a container of programme codes for matrix operations.

(3) Replace the previous deterministic inputs of c_{u1} and c_{u2} (cells H7 and I7, Figure 8.3a) with cell addresses, to read from the x^* values.

The column labelled 'n' were initially zeros. The spreadsheet's built-in *Solver* tool was then invoked, *To Minimize* the β cell, *By Changing* the two values under the 'n' column and the three cells (x_c, y_c, D) which defines a slip circle, '*Subject To*' the constraints $g(x) \leq 0$, and other constraints, as shown in the image of the *Solver Parameters* window in Figure 8.3b. The Solver option '*Use automatic scaling*' (under *Options*) can also be activated. Do not check the small box below the constraints ('*Make Unconstrained Variables Non-Negative*'), so that the Solver tool can investigate negative values of n during its constrained optimization search based on the generalized reduced gradient (GRG) nonlinear method.

For H = 6 m as in Figure 8.1a, Solver obtained a reliability index β equal to 1.81, implying a probability of failure $P_f = \Phi(-\beta)$ of 3.5%. For ultimate limit states like slope failure, acceptable target β value is normally β =2.5, for a P_f of 0.62%, or β =3.0, for a P_f of 0.13%.

By trying different values of H in cell E7, and each time let *Solver* search for the critical slip circle and associated minimum value of the β cell, one obtains a converged target β value of 2.5 when the value of *H* is 5.245 m for the statistical inputs of Figure 8.3a. The critical slip circle is again a midpoint circle, because the x_c value of 7.87 m in Figure 8.3a is on a vertical line above the mid-slope of the 5.245-m deep excavation with side slopes of 1 (vertical) to 3 (horizontal).

For the given statistical inputs, including probability distribution and parametric correlation, the excavation should not go deeper than 5.25 m in order to have a β value of 2.5, which means a probability of failure of about 0.62%. The P_f of 0.62% based on $\Phi(-\beta)$ is exact only if the variates obey normal distribution and the limit state surface (LSS) defined by $g(x) = 0$ is a plane. For comparison, Monte Carlo simulation with 800,000 realizations was conducted, obtaining a failure probability of 0.58%. The comparison is summarized in the inset table in Figure 8.3c. In principle, a search for critical slip circle should be conducted for each of the 800,000 realizations of c_u values, at the cost of long simulation time. Hence, the critical slip circle from the RBD results of Figure 8.3a was used for all the 800,000 realizations.

8.2.3 Perspectives, insights and information provided by RBD-via-FORM

The constrained optimization setting in Figure 8.3b literally instructs the *Solver* tool to find the smallest expanding *equivalent* dispersion ellipsoid that just touches the LSS, defined by $g(\mathbf{x}) = 0$. The nearest point of contact between the *equivalent* ellipsoid and the LSS is the most probable point (MPP) of failure, also called the design point when there is a target reliability index.

For the design point under the $\mathbf{x}^*$ column in Figure 8.3a, the simple Low and Tang (2004) function code EqvN(...) was used to obtain the Rackwitz-Fiessler equivalent normal transformation of the lognormally variates c_{u1} and c_{u2}. The equivalent normal mean (μ^N) of c_{u1} and c_{u2} are 27.85 and 18.01, and the equivalent normal standard deviations (σ^N) are 4.032 and 2.426. These have been used to plot, to scale, the equivalent ellipses of 1σ and $\beta = 2.5$ in the original space of the lognormal variates c_{u1} and c_{u2}, Figure 8.4a, and in the normalized (but unrotated) space of the sensitivity indicators $\mathbf{n}$, Figure 8.4b. The ellipses are rotated because of the positive correlation of $\rho = 0.5$ between c_{u1} and c_{u2}. The LSS satisfying $g(\mathbf{x}) = 0$ and $g(\mathbf{n}) = 0$, respectively, are also shown, together with the contact point (the design point) between the β ellipse and the LSS in the two plots, namely '20.36, 12.25' kPa in Figure 8.4a, and '-1.858, -2.377' in Figure 8.4b. These are the values under the $\mathbf{x}^*$ and $\mathbf{n}$ columns in Figure 8.3a, not used in plotting but automatically identified at the single contact point in the two plots.

Although the equivalent mean-value point '27.85, 18.01' (at the centre of the equivalent ellipses in Figure 8.4a) is safe (i.e., $F_s > 1$), failure condition is reached when c_{u1} *decreases* by $1.858 \times \sigma_1^N$ (i.e., 1.858×4.032) from its *equivalent* normal mean value μ_1^N of 27.85, and c_{u2} *decreases* by $2.377 \times \sigma_2^N$ (i.e., 2.377×2.426) from its *equivalent* normal mean value μ_2^N of 18.01. The result is the design point (most probable failure point) at $\mathbf{x}^* =$ '20.36, 12.25' and $\mathbf{n} =$ '-1.858, -2.377,' in Figure 8.3a, 8.4a and 8.4b.

The reliability index β is the distance from the safe mean-value point to the design point (most probable failure point), in units of directional standard deviations. For nonnormal variates, the definition and perspective remain valid in terms of equivalent normal mean values and equivalent normal standard deviations.

The plots in Figure 8.4 convey an intuitive understanding of the physical meaning of the design point as the most probable failure point in two-dimensional space. The same perspective can be visualized in the mind's eye for higher dimensions of multiple random variables, whether normal or nonnormal.

The values of the sensitivity indicators $\mathbf{n}$ in Figure 8.3 and 8.4, at '-1.858, -2.377,' indicate that c_{u2} is a more sensitive parameter than c_{u1} with respect to the design point and the target β. A significant merit of RBD-via-FORM is that its design point reflects case-specific parameter sensitivities, which are affected by probability distributions, levels of uncertainty (mean values, standard deviations, parameter correlations) and the role of the parameters in the performance function (e.g., in this case longer arc length for c_{u2} despite its smaller mean value than c_{u1}). This means that even though c_{u1}

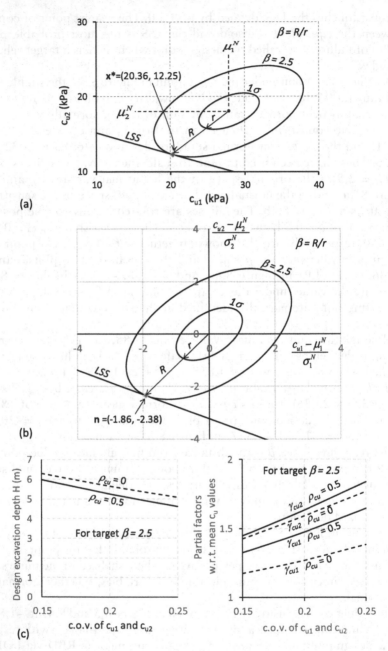

(a)

(b)

(c)

Figure 8.4 (a) Equivalent ellipse in original space, (b) equivalent ellipse in the normalized space of n, (c) permissible excavation depth H, and partial factors from the design point, for target β = 2.5.

and c_{u2} are parameters of the same nature, their sensitivities need not be the same. The partial factors of c_u, calculated as $\gamma_{cu} = \mu_{cu}/c_u^*$ (i.e., with respect to mean c_u values), are plotted in Figure 8.4c. They vary with the c.o.v. of c_u, the correlation between c_{u1} and c_{u2}, the target β value, and whether it is c_{u1} or c_{u2}.

It is informative and sometimes thought-provoking to ponder over the design point, the sensitivity indicators **n**, and implied partial factors from RBD-via-FORM, for comparison with the design point and code-recommended partial factors in a design based on EC7. It is not recommended to calibrate back-calculated partial factors for use in partial factor design, because they vary (rightly) with different cases and circumstances.

Figure 8.4c shows that the design excavation depth H varies with the c.o.v. of c_{u1} and c_{u2}, parameter correlation and target β value.

8.3 AN EXCAVATED SLOPE IN LONDON CLAY THAT FAILED DESPITE A HIGH COMPUTED FACTOR

Duncan et al. (2014) described a deep excavation in the London Clay at Bradwell, citing slope geometry and soil data from Skempton and LaRochelle (1965), Figure 8.5. Deterministic short-term stability analyses were performed for the slope by Duncan et al. (2014), using undrained shear strengths, with $\phi_u = 0$ for all of the soil strata. A vertical tension crack equal to the clay fill thickness of 3.5 m was assumed. For circular slip surface, Spencer method, Bishop simplified method and the Ordinary Method of slices all gave the same factor of safety of 1.76. The strength reduction factor (SRF) finite element analysis gave a factor of safety equal to 1.79. Despite computed factor of safety values around 1.8, the slope failed approximately 5 days after excavation was completed. Duncan et al. cited Skempton and LaRochelle on the probable causes of failure, including overestimates of the shear strength due to testing of samples of small size, strength losses due to sustained loading (creep) in the field, and the presence of fissures. Skempton and LaRochelle concluded that the opening of fissures and a low residual strength along the fissures were probable causes of failure of the slope, even though the factor of safety computed based on laboratory undrained shear strengths was relatively high. The shear strength in the field along the failure surface was much lower than small-size specimens tested in the laboratory.

We revisit this case below, for a deterministic $\phi_u = 0$ analysis with simple numerically integrated overturning moments, followed by probabilistic analysis for reliability index and probability of failure involving discretized random field. Whether deterministic or probabilistic, the safety prediction is meaningful only if the field strength input is realistic.

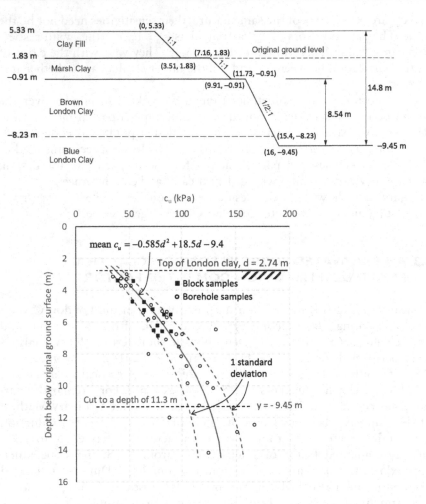

Figure 8.5 Geometry and undrained shear strength profile for the excavation slope of reactor 1 at Bradwell (after Duncan et al. 2014, from Skempton and LaRochelle, 1965). SI units and 1σ standard deviation curves added here.

8.3.1 Deterministic $\phi_u = 0$ slip circle analysis, with summation of moments from horizontal strips

For $\phi_u = 0$ slip circles, the force perpendicular to the slip surface contributes neither overturning moment nor resistance. Hence, there is no need to embark on the method of slices if the overturning moment and summation of c_u along the circular arc can be done another way. For the case in Figure 8.6 where the unit weights of soil strata are different and slope geometry is not uniform, it is more convenient to sum the overturning moments of horizontal strips than vertical slices. A simple VBA code has

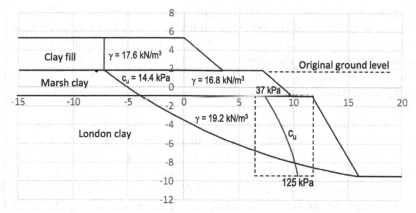

Small-specimen-size London Clay: $c_u = -0.585d^2 + 18.5d - 9.4$

(Units: m, kN/m, kN.m/m, kN/m^2, kN/m^3)

x_c	y_c	Radius	γ_{fill}	γ_{Ms}	$\gamma_{LdnClay}$	$c_{u,Ms}$	γ_w	F_s	α_{cu}
23.64	35.70	45.79	17.6	16.80	19.20	14.40	10	1.79	1

θ_{Ms}	$\theta_{LdnClay}$	θ_b	M_R	M_O	M_w	H	h_c	h_w
0.094	0.477	0.00	98401.7	54822.3	0.0	14.8	3.50	0.00

(a)

Fissured London Clay: $c_u = 0.65 \times \left(-0.585d^2 + 18.5d - 9.4 \right)$
       (Reduction factor 0.65 for c_u of fissured London clay, based on Fig. 8.7)

(Units: m, kN/m, kN.m/m, kN/m^2, kN/m^3)

x_c	y_c	Radius	γ_{fill}	γ_{Ms}	$\gamma_{LdnClay}$	$c_{u,Ms}$	γ_w	F_s	α_{cu}
22.30	32.88	42.79	17.6	16.80	19.20	14.40	10	1.18	0.65

θ_{Ms}	$\theta_{LdnClay}$	θ_b	M_R	M_O	M_w	H	h_c	h_w
0.098	0.513	0.00	61318.1	51767.1	0.0	14.8	3.50	0.00

(b)

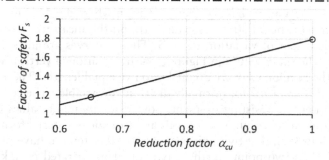

(c)

Figure 8.6 (a) Critical toe circle, (b) reduction factor $\alpha_{cu} = 0.65$, (c) effect of α_{cu} on F_s.

been written to sum the overturning moments of 800 horizontal strips from the toe to the crest of the potential sliding soil mass, similar in rationale to the code in Figure 8.2b. Another VBA code was created for integrating the depth-dependent c_u values over 100 finite arcs of equal segmental lengths in the London clay, and 20 finite arcs for the constant c_u value of the marsh clay, similar in rationale to Eq. 8.3. A dry tension crack was assumed in the compacted fill.

Initially the slip circle in Figure 8.6a were centred at $x_c = 16$ and $y_c = 20$. The Solver tool was used to search for the critical slip circle and found a factor of safety F_s equal to 1.79 (Figure 8.6a), compared with 1.76 by methods of slices and 1.79 by finite element method in Duncan et al. (2014).

That the strength of stiff fissured clay in the field could be much lower than those measured in the laboratory was discussed in Terzaghi et al. (1996, p. 366) concerning delayed slides due to gradual fissure-induced water softening of initially very stiff fissured clay over decades, and in Tomlinson (2001, pp. 138–139 and 393–395) on the need for conservative estimates of shear strength of fissured London clay for foundations and slopes.

Based on Figure 8.7 from Tomlinson, the cut slope in London Clay of Figure 8.5 is re-analyzed, with a shear strength reduction factor α_{cu} of 0.65. A lower factor of safety of 1.18 is obtained, as shown in Figure 8.6b. The effect of α_{cu} on the computed F_s is shown in Figure 8.6c.

8.3.2 Probabilistic analysis incorporating discretized random field for London Clay

The undrained shear strength values from block samples and the borehole samples in Figure 8.5 appear to follow the same trend, hence all the 45 data points were analyzed together, yielding a mean value trend described by the following equation:

$$\text{Mean } c_u = \mu_{cu} = -0.585d^2 + 18.5d - 9.4 \qquad (8.8)$$

which gives 36.9 kPa at the top surface of the London Clay where d = 2.74 m, and 124.9 kPa at the base of the excavation where d = 11.3 m. Analysis of the scatter of the 45 strength values around the mean trend of Eq. 8.8 yields a coefficient of variation of 0.15. The 1σ curves are also shown in Figure 8.5. The six c_u values in Figure 8.8a are positioned vertically so that the arc in the London Clay are divided into five equal lengths. The x^* values are simply $x_i* = \mu_i + n_i\sigma_i$ for normally distributed variables. The cell labelled M_R, which denotes resisting moment, now reads its input values of shear strength from the x^* column. Since soils are geological material formed over long durations such that two points closer together tend to have properties more alike than two points further apart (as demonstrated by Jaksa et al. 1999, for Keswick Clay in Australia, for example), the six c_u values in London Clay are assumed to be spatially autocorrelated, and modelled by

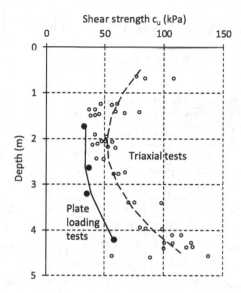

Figure 8.7 Variation of shear strength with depth in London Clay at Maldon (after Tomlinson 2001, Figure 4.1).

the exponential autocorrelation function, with a vertical autocorrelation distance of 2 m. This means that the bottom-right 6-by-6 block of the 8-by-8 correlation matrix **R** is computed from:

$$\rho_{ij} = e^{-\frac{|y_i - y_j|}{\delta}} \quad , \text{where } \delta = 2 \text{ m for this case} \tag{8.9}$$

One of the eight random variable in Figure 8.8a is the reduction factor α_{cu} for the strength of fissured clay, which we already encountered in the deterministic analyses of Figure 8.6. But in the reliability analysis of Figure 8.8, it is treated as an uncertain entity, with a mean value between 0.6 and 1.0, and a c.o.v. of 0.15.

Initially the n column values are zeros. The spreadsheet FORM procedure using Solver (as explained in connection with Figure 8.3) is then implemented, obtaining a reliability index β of 2.79 in Figure 8.8a when the mean α_{cu} is 1.0 (no reduction in shear strength), and a β value of 0.91 in Figure 8.8b when the mean α_{cu} is 0.65. The variation of β as the mean value of the reduction factor α_{cu} changes from 0.6 to 1.0 is plotted at the top right of Figure 8.8a.

In RBD for ultimate limit states, the reliability index β is usually required to be 2.5 or 3.0, corresponding to a failure probability of 0.62% and 0.13%, respectively. It is evident from the plot that, for mean α_{cu} values of 0.9 and smaller, the excavated slope of total height 14.8 m in London Clay at Bradwell-on-Sea in Essex will not satisfy the required level of reliability.

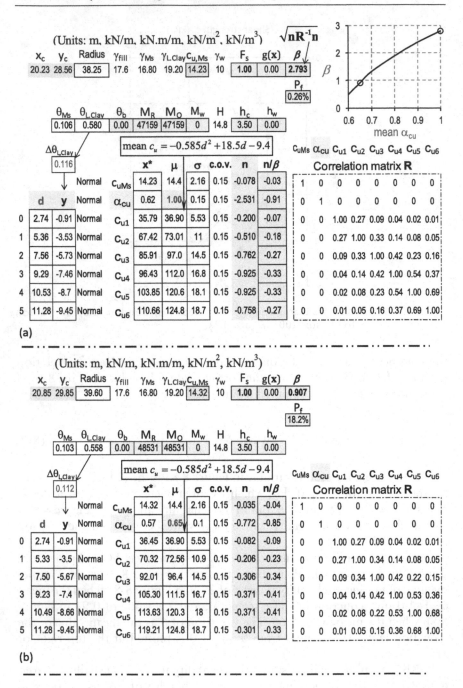

Figure 8.8 Effect of shear strength reduction factor α_{cu} on reliability index β.

For comparison, for α_{cu} value of 0.9, the factor of safety is 1.62. Even when α_{cu} is 0.8, the F_s value is still a seemingly comforting value of 1.45, as shown in Figure 8.6c. A deterministic engineer may not have basis to judge whether these F_s values of 1.62 and 1.45 signify adequate safety against slope failure.

8.4 PROBABILISTIC ANALYSES OF A SLOPE FAILURE IN SAN FRANCISCO BAY MUD

The failure of a slope excavated underwater in San Francisco Bay has been described in Duncan and Buchignani (1973), Duncan (2000) and Duncan and Wright (2005). The slope was part of a temporary excavation and was designed with an unusually low factor of safety to minimize construction costs. During construction, a portion of the excavated slope failed.

Low and Duncan (2013) analyzed the same underwater slope, first deterministically using the mean values (modelled as linear trends) of the undrained shear strength from field vane shear and laboratory triaxial tests, Figure 8.9, then probabilistically, Figure 8.10, accounting for parametric uncertainty and positive correlation of the undrained shear strength c_u and buoyant soil unit weight γ_b.

In the deterministic analysis, the factors of safety were computed with search for *critical noncircular slip surface* based on a reformulated Spencer method (described at end of chapter). The F_s values obtained were 1.20, 1.16 and 1.00, for three types of undrained shear strength data, namely field vane tests, UU triaxial tests on trimmed 35 mm specimens and UU triaxial tests on untrimmed 70 mm specimens, respectively. For comparison, the corresponding F_s values for *critical circular slip surfaces* were 1.23, 1.19 and 1.03, respectively, slightly higher than those of the critical noncircular slip surfaces. It was noted that measured strength values were affected by disturbance and rate of loading effects. Subtle errors were also caused by extrapolation of the undrained shear strength (*in situ* and lab tests data, available only for the upper 21 m of the Bay mud, from depth 6 m to depth 27 m, Figure 8.9) to the full depth of underwater excavation. Since the midpoint of a slip circular arc is at about the two-third depth, this means that in limit equilibrium slope stability analysis, half the slip surface was based on extrapolated strength.

8.4.1 Reliability analysis with correlated lognormals and statistical inputs based on *in situ* and lab tests

Figure 8.9b shows the undrained shear strength (c_u) values with depth as measured from *in situ* vane shear tests, laboratory unconsolidated undrained (UU) triaxial tests on samples trimmed to 35 mm diameter, and UU tests on untrimmed 70 mm (original diameter) samples, respectively. The mean-value trend of c_u is described by the equation $\mu_{cu} = c_0 + b \times \text{Depth}$, in

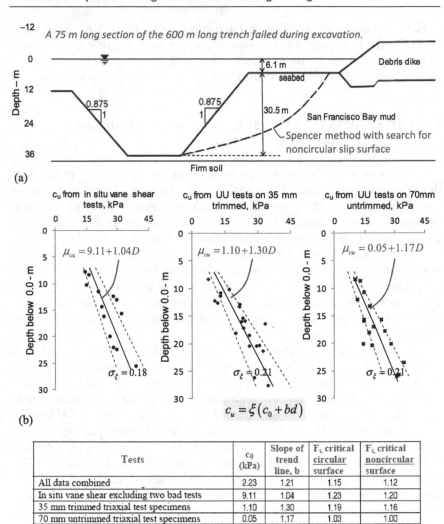

(a)

(b)

(c)

Figure 8.9 (a) An underwater slope failure in San Francisco Bay mud, (b) the average c_u profile, and the $\mu \pm 1$ standard deviation lines, (c) deterministic analysis based on mean trend of c_u, using Spencer method with search for the critical noncircular slip surface.

kPa, where the different values of c_0 and b are shown in the three μ_{cu} equations in Figure 8.9b. A parameter ξ is used to model c_u as a random variable as follows:

$$c_u = \xi \times \mu_{cu} = \xi \times (c_0 + b \times \text{Depth})$$ (8.10)

where the mean of ξ is $\mu_\xi = 1.0$, and the standard deviation of ξ is $\sigma_\xi = 0.18$, 0.21 and 0.21, based on statistical analyses of the three sets of c_u values

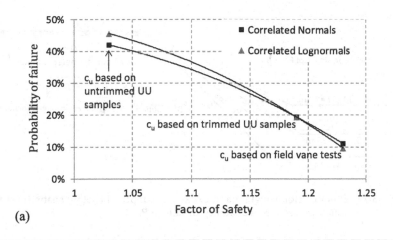

(a)

In situ vane shear excluding two bad tests

Distribution		μ	σ	x	n	Correlation matrix		β	P_f
Lognormal	ξ	1	0.18	0.806	-1.120	1	0.5	1.30	9.7%
Lognormal	γ_b	5.72	0.52	5.703	0.012	0.5	1		

Compare: P_f from three Monte Carlo simulations: 9.80%, 9.63%, 10.1%.

35 mm diameter UU triaxial trimmed specimens

Distribution		μ	σ	x	n	Correlation matrix		β	P_f
Lognormal	ξ	1	0.21	0.833	-0.778	1	0.5	0.86	19.4%
Lognormal	γ_b	5.72	0.52	5.664	-0.063	0.5	1		

Compare: P_f from three Monte Carlo simulations: 19.4%, 19.8%, 19.2%

70 mm diameter UU triaxial untrimmed specimens

Distribution		μ	σ	x	n	Correlation matrix		β	P_f
Lognormal	ξ	1	0.21	0.959	-0.100	1	0.5	0.11	45.6%
Lognormal	γ_b	5.72	0.52	5.692	-0.008	0.5	1		

Compare: P_f from three Monte Carlo simulations: 45.2%, 46.3%, 45.9%.

(b)

Figure 8.10 (a) The spread of probability of failure reflects the spread of factor of safety, (b) FORM analysis for correlated lognormal random variables ξ and γ_b.

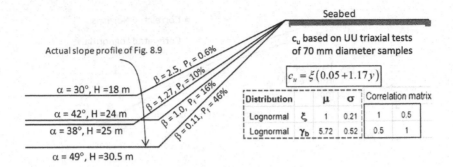

Figure 8.11 Effect of slope angle α and excavation depth H (below seabed) on reliability index and probability of failure P_f.

from vane tests, UU triaxial on trimmed specimens and UU triaxial on untrimmed specimens, respectively.

Another random variable is the buoyant unit weight γ_b of the Bay mud, with a mean value of 5.72 kN/m³ and a standard deviation of 0.52 kN/m³. Physical considerations suggest that the parameter ξ should be positively correlated with γ_b, because high c_u values are likely to occur with high soil unit weights, and vice versa. A correlation coefficient of 0.5 is assumed between ξ and γ_b.

FORM analyses can be performed easily with search for the reliability-based *critical noncircular slip surfaces*. However, obtaining probability of failure from Monte Carlo simulations for critical noncircular slip surfaces (for comparison with FORM P_f) would be very time-consuming because a search for critical noncircular slip surface is required for each set of random numbers generated in Monte Carlo simulations. In contrast, it is much simpler to obtain Monte Carlo P_f values with search for *critical circular slip surface*.

Hence, FORM analysis and Monte Carlo simulations are compared with search for the critical *circular* slip surface. Assuming lognormal distributions of ξ and γ_b, the computed FORM reliability index β are 1.30, 0.86 and 0.11 (Figure 8.10b), respectively, for the three types of c_u tests shown in Figure 8.9b. The probability of failure, $P_f \approx 1 - \Phi(\beta)$, is 9.7%, 19.4% and 45.6%, respectively. For comparison, for each of the three types of c_u tests, P_f from Monte Carlo is taken to be the average of three simulations each with 3000 trials and different initial seed, using the commercial software @ RISK. The results are as follows:

$$(9.80\% + 9.63\% + 10.1\%)/3 = 9.8\%;$$

$$(19.4\% + 19.8\% + 19.2\%)/3 = 19.5\%;$$

$$(45.2\% + 46.3\% + 45.9\%) = 45.8\%.$$

The P_f values from FORM β, equal to 9.7%, 19.4% and 45.6%, respectively, are practically the same as the Monte Carlo P_f values of 9.8%, 19.5% and 45.8%, all unacceptably high. Hence, it is not surprising that a section of the underwater excavated slope failed.

The probabilities of failure based on assuming c_u (via ξ in Eq. 8.10) and $γ_b$ as correlated lognormal variates are compared with those based on assuming c_u and $γ_b$ as normal variates, in Figure 8.10a.

This case shows that the results of both the deterministic and the probabilistic analyses are affected by biases in the strength measurements and interpretations, including extrapolating the measured strengths to greater depth. It is ironic that the test data that result in a F_s value closest to 1.0 (for a failed slope) are those that were considered to be of lowest quality, namely the tests performed on the 70 mm samples, where the most disturbed outer portion of the samples was not trimmed away. However, for these samples, the effect of the greater degree of disturbance reducing the undrained strengths was perhaps compensated closely by the effect of fast rate of lab loading increasing the undrained strengths. Nevertheless, the three β values in Figure 8.10 are much lower than the commonly required β of 2.5. Hence, a failure was not unlikely and did happen.

8.4.2 Reliability-based design of slope angle and excavation depth

The underwater slope in Figure 8.9 was designed with an unusually low factor of safety to minimize construction costs. The low consequence of slope failure and some cost–benefit considerations had presumably been considered in order to tolerate the relatively high probability of failure of about 10% (based on field vane test results) and 46% (based on laboratory UU triaxial tests on 70 mm samples), respectively. This case illustrates the principle that the target reliability index can be high or low depending on the consequence of failure. The underwater excavated slope in this case may arguably aim at a target reliability index of β that is smaller than 2.5 which is a typical target for terrestrial cases with higher consequence of failure. Figure 8.10 shows some combinations of slope angle α and the depth of excavation H (below seabed) which can reduce the probability of failure ($P_f = \Phi(-\beta)$) from 46% to 16%, 10% and 0.6%, respectively.

8.5 RELIABILITY ANALYSIS OF A NORWEGIAN SLOPE ACCOUNTING FOR SPATIAL AUTOCORRELATION

Spatial autocorrelation and spatial variability arise in geological material by virtue of its formation by natural processes acting over unimaginably long time (millions of years). This endows geomaterial with some unique statistical features (e.g., spatial autocorrelation) not commonly found in structural material manufactured under strict quality control. For example, by the nature of the slow precipitation (over many seasons) of fine-grained soil particles in water in nearly horizontal layers, two points in close horizontal or vertical proximity to one another are likely to be more positively correlated (likely to have similar undrained shear strength c_u values, for example) than two points further apart in the horizontal or vertical direction.

A clay slope in southern Norway was analyzed deterministically and probabilistically in Low et al. (2007) using the Low (2003a) spreadsheet-based reformulations of Spencer method and the intuitive first-order reliability method of Low and Tang (2004). The reformulation allows switching among the Spencer, Bishop simplified and wedge methods on the same template, by specifying different side-force inclination options and different constraints of optimization. Search for the critical circular or noncircular slip surface is possible. The deterministic procedure was extended probabilistically by implementing the first-order reliability method via constrained optimization of the equivalent dispersion ellipsoid in the original space of the random variables. The procedure was illustrated for an embankment on soft ground, and for a clay slope in southern Norway, both involving spatially correlated soil properties. The effects of autocorrelation distance on the results of reliability analysis were studied. Shear strength anisotropy and non-homogeneity were modelled via user-created simple function codes in the programming environment of the spreadsheet.

Figure 8.12 shows the results of reliability analysis involving 24 spatially correlated c_u values and 24 spatially correlated unit weight values. The size of the correlation matrix is 48 × 48. The design point obtained by the Solver tool represents the most probable combination of the 24 values of c_u and the 24 values of γ which would cause failure. As expected for resistance parameters, the 24 values of undrained shear strength c_u at the design point are all lower than their respective mean values. On the other hand, when the autocorrelation distance δ is 10 m or lower, as shown in Figure 8.12b, the design point index of γ, defined as $(\gamma_i^* - \mu_\gamma)/\sigma_\gamma$ where γ_i^* is the design point value of unit weight γ for i from 1 to 24, shows most values of unit weight γ^* above their mean value μ_γ, as expected for loading parameters, but, somewhat paradoxically, there are some design-point values of γ near the toe which are below their mean values. The implication is that the slope is less stable when the unit weights near the toes are lower. This implication can be verified by deterministic runs using higher γ values near the toe, with resulting higher factors of safety. It would be difficult for design code committee to recommend partial factors such that the design values of soil unit weight γ are above the mean along some portions of the potential sliding soil mass and below their mean along other portions. In contrast, the design point is located automatically in FORM analysis and reflects sensitivity and the underlying statistical assumptions from case to case in a way code-specified partial factors cannot.

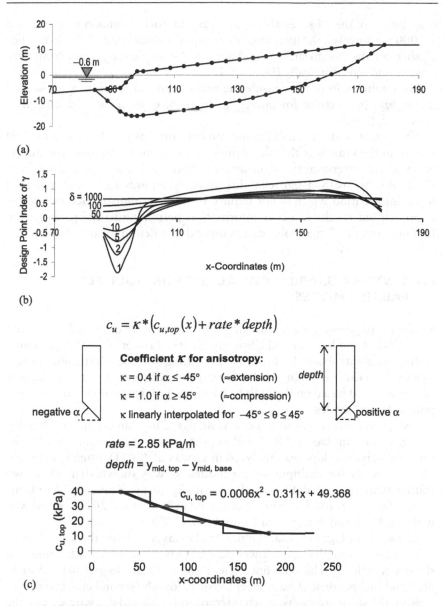

$$c_u = \kappa * \left(c_{u,top}(x) + rate * depth \right)$$

Coefficient κ for anisotropy:

$\kappa = 0.4$ if $\alpha \leq -45°$ ($\approx$extension)

$\kappa = 1.0$ if $\alpha \geq 45°$ ($\approx$compression)

κ linearly interpolated for $-45° \leq \theta \leq 45°$

negative α

positive α

depth

rate = 2.85 kPa/m

depth = $y_{mid, top} - y_{mid, base}$

$c_{u, top} = 0.0006x^2 - 0.311x + 49.368$

(c)

Figure 8.12 Reliability analysis of a Norwegian slope: (a) noncircular slip surface based on Spencer method, (b) design point index of clay unit weight as a function of horizontal autocorrelation distance δ, (c) anisotropic and non-homogeneous undrained shear strength.

At higher values of autocorrelation distance δ, the correlation coefficients approach 1.0; the design point indices of the soil unit weight γ of the 24 slices approach a common value, as shown by the nearly horizontal line in Figure 8.12b for $\delta = 1000$ m. The design point indices of c_u – defined as (c_{ui}^*

$- \mu_{cui})/\sigma_{cui}$ – of the 24 slices also approach a uniform common value when δ = 1000 m; however, the individual design-point values of c_u differ from slice to slice because the mean, μ_{cui}, and standard deviation, σ_{cui}, vary from slice to slice and anisotropically (Figure 8.12c).

The implications of not considering seabed erosion versus treating seabed as random (to account for uncertain depth of erosion) were discussed in Low et al. (2007).

The results of slope reliability analysis are only as good as the statistical inputs, in the same way that the results of deterministic analysis are only as good as the deterministic input and method used (e.g., Spencer method, FEM, or other methods). A reliability analysis requires additional statistical input information which is not required in a deterministic factor-of-safety approach but results in richer information pertaining to the performance function and the design point that is missed in a deterministic analysis.

8.6 SYSTEM RELIABILITY ANALYSIS FOR MULTIPLE FAILURE MODES

Situations requiring system reliability assessment were discussed in Wu and Kraft (1970) and Baecher and Christian (2003). Low et al. (2011) presented system reliability examples for both the sliding and the overturning failure modes of an earth retaining wall, where it was shown that system reliability can be assessed based on the FORM β indices and design points of the component failure modes.

A slope in two clayey soil layers was analyzed in Ching et al. (2009) using Monte Carlo simulation (MCS) and importance sampling (IS) methods. The same two-layered slope was analyzed in Low et al. (2011), using system reliability bounds for multiple failure modes. It was shown that when two (rather than one) reliability-based critical slip surfaces are considered, the system failure probability bounds obtained in Low et al. (2011) agreed well with the MCS and IS results of Ching et al. (2009).

As shown in Figure 8.13a, the upper clay layer is 18 m thick, with undrained shear strength c_{u1}; the lower clay layer is 10 m thick, with undrained shear strength c_{u2}. The undrained shear strengths are lognormally distributed and independent. A hard layer exists below the second clay layer.

Since the shear strengths are characterized by c_{u1} and c_{u2}, with $\phi_u = 0$, the Bishop's simplified method and the ordinary method of slices will yield the same factor of safety, and either can be used. Also, in this case where the upper clay layer is weaker than the lower clay layer, it is logical to locate two reliability-based critical slip circles, as shown in Figure 8.13a, one entirely in the upper clay layer and the other passing through both layers. The FORM reliability indices for the two modes are 2.795 and 2.893, respectively. It is interesting to note that although c_{u1} and c_{u2} are assumed to be uncorrelated, there is correlation between the two failure modes ($\rho_{12} = 0.4535$, bottom of

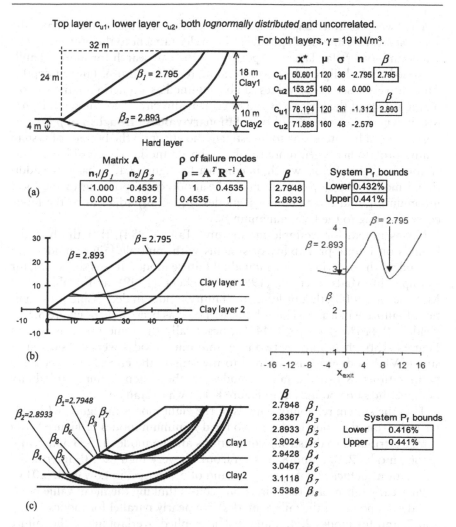

Top layer c_{u1}, lower layer c_{u2}, both *lognormally distributed* and uncorrelated.

For both layers, $\gamma = 19$ kN/m³.

	x*	μ	σ	n	β
c_{u1}	50.601	120	36	-2.795	2.795
c_{u2}	153.25	160	48	0.000	

	x*	μ	σ	n	β
c_{u1}	78.194	120	36	-1.312	2.803
c_{u2}	71.888	160	48	-2.579	

Matrix **A**		ρ of failure modes		β	System P_f bounds	
n_1/β_1	n_2/β_2	$\rho = A^T R^{-1} A$				
-1.000	-0.4535	1	0.4535	2.7948	Lower	0.432%
0.000	-0.8912	0.4535	1	2.8933	Upper	0.441%

	β	System P_f bounds	
2.7948	β_1		
2.8367	β_3	Lower	0.416%
2.8933	β_2	Upper	0.441%
2.9024	β_5		
2.9428	β_4		
3.0467	β_6		
3.1118	β_7		
3.5388	β_8		

Figure 8.13 (a) FORM results for two reliability-based critical slip circles, (b) variation of reliability indices with the x-coordinate of the left exit point of slip circles, including the two reliability-based critical slip circles with left exit points at x_{exit} of −1.5 m and +9 m, respectively, (c) system reliability analysis considering eight failure modes, including the two reliability-based critical modes of (a).

Figure 8.13a), because c_{u1} affects both slip circles. The bounds on system failure probability, computed in the two bottom-right cells in Figure 8.13a by the efficient spreadsheet implementation of the Kounias–Ditlevsen bimodal bounds for systems with multiple failure modes, are 0.432% ~ 0.441%, compared with the MCS estimated range of 0.37%~0.506% (based on MCS mean of 0.44% and c.o.v. of 15.04%, in Ching et al. 2009).

The two reliability-based critical slip circles in Figure 8.13a have the smallest β values among all possible slip circles tangent to the bottoms of the upper and lower clay layers, respectively. One can search for more reliability-based critical slip circles corresponding to different trial tangent depths. Alternatively, a series of β values can be obtained as a function of the x-coordinate values of the lower exit end of critical slip circles, as shown in Figure 8.13b, where the existence of two stationary values ('troughs') of β is obvious. It would be interesting to investigate the effect on the bounds of system failure probability when more reliability-based modes are considered. This is done in Figure 8.13c, which, in contrast to Figure 8.13b, has three additional modes (β_3, β_6, β_7) adjacent to the mode corresponding to the local minimum β_1, and three additional modes (β_4, β_5, β_8) adjacent to the mode corresponding to the local minimum β_2.

It was noted, for example in Ang and Tang (1984), that the bimodal bounds on failure probability of systems with multiple failure modes will depend on the ordering of the individual failure modes. It was suggested, for example in Madsen et al. (1986), Melchers (1999) and Haldar and Mahadevan (2000), that ordering the failure modes in decreasing probabilities of failure will lead to closer bounds. This has been done in Figure 8.13c, yielding $0.416\% \leq P_{F,sys} \leq 0.441\%$, practically the same range as that in Figure 8.13b when only the two local minimum modes were considered. A simple VBA code was also created to investigate the effects of all possible permutations (8!) of the failure modes on the system failure probability bounds: the same bounds as in Figure 8.13c was obtained.

That the system reliability bounds of the eight modes in Figure 8.13c differ little from the bounds of the two local minimum modes of Figure 8.13a can be attributed to the strong correlations among modes 1, 3, 6 and 7, and among modes 2, 4, 5 and 8, as seen from the very high (≈ 1.0) intermodal correlation coefficients of $\rho_{13}, \rho_{16}, \rho_{17}$, and of $\rho_{24}, \rho_{25}, \rho_{28}$ in Low et al. (2011).

Physically this means that direction vectors (linking the mean value point and design points of the failure modes) are nearly parallel for modes 1, 3, 6 and 7, and for modes 2, 4, 5 and 8. The implied overlapping of the failure probability contents for modes 1, 3, 6 and 7, and also for modes 2, 4, 5 and 8 means that it is sufficiently accurate to calculate the bounds for the system failure probability by considering only the two stationary values of reliability index, namely β_1 and β_2 in Figure 8.13a.

8.7 LAMBE AND WHITMAN SLOPING CORE DAM EXAMPLE, REFINED DETERMINISTICALLY AND EXTENDED PROBABILISTICALLY

Lambe and Whitman (1979, p. 368, Example 24.6) conducted a two-block stability analysis of a 198 m high sloping core dam using the wedge method of stability analysis. The strength of the clay core of the dam in Figure 8.14a

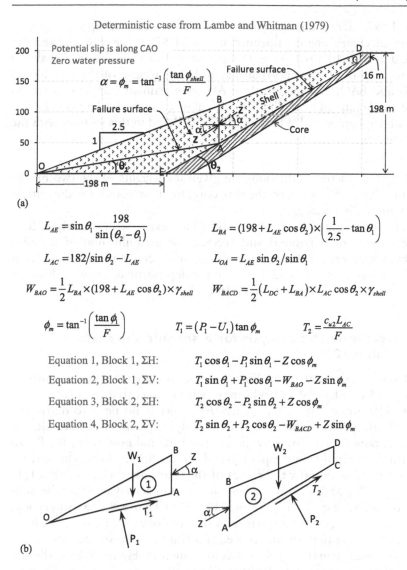

Figure 8.14 Sloping core dam example, after Lambe and Whitman (1979, p. 368, Example 24.6). Geometrical and equilibrium equations added here, to allow search for critical bilinear slip surface using constrained optimization (next figure).

was $c_u = 96$ kPa, with $\phi_u = 0$. The inclination α of the side force Z at AB was assumed to be equal to ϕ_m' where $\tan\phi_m' = (\tan\phi')/F$. That is, the ratio of the mobilized to available shear strength was the same on the vertical plane between the wedges as on the lower failure plane AO. The effective angle of friction of the granular shell material was $\phi' = 40°$. Water pressure was assumed to be zero along the bilinear slip surface CAO. A specific bilinear slip surface, comprising CA (inclined at $\theta_2 = 31.5°$) on the clay core and AO (inclined at $\theta_1 = 10°$) traversing the granular shell material, was analyzed using the graphical force polygon method. The weight of the two blocks (BACD and BAO) and the length of AC were estimated approximately in Lambe and Whitman by scaling. The calculated factor of safety was 1.65.

The above Lambe and Whitman case is analyzed in this section, with the following refinements:

(a) Exact geometrical equations are given for weights of the two blocks and the length of CA on the clay core, to allow search for the critical bilinear slip surface.
(b) The four unknowns $(P_1, P_2, Z$ and $F)$ are expressed explicitly in four equations of horizontal and vertical force equilibrium of the two blocks, Figure 8.14b, to allow solution using constrained optimization in spreadsheet. (T_1 and T_2 are not independent unknowns, because they can be calculated from P_1 and F.)

8.7.1 Deterministic analysis for a specific slip surface of $\theta_1 = 10°$

The deterministic analysis in Figure 8.15a assumes the same inputs as Example 24.6 in Lambe and Whitman, with respect to $\theta_1 = 10°$, $\phi_1' = 40°$, $c_2 = 96$ kPa and $\alpha = \phi_m = \tan^{-1}(\tan\phi_1'/F)$. There is no need to derive the factor of safety F as a function of the underlying parameters; the solution can be obtained easily by solving for the four unknowns (F, P_1, P_2, Z) from the four equilibrium equations of Figure 8.14b, using the spreadsheet *Solver* tool. Initially the values of the four unknowns were $F = 1$, $P_1 = W_1$, $P_2 = W_2$ and $Z = W_2$. These led to nonzero values in the cells labelled 'Eq. 1,' 'Eq. 2,' 'Eq. 3' and 'Eq. 4' in Figure 8.15a. Solver was then used *to change* the four values in the cells of F, P_1, P_2 and Z, *subject to the constraints* that all the four equilibrium conditions be satisfied. A factor of safety equal to 1.633 was obtained in Figure 8.15a, slightly smaller than the value of 1.65 obtained in Lambe and Whitman. The very slight difference between 1.633 and 1.65 is due to the exact calculation of the weights W_1 and W_2 (for BAO and BACD) and the length L_{AC} in Figure 8.15a, versus the approximate calculation of the same in Lambe and Whitman.

8.7.2 Search for the deterministic critical slip surface by varying the inclination angle θ_1

The slip surface with $\theta_1 = 10°$ in Figure 8.15a is only one of many possible bilinear slip surfaces. One can use the *Solver* tool to find the most critical

θ_1	θ_2	L_{AE}	L_{BA}	L_{DC}	γ_{shell}
10.00	31.5	93.81	62.18	16	17.28

W_1	W_2	L_{AC}	ϕ_1	c_1	U_1
149341.3	146581.2	254.51	40	0	0

ϕ_m	θ_1	θ_2	ϕ_2	c_2	U_2
0.4747	0.1745	0.5498	0	96	0
(radians)	(radians)	(radians)			

F	P_1	P_2	Z	T_1	T_2	F^*1
1.633	165344.9	129616.7	61798.9	84968.7	14963.6	1.633

Initially $\theta_1 = 10°$, $F = 1$, $P_1 = W_1$, $P_2 = W_2$, $Z = W_2$

Eq. 1	Eq. 2	Eq. 3	Eq. 4
$\Sigma H=0$	$\Sigma V=0$	$\Sigma H=0$	$\Sigma V=0$
0.000	0.000	0.000	0.000
Block 1		Block 2	

Solver settings:
Set Eq. 1 to zero
By Changing F, P_1, P_2, Z
Subject to Constraints:
Eqs. 2, 3 & 4 =0

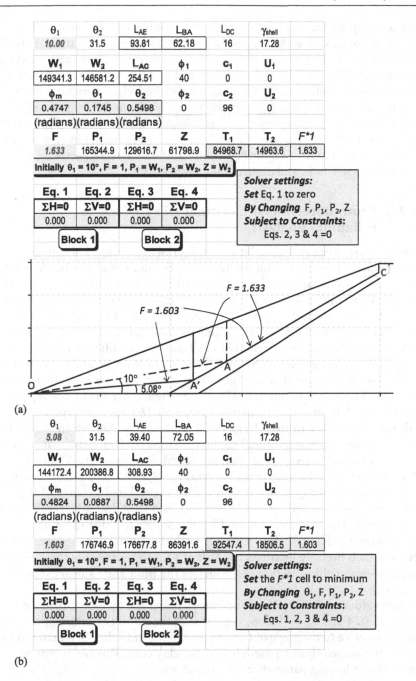

(a)

θ_1	θ_2	L_{AE}	L_{BA}	L_{DC}	γ_{shell}
5.08	31.5	39.40	72.05	16	17.28

W_1	W_2	L_{AC}	ϕ_1	c_1	U_1
144172.4	200386.8	308.93	40	0	0

ϕ_m	θ_1	θ_2	ϕ_2	c_2	U_2
0.4824	0.0887	0.5498	0	96	0
(radians)	(radians)	(radians)			

F	P_1	P_2	Z	T_1	T_2	F^*1
1.603	176746.9	176677.8	86391.6	92547.4	18506.5	1.603

Initially $\theta_1 = 10°$, $F = 1$, $P_1 = W_1$, $P_2 = W_2$, $Z = W_2$

Eq. 1	Eq. 2	Eq. 3	Eq. 4
$\Sigma H=0$	$\Sigma V=0$	$\Sigma H=0$	$\Sigma V=0$
0.000	0.000	0.000	0.000
Block 1		Block 2	

Solver settings:
Set the F^*1 cell to minimum
By Changing θ_1, F, P_1, P_2, Z
Subject to Constraints:
Eqs. 1, 2, 3 & 4 =0

(b)

Figure 8.15 Deterministic analysis for factor of safety, using the Excel *Solver* tool to solve for four unknowns in four equations: (a) lower slip surface inclined at 10°, (b) smaller F_s is found, after searching for the critical lower slip surface inclined at 5.08°.

bilinear slip surface and the corresponding minimum factor of safety. For this purpose, a cell with formula '$= F \times 1$' is created in Figure 8.15b. The *Solver* tool was then used to find the minimum value of this cell, by changing the values in the cells of θ_1, F, P_1, P_2 and Z, *subject to the constraints* that all the four equilibrium conditions be satisfied. The minimum factor of safety obtained is $F = 1.603$, Figure 8.15b, with the lower portion of the associated critical bilinear slip surface defined by $\theta_1 = 5.08°$. These values are compared with $F = 1.633$ and $\theta_1 = 10°$ of Figure 8.15a. The two slip surfaces, CA′O and CAO, for $\theta_1 = 5.08°$ and $\theta_1 = 10°$, respectively, are also shown, as solid lines and dashed lines, respectively.

8.7.3 From deterministic critical slip surface to reliability analysis

Since the clay core and granular shell material are compacted soils from different selected fill, spatial autocorrelation does not need to be modelled. It is also reasonable to assume that the angle of friction (ϕ_1) of the granular shell material and the undrained shear strength (c_2) of the clay core are independent. A third variate M_F (with mean value 1.0 and standard deviation 0.1) has been added to account for the approximate nature of the wedge method which satisfies force equilibrium but not moment equilibrium.

As shown in Figure 8.16b, only three rows need be added to go from the deterministic template of Figure 8.16a to reliability analysis based on FORM. The previous numerical inputs (Figure 8.15b) in the ϕ_1 and c_2 cells in the deterministic template of Figure 8.16a are now replaced by cell addresses pointing to the **x*** cells of Figure 8.16b. The mean values μ of ϕ_1 and c_2 are taken to be equal to the previous deterministic values of 40° and 96 kPa, respectively, and their standard deviations are 4° and 15 kPa. The performance function $g(\mathbf{x})$ is equal to $M_F F - 1$. For normal variates, x_i is simply equal to $x_i = \mu_i + n_i\sigma_i$, and for uncorrelated variates, $\beta = \sqrt{\Sigma n_i^2}$, or, using spreadsheet formula, '=sqrt(sumsq(n)).' Initially the **n** values were zeros. The solution obtained by the *Solver* tool for the **n** values and β are shown in Figure 8.16b, together with the setting of the Solver parameters, which literally instructs the *Solver* tool to find the smallest expanding dispersion ellipsoid that just touches the limit state surface defined by $g(\mathbf{x}) = 0$.

The values of the sensitivity indicators under the column labelled **n**, with $n_{\phi 1} = -2.024$ and $n_{c2} = -0.306$, suggest that the stability of the slope is much more sensitive to the angle of friction (ϕ_1) of the granular shell material than to the undrained shear strength (c_2) of the clay core. This can be appreciated from the following parametric investigations:

(a) When the value of ϕ_1 in the deterministic template of Figure 8.15b is increased by one standard deviation, to 44°, the factor safety increases from 1.60 to 1.82, an increase of about 14%.

θ_1	θ_2	L_{AE}	L_{BA}	L_{DC}	γ_{shell}
5.08	31.5	39.40	72.05	16	17.28

W_1	W_2	L_{AC}	ϕ_1	c_1	U_1
144172.4	200386.8	308.93	31.91	0	0

ϕ_m	θ_1	θ_2	ϕ_2	c_2	U_2
0.4711	0.0887	0.5498	0	91.41	0

(radians)(radians)(radians)

F	P_1	P_2	Z	T_1	T_2	F*1
1.222	174152.5	177291.3	81852.7	88704.6	23102.4	1.222

Initially $\theta_1 = 10°$, F = 1, $P_1 = W_1$, $P_2 = W_2$, Z = W_2

Eq. 1	Eq. 2	Eq. 3	Eq. 4	From deterministic to probabilistic:
$\Sigma H=0$	$\Sigma V=0$	$\Sigma H=0$	$\Sigma V=0$	Replace the ϕ_1 and c_2 cells above
0.000	0.000	0.000	0.000	with *cell addresses* pointing to the

Block 1   Block 2

x* values of ϕ_1 and c_2 cells below.

(a)

$= M_F F - 1$

		x*	μ	σ	n	g(x)	β	P_f
Normal	M_F	0.8181	1	0.1	-1.8186	0.00	2.738	0.31%
Normal	ϕ_1	31.91	40	4	-2.0237			
Normal	c_2	91.41	96	15	-0.3063			

Solver Parameters

Set the β cell to minimum,

By Changing the F, P_1, P_2 and Z cells, and the three n cells.

Subject to the constraints:

The equilibrium equation cells = 0

Performance function g(x) cell = 0

(b)

Figure 8.16 Reliability analysis of the sloping core dam at the critical bilinear slip surface. A model factor M_F accounts for the approximate nature of the wedge method.

(b) When the value of c_2 in the deterministic template of Figure 8.15b is increased by one standard deviation, to 111 kPa, the factor of safety increases from 1.60 to 1.63, an increase of about 2%.

It can also be observed in the solution in Figure 8.15b that the mobilized shear resistance T_1 (92,547 kN/m) through the granular shell material is five times the T_2 value (18,507 kN/m) based on mobilized undrained shear strength c_2 of the clay core, thereby revealing the greater influence of ϕ_1 on stability than c_2.

It is a significant merit of FORM analysis such as Figure 8.16b that it conveys information on parameter sensitivities at its most probable failure point.

8.8 SPENCER METHOD REFORMULATED FOR SPREADSHEET, FOR NON-CIRCULAR SLIP SURFACE

8.8.1 Deterministic reformulations and example analyses

Section 8.4 on a San Francisco underwater slope and Section 8.5 on a Norwegian slope both used the reformulated Spencer method in spreadsheet, for undrained analysis with search for the critical non-circular slip surfaces. This section presents the reformulation of the Spencer method, with illustrations of effective stress slope stability analysis involving noncircular slip surface, and extensions to probabilistic approach.

The derivations and procedures in this section are based on Low (2003a) and Low et al. (2007), using the notations in Nash (1987). The sketch at the top of Figure 8.17 shows the forces acting on a slice (slice i) that forms part of the potential sliding soil mass. The notations are: weight W_i, base length l_i, base inclination angle α_i, total normal force P_i at the base of slice i, mobilized shearing resistance T_i at the base of slice i, horizontal and vertical components $(E_i, E_{i-1}, \lambda_i E_i, \lambda_{i-1} E_{i-1})$ of side force resultants at the left and right vertical interfaces of slice i, where λ_{i-1} and λ_i are the tangents of the side force inclination angles (with respect to horizontal) at the vertical interfaces. Adopting the same assumptions as Spencer (1973), but reformulated for spreadsheet-based automatic constrained optimization approach, one can derive the following from Mohr-Coulomb criterion and equilibrium considerations:

$$T_i = \left[c_i' l_i + \left(P_i - u_i l_i \right) \tan \phi_i' \right] / F \quad \text{(Mohr-Coulomb criterion)} \quad (8.11)$$

$$P_i \cos \alpha_i = W_i - \lambda_i E_i + \lambda_{i-1} E_{i-1} - T_i \sin \alpha_i \quad \text{(Vertical equilibrium of slice} i) \quad (8.12)$$

$$E_i = E_{i-1} + P_i \sin \alpha_i - T_i \cos \alpha_i \quad \text{(Horizontal equilibrium of slice} i) \quad (8.13)$$

$$P_i = \frac{\begin{bmatrix} W_i - \left(\lambda_i - \lambda_{i-1} \right) E_{i-1} \\ -\dfrac{1}{F} \left(c_i' l_i - u_i l_i \tan \phi_i' \right) \left(\sin \alpha_i - \lambda_i \cos \alpha_i \right) \end{bmatrix}}{\begin{bmatrix} \lambda_i \sin \alpha_i + \cos \alpha_i \\ +\dfrac{1}{F} \tan \phi_i' \left(\sin \alpha_i - \lambda_i \cos \alpha_i \right) \end{bmatrix}} \quad \text{(From above three equations)}$$

$$(8.14)$$

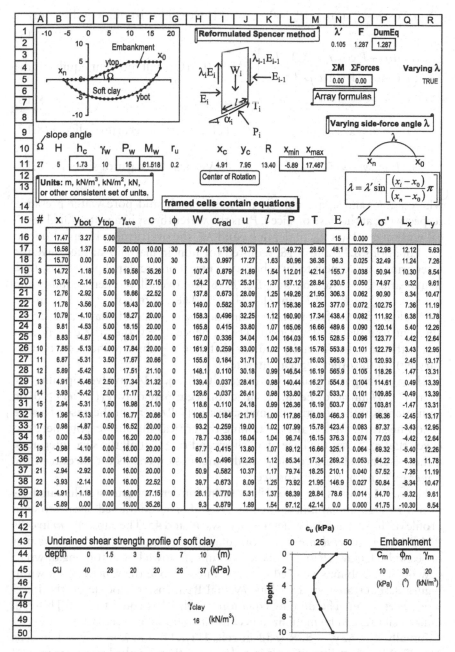

Figure 8.17 Deterministic analysis of a 5 m high embankment on soft ground with depth-dependent undrained shear strength. The limit equilibrium method of slices is based on reformulated Spencer method, with half-sine variation of side force inclination.

$$\sum[T_i\cos\alpha_i - P_i\sin\alpha_i] - P_w = 0 \qquad \text{(Overall horizontal equilibrium)}$$
$$(8.15)$$

$$\sum\left[\begin{array}{l}(T_i\sin\alpha_i + P_i\cos\alpha_i - W_i)\times L_{xi}\\ +(T_i\cos\alpha_i - P_i\sin\alpha_i)\times L_{yi}\end{array}\right] - M_w = 0 \quad \text{(Overall moment equilibrium)}$$
$$(8.16)$$

$$L_{xi} = 0.5(x_i + x_{i-1}) - x_c \qquad \text{(Horizontal lever arm of slice } i) \quad (8.17)$$

$$L_{yi} = y_c - 0.5(y_i + y_{i-1}) \qquad \text{(Vertical lever arm of slice } i) \quad (8.18)$$

where c_i', ϕ_i' and u_i are cohesion, friction angle and pore water pressure, respectively, at the base of slice i, P_w is the water thrust in a water-filled vertical tension crack (at x_0) of depth h_c, and M_w the overturning moment due to P_w. Equations 8.17 and 8.18, required for noncircular slip surface, give the lever arms with respect to an arbitrary centre. For circular slip surfaces, $L_{xi} = R\sin\alpha_i$ and $L_{yi} = R\cos\alpha_i$, and Eq. 8.16 for overall moment equilibrium reduces to: $\sum[T_iR - WR\sin\alpha_i] - M_w = 0$. The use of both λ_i and λ_{i1} in Eq. 8.14 allows for the fact that the right-most slice (slice #1) has a side that is adjacent to a water-filled tension crack, hence $\lambda_0 = 0$ (i.e., the direction of water thrust is horizontal), and for different λ values (either constant or varying) on the other vertical interfaces.

The algebraic manipulation that results in Eq. 8.14 involves opening the term $(P_i - u_il_i)\tan\phi_i'$ of Eq. 8.11, an action legitimate only if $(P_i - u_il_i) \geq 0$, or, equivalently, if the effective normal stress σ_i' $(= P_i/l_i - u_i)$ at the base of a slice is non-negative. Hence, after obtaining the critical slip surface, one needs to check that $\sigma_i' \geq 0$ at the base of all slices and $E_i \geq 0$ at all the slice interfaces. Otherwise, one should consider modelling tension cracks for slices near the upper exit end of the slip surface.

Figure 8.17 shows the spreadsheet setup for deterministic stability analysis of a 5 m high embankment on soft ground. The undrained shear strength profile of the soft ground is defined in rows 44 and 45. The subscript m in cells P44:R44 denotes *embankment*. Formulas need to be entered only in the first or second cell (row 16 or 17) of each column, followed by autofilling down to row 40. The columns labelled y_{top}, γ_{ave} and c invoke the functions shown in Figure 8.18, created via Developer/Visual Basic/Insert Module on the Excel worksheet menu. The dummy equation in cell P2 is equal to F*1. This cell, unlike cell O2, can be minimized because it contains a formula.

Initially $x_c = 6$, $y_c = 8$, $R = 12$ in cells I11:K11, and $\lambda' = 0$, $F = 1$ in cells N2:O2. Microsoft Excel's built-in Solver was then invoked to set target and constraints as shown in Figure 8.18. The Solver option 'Use Automatic Scaling' was also activated. The critical slip circle and factor of safety $F = 1.287$ shown

```
Function ytop(x, omega, H)
grad = Tan(omega * 3.14159 / 180)
If x < 0 Then ytop = 0
If x >= 0 And x < H / grad Then ytop = x * grad
If x >= H / grad Then ytop = H
End Function

Function AveGamma(ytmid, ybmid, gm, gclay)
If ybmid < 0 Then
    Sum = (ytmid * gm + Abs(ybmid) * gclay)
    AveGamma = Sum / (ytmid - ybmid)
    Else: AveGamma = gm
End If
End Function
```

(a)

```
Function Slice_c(ybmid, dmax, dv, cuv, cm)
'comment: dv = depth vector, cuv = cu vector.
If ybmid > 0 Then Slice_c = cm
  Exit Function
End If
ybmid = Abs(ybmid)
If ybmid > dmax Then    'undefined domain,
  Slice_c = 300    'hence assume hard stratum.
  Exit Function
End If
For j = 2 To dv.Count    'array size=dv.Count
  If dv(j) >= ybmid Then
    interp = (ybmid - dv(j - 1)) / (dv(j) - dv(j - 1))
    Slice_c = cuv(j - 1) + (cuv(j) - cuv(j - 1)) * interp
    Exit For
  End If
Next j
End Function
```

Solver Parameters

Set Objective: $P$2

To: ○ Max   ● Min   ○ Value Of:

By Changing Variable Cells:

$N$2:$O$2, $I$11:$K$11

Subject to the Constraints:

```
$N$2 <= 1
$N$2 >= -1
yc >= H
$N$5:$O$5 = 0
xmax >= $B$11/TAN(RADIANS($A$11))
Radius <= $J$11+MAX($C$44:$H$44)
xmin <= 0
```

☐ Make Unconstrained Variables Non-Negative

Select a Solving Method: GRG Nonlinear

Options

☑ Use Automatic Scaling

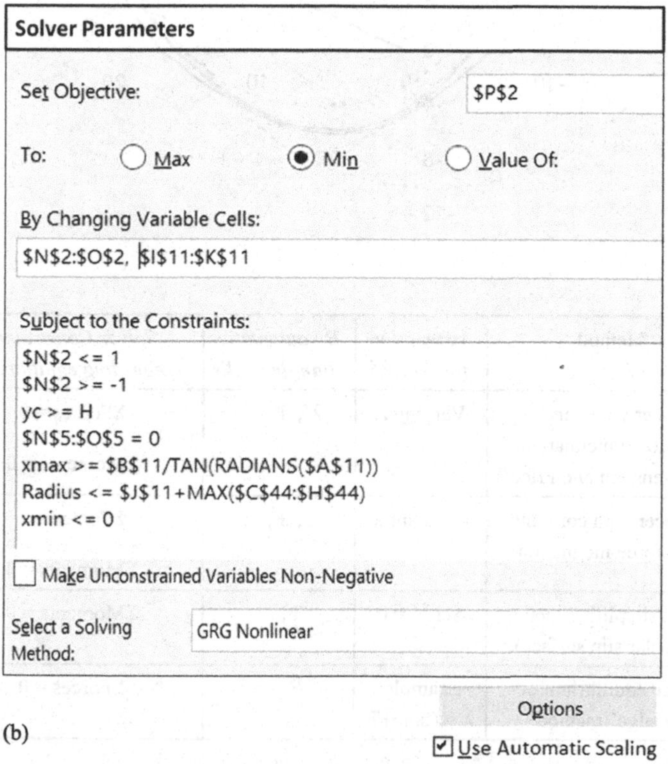

(b)

Figure 8.18 (a) User-defined VBA functions, called by columns y_{top}, γ_{ave} and c of Figure 8.17, (b) Excel Solver settings to obtain the solution of Figure 8.17.

in Figure 8.17 were obtained automatically within seconds by Solver via cell-object-oriented constrained optimization.

Noncircular critical slip surface can also be searched using Solver as in Figure 8.18, except that 'By Changing Cells' are N2:O2, B16, B18, B40, C17 and C19:C39, and with the following additional cell constraints: B16 ≥ B11/tan(radians(A11)), B16 ≥ B18, B40 ≤ 0, C19:C39 ≤ D19:D39, O2 ≥ 0.1 and P17:P40 ≥ 0.

Figure 8.19a tests the robustness of the search for noncircular critical surface. Starting from four arbitrary initial circles, the final noncircular critical surfaces (solid curves, each with 25 degrees of freedom) are close enough to each other, though not identical. Perhaps more pertinent, their factors of

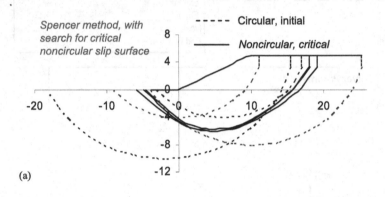

(a)

Method	Assumption for λ or λ'	By automatic changing cells	Solver Constraints regarding equilibrium
Spencer with varying side-force inclination, Morgenstern and Price	Varying λ	λ', F, ...	ΣForces = 0 ΣMoments = 0
Spencer with constant side-force inclination	Constant λ	λ', F, ...	ΣForces = 0 ΣMoments = 0
Bishop simplified method (Circular slip surface)	Set $\lambda' = 0$	F, ...	ΣMoments = 0
Force equilibrium or "wedge" method	Example: $\lambda = (\tan\phi)/F$	F, ...	ΣForces = 0

Note: λ and λ' are defined in the top inset in Fig. 8.17

(b)

Figure 8.19 (a) Testing the robustness of search for the deterministic critical noncircular slip surface, (b) implementing Spencer, Bishop Simplified and Force Equilibrium methods on the same spreadsheet template by using different optimization settings.

safety vary narrowly within 1.253–1.257. This compares with the minimum factor of safety 1.287 of the critical circular surface of Figure 8.17.

The reformulated Spencer method enables effective stress analysis with search for noncircular slip surfaces in heterogeneous soil characterized by shear strength parameters c' and φ'. Figure 8.20a shows that the factor of safety of 1.007 computed by the Low et al. (2007) reformulated Spencer method is practically the same as the 1.010 value computed by the Low et

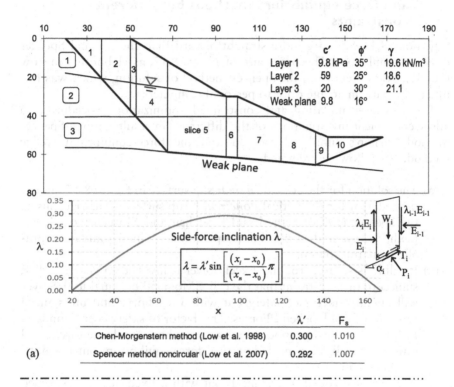

	λ'	F_s
Chen-Morgenstern method (Low et al. 1998)	0.300	1.010
Spencer method noncircular (Low et al. 2007)	0.292	1.007

(a)

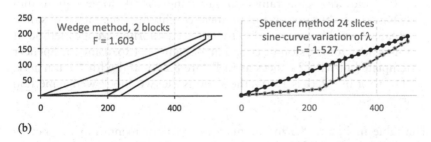

(b)

Figure 8.20 (a) Effective stress slope analysis of multilayer soil using Chen-Morgenstern method and Spencer method for non-circular slip surface, (b) comparing wedge method with Spencer method for the sloping core dam of Figure 8.15b.

al. (1998) spreadsheet method of the Chen-Morgenstern integral formulation. Figure 8.20b shows that, for the case of the sloping core dam of Figure 8.15b, the Spencer method (satisfying overall force and moment equilibriums), with varying side-force inclination λ, obtains a slightly smaller factor of safety of 1.53, compared with 1.60 using the wedge method which satisfies force equilibrium but not moment equilibrium.

8.8.2 Same template for Spencer, Bishop Simplified and force equilibrium methods but different constraints

Equations 8.11–8.18 are more straightforward than the original Spencer (1973) formulations. The procedure of Figure 8.17, as elaborated in Low (2003a), also differs from the Spencer method of solution, which was formulated for use in an age prior to personal computers.

The present formulation and constrained optimization procedure also allow convenient investigations of the difference in results among Spencer method, Bishop's simplified method and the force-equilibrium wedge method. The following are noted:

(a) The solution for the critical *circular* slip surface in Figure 8.17 (and, if desired, that for the critical *noncircular* slip surface) is as rigorous as the Spencer (1973) method and the Chen and Morgenstern (1983) method (Low et al., 1998) but is operationally simpler and conceptually more transparent.

(b) If cell λ' is set to zero, and Excel Solver is invoked to change cell F, subject to the constraint that the cell 'ΣMoment' be equal to 0, Solver will obtain a factor of safety but with a net horizontal unbalanced force in the cell labelled ΣForces. This factor of safety corresponds to Bishop's simplified method, which assumes horizontal side-forces, and satisfies overall moment and vertical force equilibrium but not horizontal equilibrium.

(c) If Solver is invoked to change F but with λ fixed at some assumed value, and with constraints on satisfying overall force equilibrium (ΣForces = 0) but not overall moment equilibrium, the factor of safety obtained would correspond to the force-equilibrium wedge method. If this is done for the sloping core dam of Figure 8.15b, using the Spencer template but with 24 slices as in Figure 8.20b, and with $\lambda = \tan \phi_1/F$, the factor of safety is 1.603, the same as the two-block wedge method analysis.

The table in Figure 8.19b summarizes the implementation of Spencer method, Bishop Simplified method and Force Equilibrium method on the same spreadsheet template by using different optimization settings of the *Solver* tool.

8.8.3 Reformulated Spencer method extended probabilistically

Reliability analyses were performed for the embankment of Figure 8.17, as shown in Figure 8.21, using the Low and Tang (2004) spreadsheet FORM procedure involving the Rackwitz-Fiessler equivalent normal transformations. The coupling between Figure 8.17 and Figure 8.21 is brought about simply by entering formulas in cells C45:H45 and P45:R45 of Figure 8.17 to read values from column x^* of Figure 8.21.

Spatial correlation in the soft ground is modelled by assuming an autocorrelation distance (δ) of 3 m in the following established negative exponential model:

$$\rho_{ij} = e^{\dfrac{|Depth(i) - Depth(j)|}{\delta}} \tag{8.19}$$

Only normals and lognormals are illustrated. Unlike the classical approach, there is no need to diagonalize the correlation matrix (which is equivalent to rotating the frame of reference). Also, the iterative computations of the

DistName		x^*	μ	σ	μ^N	σ^N	n	c_m	ϕ_m	γ_m	c_{u1}	c_{u2}	c_{u3}	c_{u4}	c_{u5}	c_{u6}	
													Correlation matrix **R**				
lognormal	c_m	10.47	10	1.50	9.873	1.561	0.381	1	-0.3	0.5	0	0	0	0	0	0	
lognormal	ϕ_m	30.95	30	3.00	29.83	3.088	0.363	-0.3	1	0.5	0	0	0	0	0	0	
lognormal	γ_m	20.76	20	1.00	19.96	1.037	0.770	0.5	0.5	1	0	0	0	0	0	0	
lognormal	c_{u1}	34.06	40	6.00	39.16	5.081	-1.003	0	0	0	1	0.61	0.37	0.19	0.1	0.04	0
lognormal	c_{u2}	22.72	28	4.20	27.22	3.390	-1.325	0	0	0	0.61	1	0.61	0.31	0.16	0.06	1.5
lognormal	c_{u3}	15.90	20	3.00	19.37	2.372	-1.463	0	0	0	0.37	0.61	1	0.51	0.26	0.1	3
lognormal	c_{u4}	15.86	20	3.00	19.36	2.366	-1.480	0	0	0	0.19	0.31	0.51	1	0.51	0.19	5
lognormal	c_{u5}	22.49	26	3.90	25.50	3.355	-0.898	0	0	0	0.1	0.16	0.26	0.51	1	0.37	7
lognormal	c_{u6}	34.84	37	5.55	36.55	5.196	-0.329	0	0	0	0.04	0.06	0.1	0.19	0.37	1	10
											0	1.5	3	5	7	10	depth

β | P_f
1.962 | 2.5%

```
Function EqvN(DistributionName, v, mean, StDev, code)
Select Case UCase(Trim(DistributionName))    'trim leading/trailing spaces & convert to uppercase
Case "NORMAL":    If code = 1 Then EqvN = mean
                  If code = 2 Then EqvN = StDev
Case "LOGNORMAL": If v < 0.000001 Then v = 0.000001
                  lamda = Log(mean) - 0.5 * Log(1 + (StDev / mean) ^ 2)
                  If code = 1 Then EqvN = v * (1 - Log(v) + lamda)
                  If code = 2 Then EqvN = v * Sqr(Log(1 + (StDev / mean) ^ 2))
End Select
End Function
```

Figure 8.21 Reliability analysis of the case in Figure 8.17, accounting for spatial variation of undrained shear strength. The two templates are easily coupled by replacing the values c_m, ϕ_m, γ_m and the c_u values of Figure 8.17 with formulas that point to the values of the x^* column of this figure.

equivalent normal mean (μ^N) and equivalent normal standard deviation (σ^N) for each trial design point are automatic during the constrained optimization search.

Starting with a deterministic critical noncircular slip surface of Figure 8.19a, the λ' and F values of Figure 8.17 were reset to 0 and 1, respectively. The $\mathbf{x}^*$ values in Figure 8.21 were initially assigned their respective mean values. The Solver tool was then invoked, to:

- *Minimize* the quadratic form (a nine-dimensional hyperellipsoid in original space), i.e., cell 'β.'
- *By changing* the nine random variables of $\mathbf{x}^*$, the λ' in Figure 8.17, and the 25 coordinate values $x_0, x_2, x_{24}, y_{b1}, y_{b3}:y_{b23}$ of the slip surface. (F remains at 1, i.e., at failure, or at limit state.)
- *Subject to the constraints* $- 1 \leq \lambda' \leq 1$, $x_0 \geq H/\tan(\text{radians}(\Omega))$, $x_0 \geq x_2$, $x_{24} \leq 0$, $y_{b3}:y_{b23} \leq y_{t3}:y_{t23}$, $\sum\text{Forces} = 0$, $\sum M = 0$, and $\sigma_1':\sigma_{24}' \geq 0$. The Solver option 'Use automatic scaling' was also activated.

The β index is 1.962 for the case with lognormal variates (Figure 8.21). The corresponding probability of failure based on the hyperplane assumption is 2.49%. The reliability-based noncircular slip surface is shown in Figure 8.22. The nine $\mathbf{x}^*$ values in Figure 8.21 define the most probable failure point, where the equivalent dispersion hyperellipsoid is tangent to the limit state surface (F = 1) of the reliability-based critical slip surface. At this tangent point, the values of c_u are, as expected, smaller than their mean values, but the c_m and ϕ_m values are slightly higher than their respective mean values. This peculiarity is attributable to c_m and ϕ_m being positively correlated to the unit weight of the embankment γ_m and also reflects the dependency of tension crack depth h_c (an adverse effect) on c_m, since the equation $h_c = 2c_m/ (\gamma_m\sqrt{K_a})$ is part of the model in Figure 8.17.

By replacing 'lognormal' with 'normal' in the first column of Figure 8.21, re-initializing column $\mathbf{x}^*$ to mean values, and invoking Solver, one obtains a β index of 1.857, with a corresponding probability of failure equal to 3.16%, compared with 2.49% for the case with lognormal variates. The reliability-based noncircular critical slip surface of the case with normal variates is practically indistinguishable from the case with lognormal variates. Both are however somewhat different (Figure 8.22a) from the deterministic critical noncircular surface from which they evolved via 25 degrees of freedom during Solver's constrained optimization search. This difference in slip surface geometry matters little, however, for the following reason. If the deterministic critical noncircular slip surface is used for reliability analysis, the β index obtained is 1.971 for the case with lognormals, and 1.872 for the case with normals. These β values are only slightly higher (<1%) than the β values of 1.962 and 1.857 obtained earlier with a 'floating' surface. Hence, for the case in hand, performing reliability analysis based on the fixed deterministic critical noncircular slip surface will yield practically the

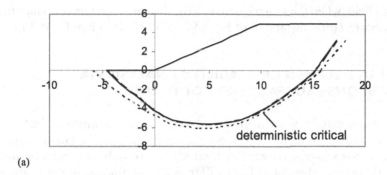

(a)

Reliability indices for the case in Figs. 8.17 and 8.21.

	[Lognormal variates]	[Normal variates]
Reliability index	1.962	1.857
	(1.971)*	(1.872)*
Prob. of failure	2.49%	3.16%
	(2.44%)*	(3.06%)*

* at deterministic critical noncircular surface

(b)

Figure 8.22 Comparison of reliability-based critical noncircular slip surfaces (the two upper curves, for normal variates and lognormal variates) with the deterministic critical noncircular slip surface (the lower dashed curve).

same reliability index as the reliability-based critical slip surface. Figure 8.22b summarizes the reliability indices for the case in hand.

8.8.4 Effect of closer discretization of random field

In the analysis above, the nonlinear undrained shear strength profile is represented by six c_u points at different depths (Figures 8.17 and 8.21). If five additional values are interpolated mid-way between the original six depths, the resulting set of 11 c_u points is an equally valid representation of the c_u profile. This 11-point c_u profile (with standard deviations equal to 15% of their respective mean values, as before), together with the other three random variables, leads to a 14-component n vector and a 14 × 14 correlation matrix, instead of the 9 × 9 matrix shown in Figure 8.21. Reliability analysis with the 14 × 14 correlation matrix, using the previously found reliability-based critical slip surface, obtains a β index of 1.95, instead of 1.96 in Figure 8.21. Another discretization interval was also investigated, with the same c_u profile represented by 21 c_u points, a 24-component n vector and a 24 × 24 correlation matrix instead of the 9 × 9 matrix shown in Figure 8.21. The β index was found to be 1.95. Thus, the computed β index is not affected by the number of discrete points used to represent the one-dimensional c_u

random field, when the spacing between the discrete c_u points is smaller than the autocorrelation distance δ of Eq. 8.9. For the case in hand, δ is 3 m.

8.9 FINITE ELEMENT RELIABILITY ANALYSIS VIA RESPONSE SURFACE METHODOLOGY

Programmes can be written in spreadsheet to handle implicit limit state functions (e.g., Low et al. 1998; Low and Tang 2004, p. 87). However, there are situations where serviceability limit states can only be evaluated using stand-alone finite element or finite difference programmes. In these circumstances, reliability analysis and reliability-based design can still be performed, provided one first obtains a response surface function (via the established response surface methodology) which closely approximates the outcome (near the LSS) of the stand-alone finite element or finite difference programmes. Once the closed-form response function of the LSS has been obtained, performing reliability-based design for a target reliability index is straightforward and fast. Performing Monte Carlo simulation on the closed form approximate response surface function of the LSS also takes little time, but without the useful information of the design point which is available in FORM. Xu and Low (2006) illustrates the finite element reliability analysis of embankments on soft ground via the response surface methodology.

Part III

Reliability-based design applied to rock engineering

Chapter 9

Plane sliding in rock slopes

9.1 INTRODUCTION

Deterministic solution procedures are presented for two-dimensional slopes in rock mass containing one or multiple joint planes along which rock blocks may slide, prior to extending them to probability-based design and analysis. Deterministic cases, and some Monte Carlo simulation design cases, from established publications, are summarized and solved (including using alternative ways), before extending them into RBD accounting for uncertainty in input variables. The first-order reliability method (FORM) and RBD-via-FORM are conducted using the Low and Tang (2007) spreadsheet procedure, with lucidly presented outcomes. The discussions on insights and information attainable from the design point of RBD-via-FORM, and connections with partial factor design methods, constitute the gist which the author hopes to share with readers. It does not matter even if readers are used to conducting FORM and RBD-via-FORM on other platforms or using stand-alone packages. Comparisons are made with second-order reliability method (SORM) and Monte Carlo simulations. The cases include two-block mechanisms from Goodman (1989), a Hong Kong slope from Hoek (2007), and reliability analysis for comparison with a Monte Carlo simulation study by Wyllie (2018) of a failed slope in a limestone quarry.

9.2 PLANE SLIDING STABILITY ANALYSIS, AN EXAMPLE FROM GOODMAN (1989)

The deterministic analysis will be conducted first, using a procedure different from the cited source, before extending the case to reliability-based design (RBD).

DOI: 10.1201/9781003112297-12

9.2.1 Deterministic analysis using an alternative procedure

Figure 9.1 shows a problem from Goodman (1989, Problem 8.3), in which a discontinuity plane P daylights into a cut. The orientation of the plane in strike and dip convention is: strike N 30° W, dip 50° NE. The weight of a potential sliding mass on plane P is 400 metric tons on an area of 200 m². The friction angle is estimated to be 30°. It is required to find the rock bolt force T that will increase the factor of safety F_s to 1.0 and 1.5 when there is no water pressure, and the water pressure u on the discontinuity plane that will cause slide when the rock bolt force for F_s = 1.5 based on dry slope assumption is acting.

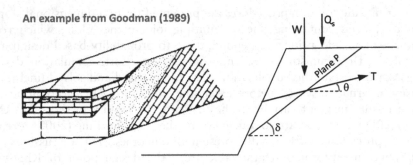

A plane P, with strike N 30° W and dip 50° NE, daylights into a cut. The weight of a potentially sliding mass on plane P is 400 metric tons on an area of 200 m². The friction angle is believed to be 30°.

(a) Find the direction and magnitude of the minimum rock bolt force to achieve a factor of safety of (i) 1.0, and (ii) a factor of safety of 1.5.

(b) What water pressure acting on plane P could cause failure after rock bolts are installed for a safety factor of 1.5?

(c) Is the direction of rock bolting for minimum required support force necessarily the best direction in which to install the rock bolts in this problem?

Answers in Goodman (1989), using the neat lower hemispherical projection technique and the friction circle concept:

(a) (i) 137 tons (1343 kN), in a direction rising 20° above horizontal to the S 60° W.

   (ii) 194 tons ((1901 kN), in a direction rising 29° above horizontal to S 60° W.

(b) The water pressure to initiate slip is 112 tons/200 m² = 0.56 tons/m² = 5.5 kPa.

(c) The minimum force direction is not the direction for shortest bolts. The latter is perpendicular to plane P. The optimum direction depends on the relative costs of steel and drill holes and lies somewhere between these two extremes.

Figure 9.1 A deterministic plane slide analysis involving rock bolt force T and water pressure on discontinuity plane P (after Goodman 1989, Problem 8.3).

Some formulas of single-block plane slides were given in Goodman, based on equating the shear force directed down the sliding surface with the shear strength along the sliding surface, when the condition for limit equilibrium is reached. For the problem in Figure 9.1, Goodman's solution was obtained using the neat lower hemispherical projection technique and the friction circle concept.

This section solves, deterministically then probabilistically, the same problem using analytical formulation and the Excel Solver tool.

Force equilibrium along the discontinuity plane leads to the following equation:

$$N'\frac{\tan\phi}{F_s} + T\cos(\delta-\theta) - (W+Q_s)\sin\delta = 0 \tag{9.1}$$

where

$$N' = (W+Q_s)\cos\delta + T\sin(\delta-\theta) - uA \tag{9.2}$$

Rearranging, one gets:

$$F_s = \frac{\left[(W+Q_s)\cos\delta + T\sin(\delta-\theta) - uA\right]\tan\phi}{(W+Q_s)\sin\delta - T\cos(\delta-\theta)} \tag{9.3}$$

Equation 9.3 means that the traditional lumped factor of safety, F_s, is the ratio of the maximum available upslope resisting force (the numerator) to the downslope sliding force (the denominator).

Another perspective is possible by rearranging Eq. 9.1 as follows:

$$N'\frac{\tan\phi}{F_s} = (W+Q_s)\sin\delta - T\cos(\delta-\theta) \tag{9.4}$$

which means *mobilized resistance = net sliding force along discontinuity plane*. This perspective is useful for EC7 and LRFD, where there are several partial factors instead of a lumped F_s.

Note that Eq. 9.3 is valid when T is a mobilized force at working condition, of the same nature as the first term in the denominator of Eq. 9.3 which is the actual downslope sliding force at working condition.

When there is no rock bolt force (i.e., T = 0), and water pressure u is also zero, the above expression reduces to:

$$F_s = \frac{\left[(W+Q_s)\cos\delta\right]\tan\phi}{(W+Q_s)\sin\delta} = \frac{\tan\phi}{\tan\delta} = \frac{\tan 30°}{\tan 50°} = 0.48 \tag{9.5}$$

which means that the maximum available frictional resistance on the discontinuity plane P is only 48% of that needed to maintain equilibrium; the block will slide if not reinforced by rock bolt. The effect of rock bolt force T and water pressure u on the factor of safety is investigated next.

Equations 9.2 and 9.3 have been entered in a spreadsheet as shown in Figure 9.2, in cells with headings N' and F_s, respectively. The solution for part (a) of the question in Figure 9.1 is obtained easily as follows in Figure 9.2a(i) and (ii):

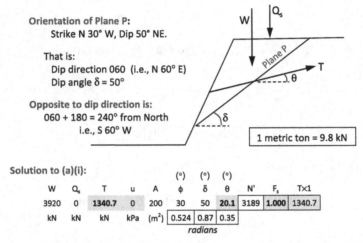

Orientation of Plane P:
  Strike N 30° W, Dip 50° NE.

That is:
  Dip direction 060 (i.e., N 60° E)
  Dip angle δ = 50°

Opposite to dip direction is:
  060 + 180 = 240° from North
    i.e., S 60° W

1 metric ton = 9.8 kN

Solution to (a)(i):

W	Q_s	T	u	A	ϕ (°)	δ (°)	θ (°)	N'	F_s	T×1
3920	0	1340.7	0	200	30	50	20.1	3189	1.000	1340.7
kN	kN	kN	kPa	(m²)	0.524	0.87	0.35			
						radians				

Use the Solver tool, to minimize the"T×1" cell, by changing cells T and θ, subject to the constraint F_s = 1.0.
Get T = 1340.7 kN = 136.8 tons, and θ = 20°, against the dip direction, i.e. S 60° W.

Solution to (a)(ii):

W	Q_s	T	u	A	ϕ (°)	δ (°)	θ (°)	N'	F_s	T×1
3920	0	1897.4	0	200	30	50	28.9	3202	1.500	1897.4
kN	kN	kN	kPa	(m²)	0.524	0.87	0.5			
						radians				

Use the Solver tool, to minimize the"T×1" cell, by changing cells T and θ, subject to the constraint F_s = 1.5.
Get T = 1897.4 kN = 193.6 tons, and θ = 29°, against the dip direction, i.e. S 60° W.

Solution to (b):

W	Q_s	T	u	A	ϕ (°)	δ (°)	θ (°)	N'	F_s	T×1
3920	0	1897.4	5.34	200	30	50	28.9	2135	1.000	1897.4
kN	kN	kN	kPa	(m²)	0.524	0.87	0.5			
						radians				

With T and θ as found in a(ii) above, use Solver to set cell F_s to 1.0, by changing u.
Get u = 5.34 kPa = 0.54 ton/m². (Can also use Excel GoalSeek tool in this case when there is only one input value, u, to be changed.)

Figure 9.2 An alternative deterministic analysis of the plane slide example, using force equilibrium equation and the Excel Solver optimization tool.

Figure 9.2a(i): Initially T = 0, and θ = 0. Excel Solver was invoked, to minimize the 'T × 1' cell, by changing (automatically) cells T and θ, subject to the constraint F_s = 1.0. The solution obtained by the Solver tool is T = 1340.7 kN = 136.8 tons, and θ = 20.1°, practically the same as those (137 tons and 20°, Figure 9.1) obtained by stereographic projection and friction circle analysis.

Figure 9.2a(ii): Invoke Excel Solver, to minimize the 'T × 1' cell, by changing (automatically) cells T and θ, subject to the constraint F_s = 1.5. Get T = 1897.4 kN = 193.6 tons, and θ = 28.9°, again in agreement with those (194 tons, and 29°, Figure 9.1) obtained by stereographic projection and friction circle analysis.

The solution for part (b) of the question in Figure 9.1 is obtained easily as follows in Figure 9.2b:

Figure 9.2b: With T and θ as found in a(ii) above, invoke Solver to set cell F_s to 1.0, by changing u. Get u = 5.34 kPa = 0.54 ton/m², compared with 0.56 ton/m² obtained approximately in Figure 9.1 from scaling the water force U in a force polygon.

Note that the weight of 400 tons on an area of 200 m² (mentioned in the question in Figure 9.1) means that the potential sliding mass has an average thickness of only about 0.75 m. Hence, even a seemingly small water pressure of 5.34 kPa can reduce the F_s from 1.5 (dry case, Figure 9.2a(ii)) to 1.0, Figure 9.2b, where the water force U is 5.34 kPa × 200 m² = 1066 kN. One can already anticipate the context-dependent sensitivity of slope instability to water pressure: The safety of heavier potential sliding block (or same weight W but smaller area A) will be less sensitive to the same water pressure than a lighter potential sliding block (or same weight W but bigger area A) to the same water pressure. Such context-dependent sensitivities of water pressure and other parameters will be automatically reflected in the outcome of RBD-via-FORM, as demonstrated later.

Goodman (1989) aptly noted that the minimum force direction is not the direction for shortest bolts. The latter is perpendicular to the sliding plane. The optimum direction depends on the relative costs of steel and drill holes and lies somewhere between these two extremes.

We next consider a case similar to Figure 9.2, but with the aim to determine the mean rock bolt force T which will achieve a reliability index β of 2.5 against sliding of the rock block. The uncertainties of five of the inputs, namely W, A, u, T and φ, are accounted for.

9.2.2 RBD-via-FORM of potential plane sliding

The random variables are assumed to be the weight of rock W above the discontinuity plane, the discontinuity surface area A under the potential sliding block, mobilized rock bolt force T at working condition, water pressure u on the discontinuity plane and friction angle φ of the discontinuity plane. Weight W and area A are assumed to be positively correlated with ρ_{WA} = 0.5, on the ground that larger area A tends to occur with bigger weight.

The mean values of W, A and φ are 3920 kN, 200 m^2 and 30°, respectively, same as the deterministic inputs in Figures 9.1 and 9.2. The mean value of T is the design value to be determined, such that the reliability is 99.38% against sliding failure, corresponding to a reliability index β of 2.5, that is, failure probability of 0.62%. The standard deviations of W, A, T and φ are assumed to be 10% of their respective mean values. These four random variables are assumed to obey the normal distribution. The pore water pressure u is assumed to follow the gamma distribution, Gamma(5, 0.5). The two parametric values (5, 0.5) correspond to mean μ = 2.5 kPa and standard deviation σ ≈ 1.12 kPa for the water pressure u.

The two columns labelled 'Para1' and 'Para2' and the 5-by-5 correlation matrix show the statistical inputs described above. The transition from the deterministic setup of Figure 9.2 to the RBD set-up in Figure 9.3 is straight-forward: replace the numerical values of W, A, u, T and φ in the first row by cell addresses so that they read their values from the column labelled 'x*.' The five cells under the x* column invoke simple Visual Basic for Applications (VBA) programme codes in Microsoft Excel to calculate x* values from the corresponding values under the column labelled 'n,' as explained in the Appendix and in Low and Tang (2007). The performance function g(x) in Figure 9.3a is Eq. 9.1 with F_s = 1.0:

$$g(\mathbf{x}) = N' \tan\phi + T \cos(\delta - \theta) - (W + Q_s)\sin\delta \qquad (9.6)$$

The equation for the FORM reliability index β in Figure 9.3a is:

$$\beta = \min_{\mathbf{x} \in F} \sqrt{\mathbf{n}^T \mathbf{R}^{-1} \mathbf{n}} \qquad \left(\text{entered as an array formula}\right) \qquad (9.7)$$

The physical meaning of and the symbols in Eq. 9.7 are explained and defined in Chapter 2. Equation 9.7 literally means finding the smallest equivalent 5D dispersion ellipsoid that just touches the limit state surface (LSS) at the most probable failure point.

Initially the n column values were zeros. Then:

a) A trial value (e.g., 1800 kN) was input for the mean value of T (under the 'Para1') column, such that the cell labelled g(x) displays a positive value.
b) Microsoft Excel built-in routine *Solver* was invoked, to *minimize* the cell β, by automatically changing the five values under the n column cells, subject to the constraint that the performance function cell 'g(x)' be equal to 0.0.
c) Repeat the above steps until a mean value of T achieves the target reliability index (2.5 in this case).

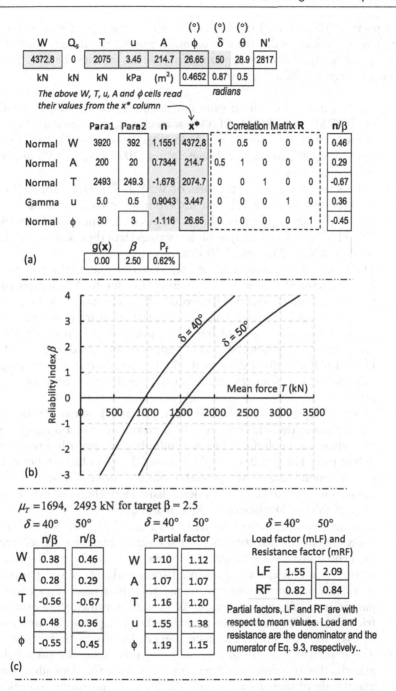

W	Q_s	T	u	A	φ (°)	δ (°)	θ (°)	N'
4372.8	0	2075	3.45	214.7	26.65	50	28.9	2817
kN	kN	kN	kPa	(m²)	0.4652	0.87	0.5	

The above W, T, u, A and φ cells read their values from the x* column —— radians

		Para1	Para2	n	x*	Correlation Matrix R					n/β
Normal	W	3920	392	1.1551	4372.8	1	0.5	0	0	0	0.46
Normal	A	200	20	0.7344	214.7	0.5	1	0	0	0	0.29
Normal	T	2493	249.3	-1.678	2074.7	0	0	1	0	0	-0.67
Gamma	u	5.0	0.5	0.9043	3.447	0	0	0	1	0	0.36
Normal	φ	30	3	-1.116	26.65	0	0	0	0	1	-0.45

(a)

g(x)	β	P_f
0.00	2.50	0.62%

(b)

$\mu_T = 1694$, 2493 kN for target β = 2.5

	δ = 40°	50°		δ = 40°	50°		δ = 40°	50°
	n/β	n/β		Partial factor			Load factor (mLF) and Resistance factor (mRF)	
W	0.38	0.46	W	1.10	1.12			
A	0.28	0.29	A	1.07	1.07			
T	-0.56	-0.67	T	1.16	1.20			
u	0.48	0.36	u	1.55	1.38			
φ	-0.55	-0.45	φ	1.19	1.15			

	δ = 40°	50°
LF	1.55	2.09
RF	0.82	0.84

Partial factors, LF and RF are with respect to mean values. Load and resistance are the denominator and the numerator of Eq. 9.3, respectively..

(c)

Figure 9.3 (a) RBD of mean T for β of 2.5, (b) reliability index β varies with μ_T and dip angle δ of discontinuity plane, (c) sensitivities, partial factors and LF and RF are context-dependent.

A mean value of T of 2493 kN achieves the target reliability index β of 2.5. The five values under the $\mathbf{x}^*$ column represent the design point where the expanding *equivalent* dispersion ellipsoid first touches the LSS. The following in Figure 9.3a are noteworthy:

(1) The negative sensitivity **n** values of uncorrelated T and ϕ from RBD-via-FORM, –1.678 and – 1.116, respectively, indicate that these two are resistance parameters, which decrease from their safe mean values (2493 kN, 30°) to smaller design values (2075 kN, 26.65°) under the column labelled $\mathbf{x}^*$.

(2) In contrast, the positive **n** values of W, A and u, at 1.155, 0.734 and 0.904, respectively, reveal their unfavourable effects on stability (they are 'loads'), in having their design values under the $\mathbf{x}^*$ column (4373 kN, 214.7 m², 3.447 kPa) at higher values than their safe mean values of 3920 kN, 200 m² and 2.50 kPa.

(3) The unfavourable character of area A arises due to the product uA being a destabilizing water force, and also due to A being positively correlated with weight W. However, if shear resistance along discontinuity plane derives from cohesion c apart from friction angle ϕ, the random variable A would have stabilizing–destabilizing duality, in contributing resistance cA and unfavourable load uA. RBD-via-FORM will automatically reveal whether decreasing or increasing the value of area A is more critical in reaching the most probable failure point.

(4) In summary, the mean values of W, A, T, u and ϕ constitute the mean-value point in the safe domain if the performance function g($\mathbf{x}$) is positive when the **n** column values are initially zeros prior to invoking the Solver tool. The five design values under the $\mathbf{x}^*$ column represent the design point (the most probable failure point) where the five-dimensional expanding equivalent dispersion ellipsoid first touches the LSS in the 5D space of W, A, T, u and ϕ.

(5) It is an important merit of RBD-via-FORM (Figure 9.3a) that the context-dependent sensitivities of the random variables are reflected in the probabilistic design outcome. Figure 9.3a reveals that the stability of the rock block, for the adopted statistical inputs and parametric correlations, is most sensitive to the rock bolt force T (a resistance parameter), followed by weight W (a load parameter), discontinuity friction angle ϕ (a resistance parameter), water pressure u and area A (uA is a load parameter), in decreasing order of sensitivity.

(6) The sensitivities of random variable (in their influence on reliability and the design point) can vary from case to case, depending on the magnitude of their mean values, the level of uncertainty represented by their standard deviations, the correlations among random variables as justified by physical considerations, the probability distributions of

the random variables, and the roles played by the random variables in the performance function. RBD-via-FORM will automatically reveal context-dependent and case-specific sensitivities of random variables.

From the perspective of the load and resistance factor design (LRFD), it is possible to back-calculate load factors (LF) and resistance factor (RF) from the outcome of RBD-via-FORM. To do this, one needs to first define loads Q_i and resistance R, and nominal values of loads and resistance.

Based on Eq. 9.3, one possible way of defining load Q and resistance R is as follows:

$$Q = (W + Q_s)\sin\delta - T\cos(\delta - \theta) \tag{9.8}$$

$$R = N'\tan\phi = \left[(W + Q_s)\cos\delta + T\sin(\delta - \theta) - uA\right]\tan\phi \tag{9.9}$$

Partial factors LF and RF can then be back-calculated from:

$$LF = Q^* / Q_n \tag{9.10a}$$

$$\left(\text{or}\, mLF = Q^* / \mu_Q\right) \tag{9.10b}$$

and

$$RF = R^* / R_n \tag{9.11a}$$

$$\left(\text{or}\, mRF = R^* / \mu_R\right) \tag{9.11b}$$

where Q^* and R^* are calculated from the design values of the underlying random variables, Q_n is nominal load (typically greater than mean load) and R_n is nominal resistance (typically smaller than mean resistance). Such back-calculated LF and RF will vary from case to case. For example, those for $\delta = 40°$ will be different from those for $\delta = 50°$, and those for target β equal to 3.0 will be different from those for target β equal to 2.5.

Figure 9.3b shows that the required mean value of rock bolt force T depends on the target reliability level and the dip angle δ of the discontinuity plane. Figure 9.3c shows the context-dependent sensitivity indicators (whether n, or n/β) on the left panel, the back-calculated partial factors of the underlying five parameters in the middle panel, and the LF and RF on the right (based on Eqs. 9.8 and 9.9). The partial factors of the five random variables and the LF and RF are based on mean values (including Eqs. 9.10b and 9.11b). Other values of partial factors and LF and RF are possible, depending on the level of conservatism in the adopted characteristic values (in Eurocode 7) and nominal values (in LRFD).

Other definitions of Q and R are possible apart from Eqs. 9.8 and 9.9. For example, a designer may decide simply to let Q = W, and R = T, and obtain the back-calculated LF and RF as LF = W^*/W_n, and RF = T^*/T_n, and so on.

It is obvious from Figure 9.3 and the above discussions that the values of partial factors and LF and RF back-calculated from the results of RBD-via-FORM are affected by the following:

(i) The target level of reliability in RBD-via-FORM, whether β = 2.5, 3.0 or 3.5.
(ii) The level of conservatism in selecting the characteristic or nominal values.
(iii) The definition of loads and resistance in LRFD.
(iv) The relative magnitudes of the mean values and COV of the random variables in RBD-via-FORM.
(v) Other details which may cause the sensitivity of a random variable to vary from case to case. For example, for the case in hand, different dip angle of the discontinuity plane may change the sensitivities (and hence back-calculated partial factors and LF and RF) of W, T, u and ϕ.

Sensitivity indicators $\mathbf{n}$ (or $\mathbf{n}/\beta$), and back-calculated partial factors and L_F and R_F from RBD-via-FORM, are meaningful for enhancing EC7 and LRFD, in a case-specific (i.e., context-dependent) manner. It is not meaningful to calibrate back-calculated partial factors and LF and RF because the calibrated partial factors do not have general applicability.

To illustrate, Figure 9.4a shows a case identical to Figure 9.3a except there is an additional cohesive resistance with mean cohesion μ_c = 10 kPa and standard of deviation σ_c = 2 kPa, which means a mean cohesive resistance of 10 kPa × 200 m^2 = 2 MN. This makes the cohesion by far the most pivotal random variable, with a n_c value of −1.92, or n_c/β of −0.77, as shown in the last row of Figure 9.4a. The n/β values from Figure 9.3a (with no cohesion) is shown in the rightmost column in Figure 9.4a, for comparison. In contrast, Figure 9.4b shows a case with smaller mean discontinuity area A (μ_A = 60 m^2) than Figure 9.4a, where mean area μ_A was 200 m^2. This reduces both the mean cohesive force cA (a resistance) and mean normal water force uA (an unfavourable load) to only 30% of their mean values in Figure 9.4a and greatly diminishes the values of the sensitivity indicators n/β of cohesion c and water pressure u: −0.07 and 0.12 in Figure 9.4b versus −0.77 and 0.41 in Figure 9.4a when the area A was 200 m^2. The random variables W, T and ϕ in Figure 9.4a, with small n/β values of 0.26, −0.20 and 0.10, respectively, became the three most important players in Figure 9.4b, with n/β values of 0.51, −0.58 and − 0.49, respectively. One may also note that the less-sensitive random variable ϕ in Figure 9.4a has its design value (30.76°) slightly above its mean value of 30°, due to its negative correlation with the most-sensitive cohesion c there where discontinuity area was

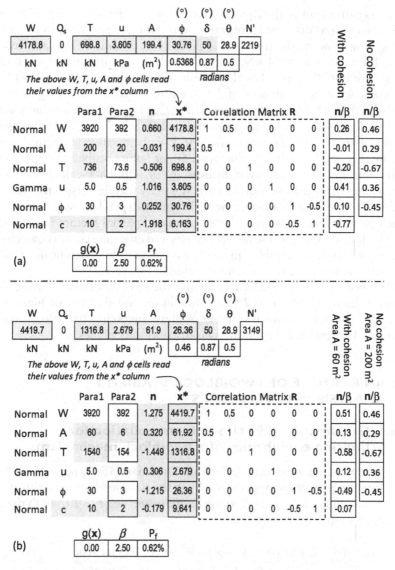

W	Q_s	T	u	A	ϕ (°)	δ (°)	θ (°)	N'
4178.8	0	698.8	3.605	199.4	30.76	50	28.9	2219
kN	kN	kN	kPa	(m²)	0.5368	0.87	0.5	

The above W, T, u, A and ϕ cells read their values from the x* column — *radians*

		Para1	Para2	n	x*	Correlation Matrix R						n/β (With cohesion)	n/β (No cohesion)
Normal	W	3920	392	0.660	4178.8	1	0.5	0	0	0	0	0.26	0.46
Normal	A	200	20	-0.031	199.4	0.5	1	0	0	0	0	-0.01	0.29
Normal	T	736	73.6	-0.506	698.8	0	0	1	0	0	0	-0.20	-0.67
Gamma	u	5.0	0.5	1.016	3.605	0	0	0	1	0	0	0.41	0.36
Normal	ϕ	30	3	0.252	30.76	0	0	0	0	1	-0.5	0.10	-0.45
Normal	c	10	2	-1.918	6.163	0	0	0	0	-0.5	1	-0.77	

(a)

g(x)	β	P_f
0.00	2.50	0.62%

W	Q_s	T	u	A	ϕ (°)	δ (°)	θ (°)	N'
4419.7	0	1316.8	2.679	61.9	26.36	50	28.9	3149
kN	kN	kN	kPa	(m²)	0.46	0.87	0.5	

The above W, T, u, A and ϕ cells read their values from the x* column — *radians*

		Para1	Para2	n	x*	Correlation Matrix R						n/β (With cohesion Area A = 60 m²)	n/β (No cohesion Area A = 200 m²)
Normal	W	3920	392	1.275	4419.7	1	0.5	0	0	0	0	0.51	0.46
Normal	A	60	6	0.320	61.92	0.5	1	0	0	0	0	0.13	0.29
Normal	T	1540	154	-1.449	1316.8	0	0	1	0	0	0	-0.58	-0.67
Gamma	u	5.0	0.5	0.306	2.679	0	0	0	1	0	0	0.12	0.36
Normal	ϕ	30	3	-1.215	26.36	0	0	0	0	1	-0.5	-0.49	-0.45
Normal	c	10	2	-0.179	9.641	0	0	0	0	-0.5	1	-0.07	

(b)

g(x)	β	P_f
0.00	2.50	0.62%

Figure 9.4 (a) With mean cohesion μ_c = 10 kPa, which results in a mean cohesive resistance of 10 × 200 = 2 MN; cohesion becomes by far the most pivotal random variable, (b) with cohesion, but smaller mean area A of 60 m² instead of 200 m² in (a); cohesion no longer a pivotal parameter.

200 m². This is an example of correlated sensitivities which was explained in Section 2.5 in the context of rotational ULS of a spread foundation.

In RBD-via-FORM, it is important to distinguish positive from negative reliability index. For example, for the case in Figure 9.4b, a mean value of T = 446 kN will also achieve a computed β of 2.50, which must be regarded as the negative root of the equation for β (Eq. 9.7), because the performance function cell $g(\mathbf{x})$ displayed a negative value when the $\mathbf{n}$ values were initially zeros, which means the mean-value point is in the unsafe domain. In contrast, for mean $T > 902.1$ kN, the $g(\mathbf{x})$ displays a positive value when the $\mathbf{n}$ values are initially zeros. For the case in Figure 9.4b, a mean value of $T = 1540$ kN achieves the target β of 2.5 (the positive root of Eq. 9.7), or failure probability P_f of 0.62%. For mean $T = 902.1$ kN, the mean-value point is right on the limit state surface $g(\mathbf{x}) = 0$ ($F_s = 1.0$), which means a failure probability of about 50%. For mean $T = 446$ kN, the mean-value point is inside the unsafe domain, with a β value of -2.50, or failure probability of 99.38%.

To sum up, the sensitivity indicators ($\mathbf{n}$, or $\mathbf{n}/\beta$), back-calculated partial factors and other information in FORM offer valuable perspective for EC7 and LRFD design in a case-specific and context-dependent way. There is not much point trying to calibrate back-calculated partial factors from FORM for LRFD and EC7, because different cases and changing context can change important random variables in one case to become insignificant random variables in another case, as demonstrated in Figure 9.4.

The above examples and figures involve the stability of a single rock block above a discontinuity plane P, as shown in the top sketches of Figures 9.1 and 9.2. We next consider stability analysis involving two-block mechanisms and potential sliding along two discontinuity planes.

9.3 AN EXAMPLE OF TWO-BLOCK STABILITY ANALYSIS IN ROCK SLOPES

9.3.1 Goodman (1989) closed form equation for limiting equilibrium of a two-block mechanism

The following equation was derived in Goodman (1989, pp. 465–468), to calculate the required support force R_b in the passive block, in a direction θ below horizontal, *to achieve limiting equilibrium* (i.e., $F_s = 1.0$) with the friction angles input in the equation:

$$R_b = \frac{W_1 \sin(\delta_1 - \phi_1)\cos(\delta_2 - \phi_2 - \phi_3) + W_2 \sin(\delta_2 - \phi_2)\cos(\delta_1 - \phi_1 - \phi_3)}{\cos(\delta_2 - \phi_2 + \theta)\cos(\delta_1 - \phi_1 - \phi_3)} \quad (9.12)$$

where, as shown in the top sketch of Figure 9.5a,

   ϕ_1, ϕ_2, ϕ_3 are the friction angles applicable to sliding along the upper, lower and vertical slide surfaces, respectively;
   δ_1 and δ_2 are the inclinations of the upper and lower slide surface, respectively;
   W_1 and W_2 are the weights of the (upper) active and the (lower) passive blocks per unit of slide width.

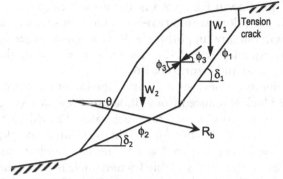

A problem of two-block stability (after Goodman, 1980, problem 8.9)

A creeping rock slide above a road has moved a total of 3 m. The direction of resultant displacement at the top of the slope is 60° below the horizontal while the direction of the resultant displacement in the lower portion of the slope is 25° below the horizontal. A cross section of the slope with these angles (δ_1 and δ_2) and the known position of the tension crack above the slide determines the area of the active upper block to be 10,000 m³ and the area of the passive lower block to be 14,000 m³. The rock weighs 0.027 MN/m³. There are no rock anchors in the slope.

(a) Assuming that the factor of safety of the slide is now 1.0 and that all the friction angles are the same (i.e., $\phi_1 = \phi_2 = \phi_3$), calculate the available friction angle.

(b) Calculate the increase in the factor of safety if 4000 m³ are excavated from the active block and removed from the slide.

(c) Calculate the horizontal anchorage force required to achieve the same factor of safety as the excavation in (b).

Equation 8.13 from Goodman (1980):

$$R_b = \frac{W_1 \sin\left(\delta_1 - \phi_1\right)\cos\left(\delta_2 - \phi_2 - \phi_3\right) + W_2 \sin\left(\delta_2 - \phi_2\right)\cos\left(\delta_1 - \phi_1 - \phi_3\right)}{\cos\left(\delta_2 - \phi_2 + \theta\right)\cos\left(\delta_1 - \phi_1 - \phi_3\right)}$$

Answers from Goodman (1980):

(a) Assume $\phi_1 = \phi_2 = \phi_3$.
   With $\theta = 0$, try different ϕ_j values,
   get $\phi = 36.4°$ for $R_b = 0$ by the above equation.

(b) Assume $\phi_1 = \phi_2 = \phi_3$. With $\theta = 0$, and W_1 based on 6000 m³, try different ϕ values, get $\phi_j = 33.3°$ for $R_b = 0$. Hence factor of safety is equal to
   $\tan36.4°/\tan33.3° = 1.12$.

(c) With W_1 based on 10,000 m³ as originally given, and $\phi_j = 33.3°$, the required anchorage force R_b is, from the above equation, equal to 37.1 MN.

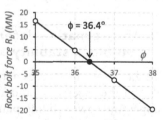

Figure 9.5 Deterministic two-block stability analysis based on Goodman (1989) closed form equation.

Part (a) of the question in Figure 9.5 is about back-calculating the available friction angle for a slope with-a two-block mechanism that failed. Since failure occurred without any support force, R_b is set equal to zero on the left of the closed form equation (Eq. 9.12). Assuming $\phi_1 = \phi_2 = \phi_3$, the solution of ϕ is obtained by trial and error to be 36.4°.

Part (b) of the question concerns evaluating the factor of safety if the volume of the upper block is reduced from 10,000 to 6000 m³. Assuming $\phi_1 = \phi_2 = \phi_3$, the solution is first obtained for the ϕ value that will just maintain equilibrium at the reduced destabilizing weight of the upper block, based on Eq. 9.12 with $R_b = 0$. The required ϕ for limiting equilibrium is 33.3°. Knowing from part (a) that the available friction angle is 36.4°, the factor of safety is therefore calculated to be $F_s = \tan(36.4°)/\tan(33.3°) = 1.12$.

Part (c) of the question concerns calculating the support force R_b required to achieve the same F_s as the excavation in part (b). Equation 9.12 for R_b is for limiting equilibrium, i.e., $F_s = 1.0$. Nevertheless, the required R_b can be calculated by input of $\phi_1 = \phi_2 = \phi_3 = 33.3°$ in Eq. 9.12, because 33.3° corresponds to an F_s of 1.12 form part (b). Therefore, a support force of 37.1 MN is required to achieve the same F_s as the excavation in Part (b).

Using the same principle as that in the solution of Figure 9.5c, the support force R_b to achieve any F_s can be calculated using the above Eq. 9.12 from Goodman (1989). For example, for the case in hand, based on the original volumes $V_1 = 10,000$ m³ and $V_2 = 14,000$ m³, if the desired F_s is 1.20, one can input $\phi_1 = \phi_2 = \phi_3 = \text{atan}(\tan(36.4°)/1.2)) = 31.57°$ in the Goodman equation of R_b and obtain the required support force of $R_b = 58.4$ MN.

In engineering analysis, if the same solution/s can be obtained by different procedures, it will often enhance understanding to explore and compare the different procedures. In this spirit, an alternative deterministic solution procedure to the three-part problem of Figure 9.5 is presented next, before extending it to reliability-based design.

9.3.2 An alternative deterministic procedure for two-block stability analysis

Figure 9.6 shows the forces acting on the two blocks of the slope, namely weights of the blocks W_1 and W_2, mobilized shear forces T_1 and T_2 along the upper and lower sliding planes, total normal forces N_1 and N_2 perpendicular to the sliding planes, external rock anchor force R_b, and internal resultant force Z inclined at an angle α with the normal to the vertical interface. Water pressure and cohesion on the discontinuity planes can be modelled if appropriate. In this case, there is no water force, and shear resistances along discontinuity planes are frictional. Hence, U_1, U_2, c_1 and c_2 are zeros. The horizontal and vertical equilibriums of the two blocks are satisfied when the four simple equations of ΣH and ΣV shown in Figure 9.6 are equal to zeros.

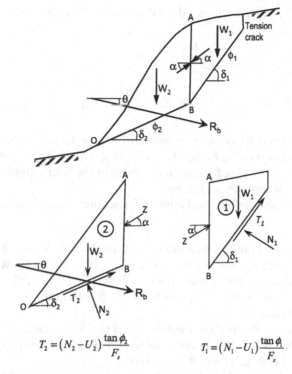

$$T_2 = \left(N_2 - U_2\right)\frac{\tan\phi_2}{F_s} \qquad\qquad T_1 = \left(N_1 - U_1\right)\frac{\tan\phi_1}{F_s}$$

Equation 1, Block 1, ΣH: $T_1\cos\delta_1 - N_1\sin\delta_1 + Z\cos\alpha$

Equation 2, Block 1, ΣV: $T_1\sin\delta_1 + N_1\cos\delta_1 - W_1 + Z\sin\alpha$

Equation 3, Block 2, ΣH: $T_2\cos\delta_2 - N_2\sin\delta_2 - Z\cos\alpha + R_b\cos\theta$

Equation 4, Block 2, ΣV: $T_2\sin\delta_2 + N_2\cos\delta_2 - W_2 - Z\sin\alpha - R_b\sin\theta$

Make assumption on α, then use the spreadsheet Solver tool to find the
values of four unknowns which satisfy the above four equilibrium equations.

Figure 9.6 An alternative deterministic procedure for two-block stability analysis
that obtains the same solution as the Goodman (1989) closed form
equation.

There are more unknowns than can be solved by the four equilibrium equations. One assumption commonly made to reduce the number of unknowns
is that the inclination angle α of the side force Z at the vertical interface is
equal to the mobilized friction angle ϕ_m on the discontinuity planes:

$$\alpha = \phi_m = \tan^{-1}\left(\tan\phi\,/\,F\right) \qquad\qquad (9.13)$$

as assumed in the Goodman solutions of Figure 9.5.

For the case in hand, the input parameters for the four equilibrium equa-
tions in Figure 9.6 are δ_1, δ_2, α, R_b, θ, W_1 and W_2. The mobilized shear force

T_1 and T_2 can be calculated once the values of N_1, N_2 and F_s are determined, as follows:

$$T_1 = N_1 \frac{\tan \phi_1}{F_s} \qquad (9.14a)$$

and

$$T_2 = N_2 \frac{\tan \phi_2}{F_s} \qquad (9.14b)$$

There is no need to derive the factor of safety F_s as a function of the underlying parameters (δ_1, δ_2, R_b, θ, W_1, W_2, ϕ_1, ϕ_2, ...). The solution can be obtained easily by solving for four unknowns in the four equilibrium equations shown in Figure 9.6, as follows:

For part (a) of the problem in Figure 9.5, the solution is obtained in Figure 9.7a as follows:

  (i) The value of F_s is 1.0. The unknowns are N_1, N_2, Z and ϕ (the same for ϕ_1, ϕ_2 and α). Initially $\phi_1 = 30°$, $F = 1$, $N_1 = W_1$, $N_2 = W_2$, $Z = W_2$. These led to nonzero values in the cells labelled 'Eq. 1,' 'Eq. 2,' 'Eq. 3' and 'Eq. 4' in Figure 9.7a, which means force equilibrium conditions were not satisfied.
 (ii) The *Solver* tool was then invoked, *to set* the 'Eq. 1' cell to 0, *by changing* the N_1, N_2, Z and ϕ_1 cells, *subject to the constraints* that the Eqs. 2, 3 and 4 cells in Figure 9.7a be zeros.
(iii) A solution of $\phi = 36.38°$ was obtained (same as that in Figure 9.5a), together with those of N_1, N_2 and Z. All the four 'Eq.' cells are zeros, indicating satisfaction of force equilibrium conditions.

For part (b) of the problem in Figure 9.5, a volume value of "6000" is entered in cell V_1 of Figure 9.7b, the solution is then obtained as follows:

  (i) The unknowns are F_s, N_1, N_2 and Z. Initially $F = 1$, $N_1 = W_1$, $N_2 = W_2$ and $Z = W_2$.
 (ii) The *Solver* tool was invoked, *to set* the 'Eq. 1' cell to 0, *by changing* the F_s, N_1, N_2 and Z cells, *subject to the constraints* that the Eqs. 2, 3 and 4 cells in Figure 9.7b be zeros. (It is good to tick the 'Use Automatic Scaling' option of Solver.)
(iii) A solution of $F_s = 1.121$ was obtained (same as that in Figure 9.5b), together with those of N_1, N_2 and Z.

For part (c) of the problem in Figure 9.5, a volume value of "10000" is entered in cell V1 of Figure 9.7c, the solution is then obtained as follows:

  (i) The unknowns are R_b, N_1, N_2 and Z. Initially $F = 1.121$, $R_b = 0$, $N_1 = W_1$, $N_2 = W_2$ and $Z = W_2$.
 (ii) The *Solver* tool was invoked, *to set* the 'Eq. 1' cell to 0, *by changing* the R_b, N_1, N_2 and Z cells, *subject to the constraints* that the Eq. 2, 3 and 4 cells in Figure 9.7c be zeros.

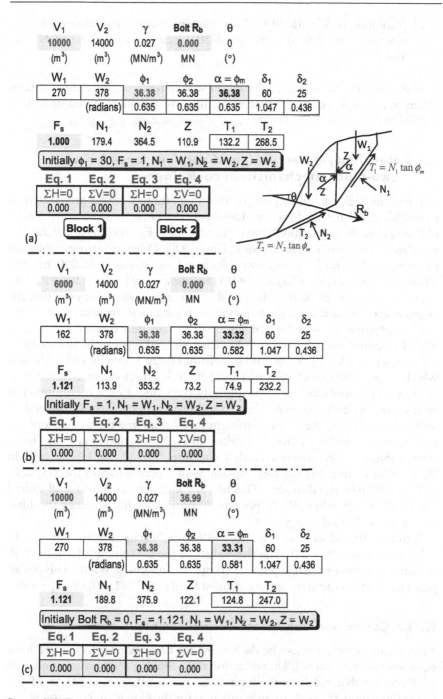

(a)

V_1	V_2	γ	Bolt R_b	θ
10000	14000	0.027	0.000	0
(m^3)	(m^3)	(MN/m^3)	MN	$(°)$

W_1	W_2	ϕ_1	ϕ_2	$\alpha = \phi_m$	δ_1	δ_2
270	378	36.38	36.38	36.38	60	25
	(radians)	0.635	0.635	0.635	1.047	0.436

F_s	N_1	N_2	Z	T_1	T_2
1.000	179.4	364.5	110.9	132.2	268.5

Initially $\phi_1 = 30$, $F_s = 1$, $N_1 = W_1$, $N_2 = W_2$, $Z = W_2$

Eq. 1	Eq. 2	Eq. 3	Eq. 4
$\Sigma H = 0$	$\Sigma V = 0$	$\Sigma H = 0$	$\Sigma V = 0$
0.000	0.000	0.000	0.000

Block 1   Block 2

(b)

V_1	V_2	γ	Bolt R_b	θ
6000	14000	0.027	0.000	0
(m^3)	(m^3)	(MN/m^3)	MN	$(°)$

W_1	W_2	ϕ_1	ϕ_2	$\alpha = \phi_m$	δ_1	δ_2
162	378	36.38	36.38	33.32	60	25
	(radians)	0.635	0.635	0.582	1.047	0.436

F_s	N_1	N_2	Z	T_1	T_2
1.121	113.9	353.2	73.2	74.9	232.2

Initially $F_s = 1$, $N_1 = W_1$, $N_2 = W_2$, $Z = W_2$

Eq. 1	Eq. 2	Eq. 3	Eq. 4
$\Sigma H = 0$	$\Sigma V = 0$	$\Sigma H = 0$	$\Sigma V = 0$
0.000	0.000	0.000	0.000

(c)

V_1	V_2	γ	Bolt R_b	θ
10000	14000	0.027	36.99	0
(m^3)	(m^3)	(MN/m^3)	MN	$(°)$

W_1	W_2	ϕ_1	ϕ_2	$\alpha = \phi_m$	δ_1	δ_2
270	378	36.38	36.38	33.31	60	25
	(radians)	0.635	0.635	0.581	1.047	0.436

F_s	N_1	N_2	Z	T_1	T_2
1.121	189.8	375.9	122.1	124.8	247.0

Initially Bolt $R_b = 0$, $F_e = 1.121$, $N_1 = W_1$, $N_2 = W_2$, $Z = W_2$

Eq. 1	Eq. 2	Eq. 3	Eq. 4
$\Sigma H = 0$	$\Sigma V = 0$	$\Sigma H = 0$	$\Sigma V = 0$
0.000	0.000	0.000	0.000

Figure 9.7 Deterministic analysis based on horizontal and vertical equilibrium of the two rock blocks, for comparison with the three solutions of Goodman (1989) in Figure 9.5.

(iii) A solution of $R_b = 36.99$ MN was obtained (very slight difference with that in Figure 9.5c, due to round-off), together with those of N_1, N_2 and Z.

The deterministic procedure based on simple force equilibrium considerations in Figure 9.7 can be extended readily into reliability-based design, as shown next.

9.3.3 Reliability-based design of support force for a two-block mechanism in rock slope

The friction angles ϕ_1 and ϕ_2 on the two discontinuity planes are treated as independent normal random variables, with the same mean value of 36.4°, obtained from the deterministic back-calculations in Figures 9.5a and 9.7a. The standard deviation is 3.5°, as shown in Figure 9.8b. Other random variables are the volumes V_1 and V_2 of the two blocks, of mean values 10,000 m³ and 14,000 m³, respectively, and a c.o.v. of 0.1, and the external support force R_b with a c.o.v. 0.1. It is assumed that the variables are independent. Suppose that the engineer wants to determine the required mean value of R_b for a target β of 2.5.

The deterministic template of Figure 9.8a can be extended easily into RBD by adding the part shown in Figure 9.8b, and replace the five numerical inputs in the V_1, V_2, R_b, ϕ_1 and ϕ_2 cells of Figure 9.8a with cell addresses which refer to the five $\mathbf{x}^*$ cells of Figure 9.8b. The performance function in the $g(\mathbf{x})$ is the formula '= F_s − 1.' Initially the values in the n column cells were zeros. For different trial values of the mean support force R_b, the Solver tool was used *to set the* β cell to minimum, *by changing* the five cells of the n column and the four numerical values of F_s, N_1, N_2 and Z cells in Figure 9.8a, *subject to the constraints* that the four 'Eq. 1' ..., 'Eq. 4' cells in Figure 9.8a (containing the four equations of Figure 9.6) and the $g(\mathbf{x})$ cells in Figure 9.8b be equal to zero. This literally instructs the *Solver* tool to find the smallest expanding 5D dispersion ellipsoid that just touches the limit state surface defined by $g(\mathbf{x}) = 0$.

A mean value of support force $R_b = 89.45$ MN (Figure 9.8b) is found to achieve the target β of 2.5. The five values in the $\mathbf{x}^*$ column of Figure 9.8b constitute the most probable point of failure (i.e., the design point). This design point lies on the limit state surface defined by $g(\mathbf{x}) = F_s − 1 = 0$, i.e., $F_s = 1.0$.

9.3.3.1 Comparison with Monte Carlo simulation

Monte Carlo simulation can be done using the Goodman (1989) closed form equation (Eq. 9.12 above). That equation is for limiting equilibrium condition, which means that it defines the limit state surface where $F_s = 1.0$. In Monte Carlo simulation, the performance function for the Goodman equation is:

$$g(\mathbf{x}) = R_{b,MCarlo} − R_{b,Goodman} \qquad (9.15)$$

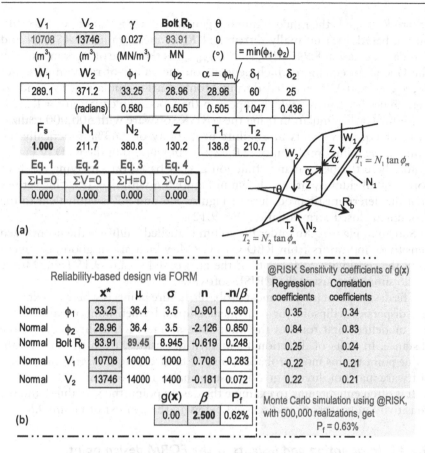

V_1	V_2	γ	Bolt R_b	θ	$= \min(\phi_1, \phi_2)$	
10708	13746	0.027	83.91	0		
(m^3)	(m^3)	(MN/m^3)	MN	$(°)$		

W_1	W_2	ϕ_1	ϕ_2	$\alpha = \phi_m$	δ_1	δ_2
289.1	371.2	33.25	28.96	28.96	60	25
	(radians)	0.580	0.505	0.505	1.047	0.436

F_s	N_1	N_2	Z	T_1	T_2
1.000	211.7	380.8	130.2	138.8	210.7

Eq. 1	Eq. 2	Eq. 3	Eq. 4
$\Sigma H=0$	$\Sigma V=0$	$\Sigma H=0$	$\Sigma V=0$
0.000	0.000	0.000	0.000

(a)

$T_1 = N_1 \tan \phi_m$
$T_2 = N_2 \tan \phi_m$

Reliability-based design via FORM

		x^*	μ	σ	n	$-n/\beta$
Normal	ϕ_1	33.25	36.4	3.5	-0.901	0.360
Normal	ϕ_2	28.96	36.4	3.5	-2.126	0.850
Normal	Bolt R_b	83.91	89.45	8.945	-0.619	0.248
Normal	V_1	10708	10000	1000	0.708	-0.283
Normal	V_2	13746	14000	1400	-0.181	0.072

$g(x)$	β	P_f
0.00	2.500	0.62%

@RISK Sensitivity coefficients of $g(x)$

Regression coefficients	Correlation coefficients
0.35	0.34
0.84	0.83
0.25	0.24
-0.22	-0.21
0.22	0.21

Monte Carlo simulation using @RISK, with 500,000 realizations, get $P_f = 0.63\%$

(b)

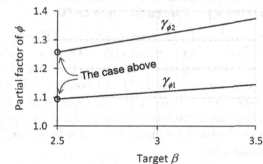

(c)

Figure 9.8 (a) Deterministic template as in Figure 9.7c, (b) RBD-via-FORM indicate that a mean support force R_b of 89.45 MN is required to achieve a target β of 2.5, (c) greater partial factor of ϕ_2 than ϕ_1 suggested by RBD-via-FORM.

where $R_{b,MCarlo}$ is the random number generated during Monte Carlo simulation, based on a normally distributed R_b of mean value 89.45 MN and standard deviation 8.945 MN, and $R_{b,Goodman}$ is the value of R_b computed by the Goodman equation which depends on the values of random variables ϕ_1, ϕ_2, V_1 and V_2 generated during Monte Carlo simulation. (V_1 and V_2 determines W_1 and W_2, which appear in the Goodman equation for R_b.)

Monte Carlo simulation using the @RISK software with 500,000 realizations of Eq. 9.15 results in a failure probability of 0.63%, practically the same as the failure probability of 0.62% (from $\Phi(-\beta)$) in the RBD of Figure 9.8b where the performance function g(x) is based on satisfying the four force equilibrium equations shown in Figure 9.6 with $F_s = 1.0$. This verifies that the deterministic procedure of Figures 9.6 and 9.7 is equivalent to the Goodman closed form equation (Eq. 9.12).

Shown in Figure 9.8b under the column labelled '$-n/\beta$' are the normalized sensitivity indicators from RBD-via-FORM, which are in good agreement with the sensitivity coefficients of the g(x) based on Eq. 9.15 from Monte Carlo simulation using the @RISK software.

The design point (i.e., the most probable failure point) is where an expanding dispersion ellipsoid (or equivalent ellipsoid when nonnormal variates are modelled) first touches the limit state surface. Reliability index is the distance, in units of directional standard deviations, from the safe mean-value point to this most probable point of failure. (Monte Carlo simulation software may not directly give a most probable failure point.)

It is often enlightening to examine the design point (the $\mathbf{x}^*$ values) and the sensitivity indicators $\mathbf{n}$, as discussed below in the context of Figure 9.8.

9.3.3.2 Information and insights at the FORM design point, and implications for Eurocode 7 and LRFD

The partial factors γ_i and load and resistance factors LF and RF, in the discussions below, are with respect to mean values. This is to avoid ambiguity that arises from back-calculated partial factors and LF and RF when different characteristic values (for Eurocode 7) and nominal values (for LRFD) are adopted. RBD-via-FORM needs statistical inputs, but not partial factors. Nevertheless, the partial factors implied at the design point of RBD-via-FORM are back-calculated for discussions with Eurocode 7 and LRFD in the following paragraphs.

(1) The values of the sensitivity indicators under the column labelled $\mathbf{n}$, with $n_{\phi2} = -2.126$ and $n_{\phi1} = -0.901$, means that the most probable failure value of ϕ_2 (28.96°, under the $\mathbf{x}^*$ column) is 2.126 × $\sigma_{\phi2}$ smaller than its mean value of 36.4°, while the most probable failure value of ϕ_1 (33.25°) is 0.901 × $\sigma_{\phi1}$ smaller than its mean value of 36.4°. From the Eurocode 7 perspective, the implied partial factors are $\gamma_{\phi2} = 36.4/28.96 \approx 1.26$, and $\gamma_{\phi2} = 36.4/33.25 \approx 1.10$. This suggests that the design is

more sensitive to the friction angle ϕ_2 of the lower discontinuity plane than to the friction angle ϕ_1 of the steeper upper discontinuity plane. This conclusion is valid even when partial factors are calculated from $\gamma_\phi = \tan\phi_k/\tan\phi^*$, where ϕ_k is the characteristic value of friction angle.

(2) That two parameters (ϕ_1 and ϕ_2 in this case) of the same physical nature can have different sensitivity indicators ($n_1 = -0.901$, $n_2 = -2.126$, Figure 9.8b) is a manifestation of context-dependent sensitivity, which is accounted for automatically in RBD-via-FORM, but difficult to deal with in partial factor design approach. Hence, conducting RBD-via-FORM in tandem with partial factor design can throw much lights on the latter (e.g., EC7) and provide guidance in its continuing evolvement.

(3) From the LRFD and EC7-DA2 perspective, the mobilized resistance for the case in Figure 9.8a consists of three components in different directions, namely (i) mobilized frictional resistance T_1 along the steeper upper discontinuity plane, (ii) mobilized frictional resistance T_2 along the lower discontinuity plane and (iii) mobilized support force R_b, which is horizontal when $\theta = 0$ in this case. The prevailing LRFD method allows for multiple loads but only one resistance. Engineers may think differently on how to define a single resistance from the multi-directional three components of resistance. Even the definition of loads for use in LRFD may vary from one designer to another: should it be W_1 and W_2, or their sliding effects along discontinuity planes 1 and 2? If W_1 and W_2 are regarded as loads, the n_{V1} value of 0.708 and n_{V2} value of -0.181 in Figure 9.8b suggest that W_1 should be factored up, but not W_2. In contrast, if sliding forces Q_1 and Q_2 along discontinuity planes 1 and 2 are regarded as loads, both should of course be factored up.

It is a significant merit of RBD-via-FORM such as Figure 9.8b that it conveys context-dependent information on parameter sensitivities at its most probable failure point. It is likely that conducting RBD-via-FORM in tandem with EC7 and LRFD designs can reveal issues that require attention and improve design rationale and guide EC7 and LRFD in their continuing evolution.

9.4 DETERMINISTIC FORMULATIONS AND RELIABILITY-BASED DESIGN OF SAU MAU PING SLOPE IN HONG KONG

We now consider a slope in Hong Kong, with frictional and cohesive resistance along a rock joint. The destabilizing forces are the weight of potential sliding rock, water pressure in tension crack and joint and earthquake-induced horizontal force. It will be shown that in this case where two sensitive parameters are modelled by the highly skewed truncated exponential distribution, a more precise estimate of failure probability is achievable by performing second-order reliability method (SORM) in spreadsheet, based on the design point and reliability index obtained in FORM. This section is partly based on Low (2007), Low and Phoon (2015b), and Low (2019), with substantial refinements and new contents added here.

The following limit equilibrium formulation for stability analysis of a two-dimensional rock slope in Hong Kong was presented in Hoek (2007), who also analyzed it probabilistically using Monte Carlo simulation. The deterministic formulations for the factor of safety, from Hoek (2007), are as follows (notations as shown in Figure 9.9a):

$$F_s = \frac{cA + N' \tan\phi}{W \left(\sin\psi_p + \alpha \cos\psi_p \right) + V \cos\psi_p - T \sin\theta} \tag{9.16}$$

where

$$A = \left(H - z \right) / \sin\psi_p, \quad \text{area of the sheet joint} \tag{9.16a}$$

$$z = H \left(1 - \sqrt{\cot\psi_f \tan\psi_p} \right), \quad \text{critical } z \text{ for dry slope} \tag{9.16b}$$

$$N' = W \left(\cos\psi_p - \alpha \sin\psi_p \right) - U - V \sin\psi_p + T \cos\theta \tag{9.16c}$$

$$W = 0.5\gamma H^2 \left(\left[1 - \left(\frac{z}{H} \right)^2 \right] \cot\psi_p - \cot\psi_f \right) \tag{9.16d}$$

$$U = 0.5\gamma_w z_w A \tag{9.16e}$$

$$V = 0.5\gamma_w z_w^{\,2} \tag{9.16f}$$

Figure 9.9a shows the reliability-based design of reinforcing force T, using the Low and Tang (2007) FORM procedure (Eq. 9.7). The five random variables modelled for this two-dimensional rock slope are: the shear strength parameters ϕ (degrees) and c (tonne/m^2) of the failure surface (the joint plane with inclination angle ψ_p), tension crack depth z, the ratio z_w/z where z_w is the height of water in the tension crack and the coefficient of horizontal earthquake acceleration α. From physical considerations, the minimum value for z_w/z is 0, when the tension crack is dry. The maximum value is 1, when the tension crack is filled with water. For RBD-via-FORM below, and for the Monte Carlo simulations in Chapter 8 of Hoek (2007), both z and z_w will change from their mean values, but both must be restricted to the domain $0 \le z_w/z \le 1.0$.

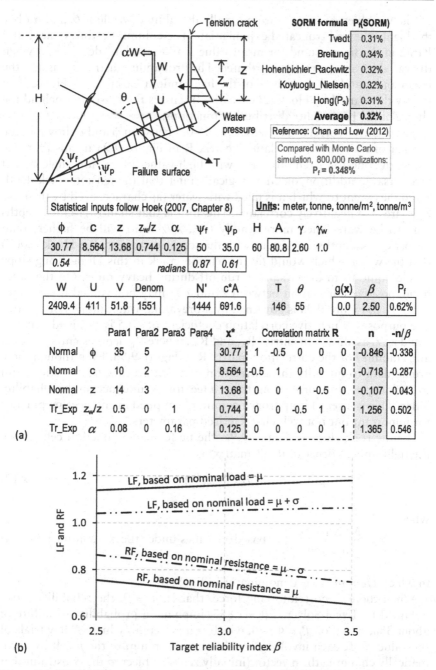

Statistical inputs follow Hoek (2007, Chapter 8) | **Units:** meter, tonne, tonne/m², tonne/m³

ϕ	c	z	z_w/z	α	ψ_f	ψ_p	H	A	γ	γ_w
30.77	8.564	13.68	0.744	0.125	50	35.0	60	80.8	2.60	1.0
0.54				radians	0.87	0.61				

W	U	V	Denom		N'	c*A		T	θ		g(x)	β	P_f
2409.4	411	51.8	1551		1444	691.6		146	55		0.0	2.50	0.62%

		Para1	Para2	Para3	Para4	x*	Correlation matrix R					n	-n/β
Normal	ϕ	35	5			30.77	1	-0.5	0	0	0	-0.846	-0.338
Normal	c	10	2			8.564	-0.5	1	0	0	0	-0.718	-0.287
Normal	z	14	3			13.68	0	0	1	-0.5	0	-0.107	-0.043
Tr_Exp	z_w/z	0.5	0	1		0.744	0	0	-0.5	1	0	1.256	0.502
Tr_Exp	α	0.08	0	0.16		0.125	0	0	0	0	1	1.365	0.546

(a)

(b)

Target reliability index β

Figure 9.9 RBD of Hong Kong slope. (a) RBD using FORM, followed by spreadsheet SORM of Chan and Low (2012a), (b) load and resistance factors inferred from RBD-via-FORM.

The variables ϕ, c and z are normally distributed, while z_w/z and α obey the highly skewed truncated exponentials. The column headings Para1 and Para2 in the figure stand for mean value μ and standard deviation σ when the variables are normally distributed. The probability distributions and the mean and standard deviations of the five random variables in Figure 9.9a follow those used by Hoek (2007), who discusses the reasoning behind the choice of the probability distribution functions.

The negative correlation coefficient of –0.5 between ϕ and c shown in the top left of the 5-by-5 correlation matrix $\mathbf{R}$ in Figure 9.9a means that low values of cohesion c tend to occur with high values of friction angle ϕ, and vice versa. In addition, one may logically infer that the tension crack depth z and the extent to which it is filled with water (as characterized by the ratio z_w/z) are also negatively correlated. This means that shallower crack depths tend to be water-filled more readily (i.e., z_w/z ratio will be higher) than deeper crack depths, consistent with the scenario suggested in Hoek (2007) that the water which would fill the tension crack in this Hong Kong slope would come from direct surface run-off during heavy rains. On the other hand, positive correlation between z and z_w/z may be justified if the water depth z_w in Figure 9.9a is governed by a prevailing water table. For illustrative purposes, a negative correlation coefficient of –0.5 is assumed between z and z_w/z, as shown in entries R_{43} and R_{34}, where R_{ij} denotes entry in row i and column j of the correlation matrix $\mathbf{R}$ in Figure 9.9a. Even though there are no data to quantify this correlation between z and z_w/z, it is still useful to explore possible correlations to get a feel for its influence on the reliability index. This is a sensible approach commonly applied in engineering practice for important but not well-characterized parameters.

For the rock slope of Figure 9.9a, the performance function cell (above the right upper corner of the R matrix) is:

$$g(\mathbf{x}) = F_s - 1 \qquad (9.17)$$

where

$$F_s = \frac{N'\tan\varphi + cA}{Denom} \quad \text{(based on values under the } \mathbf{x} * \text{column)} \quad (9.17\text{a})$$

in which $Denom$ is the denominator in Eq. 9.16.

When there is no reinforcing force, that is, $T = 0$, the reliability index obtained by Excel Solver is $\beta = 1.887$, implying a probability of failure of about 3%, based on $P_f = \Phi(-\beta)$, which is unacceptably high. A few trials of the value of T, each invoking Excel Solver to minimize the β cell by automatically changing the $\mathbf{n}$ vector (initially zeros) subject to the constraint that the g($\mathbf{x}$) cell be equal to 0, lead to the solution as shown, namely a horizontal (for which $\theta = 55°$) T force of 146 tonnes/m (i.e., 1432 kN, per m length of slope) is required to achieve a reliability index β of 2.5, corresponding to a failure probability of about 0.62%, from $P_f = \Phi(-\beta)$.

For the case in hand, the RBD is most sensitive to the coefficient of horizontal earthquake acceleration α and the ratio z_w/z. The values of the sensitivity indicators of α and z_w/z, under the column labelled **n** to the right of the **R** matrix, are 1.365 and 1.256, respectively, higher than the absolute values of the other three **n** values.

A reliability-based design via FORM is able to locate the design point case by case and in the process reflect parametric sensitivities as affected by case-specific limit state surface, statistical inputs and correlation structure in a way that design based on prescribed partial factors cannot.

The equation $P_f = \Phi(-\beta)$ is exact when the LSS is planar and the parameters follow normal distributions. Inaccuracies in P_f estimation may arise when the LSS is curved and/or nonnormal distributions are modelled. More refined alternatives are available, for example, the established second-order reliability method (SORM). SORM analysis requires the FORM β value and design point values as inputs, and therefore is an extension of FORM. In general, the SORM attempts to assess the curvatures of the LSS near the FORM design point in the dimensionless and rotated u-space. The failure probability is calculated from the FORM reliability index β and estimated principal curvatures of the LSS using established SORM equations of the following form:

$$P_f(SORM) = f(\beta_{FORM}, \text{Curvatures at the FORM design point}) \quad (9.18)$$

These SORM formulas have been used by Chan and Low (2012a), who presented a practical and efficient spreadsheet-automated approach of implementing SORM using an approximating paraboloid (Der Kiureghian et al., 1987) fitted to the LSS in the neighborhood of the FORM design point. Complex mathematical operations associated with Cholesky factorization, Gram-Schmidt orthogonalization and inverse transformation are relegated to relatively simple short function codes in the Microsoft Excel spreadsheet platform.

The $P_f(SORM)$ value of 0.32% shown in the table at the top right of Figure 9.9a was obtained using the Chan and Low (2012a) Excel spreadsheet SORM approach. This is smaller than $P_f(FORM)$ value of 0.62% based on $P_f = \Phi(-\beta)$. The eight sample points for estimating the four components of curvature at the design point in the five-dimensional random variable u-space are based on sampling grid coefficient $k = 1$ if $\beta \leq 3$, and $k = 3/\beta$ if $\beta > 3$. The accuracy of this FORM-then-SORM P_f estimation of 0.32% is adequate when compared with Monte Carlo simulations with 800,000 Latin Hypercube sampling, yielding a failure probability of 0.348%. The number of sampling points (eight) for estimating curvature at the FORM design point is five orders of magnitude smaller than the 800,000 in a Monte Carlo simulation.

The above RBD of a reinforced rock slope illustrates that RBD can be used when (i) there are no EC7 recommended partial factors yet (e.g., the parameters z_w/z and α in Figure 9.9a), (ii) when the design show context-dependent sensitivities to the underlying parameters which cannot be reflected by fixed partial factors, (iii) when the parameters are statistically correlated based on physical considerations (e.g., between c and ϕ, and between z and z_w/z in Figure 9.9a) and (iv) when there is a target reliability index and associated failure probability which can be higher or lower depending on the consequence of failure, or, for serviceability limit states, non-performance.

One may note the following connections between EC7 and the RBD example in Figure 9.9a:

(1) The design point (the five values under the column labelled x_i^*) has the same qualitative meaning as the design point in EC7. However, the FORM design point is the most probable point of failure at the contact point of an expanding equivalent dispersion ellipsoid with the limit state surface (defined by $F_s = 1$) in 5D space and reflects context-sensitivity and parameter correlations in a way the design point of EC7 cannot; this is because the design point in EC7 is obtained by applying code-specified partial factors to conservative characteristic values.

(2) Figure 9.9a uses ϕ (instead of $\tan\phi$) as a random variable, to match the statistics for ϕ in Hoek (2007, Chapter 8). This does not prevent comparisons with EC7 in which a partial factor is applied to the characteristic value of $\tan\phi_k$ instead of ϕ_k. After deciding on the characteristic value of ϕ_k, the design value for ϕ is $\phi_d = \tan^{-1}[(\tan\phi_k)/1.25]$, which can then be compared with the design value of ϕ_d in this example.

(3) The sensitivity indicator values of ϕ and c, equal to –0.85 and –0.72 under the n column in Figure 9.9a, mean that their influence on the design is similar. This outcome is opposite to the retaining wall foundation case in Chapter 5 (Figure 5.10), where the design is much more sensitive to ϕ than c. This context-dependent sensitivity is attributable to:

    (a) The roles played by ϕ and c in the bearing capacity equation of Chapter 5 are different from their roles in the shear resistance of rock joint in Eq. 9.16.
    (b) The statistics (mean values, standard deviations, correlations) of the random variables, which determine the orientation of the equivalent dispersion ellipsoid.

9.4.1 LRFD considerations: Non-uniqueness of inferred LF and RF from RBD-via-FORM

The numerator in Eq. 9.16 is the resistance R against slide failure in Figure 9.9a, and the denominator is the net downslope force Q tending to cause sliding. Monte Carlo simulations can be used to estimate the mean values and standard deviations of Q and R, namely μ_Q, σ_Q, μ_R and σ_R. Nominal load Q_n and nominal resistance R_n can then be chosen, and LF and RF calculated from the ratios $Q*/Q_n$ and $R*/R_n$, where the $Q*$ and $R*$ values are based on the design values under the **x*** column in Figure 9.9a. Figure 9.9b shows that back-calculated LF and RF values change with the target β value, and with how the nominal load and resistance values are defined. Besides back-calculating LF and RF from RBD-via-FORM, it is enlightening to conduct RBD-via-FORM in tandem with LRFD. Note that LRFD and EC7 (2004) Design Approach 2 (EC7-DA2) are similar in principle, except that in LRFD the nominal resistance is multiplied by RF (typically <1.0), and in EC7-DA2, the conservative characteristic resistance is divided by RF (typically >1.0), to arrive at the design value for resistance.

9.4.2 Combined methods of drainage, slope reprofiling and reinforcement to improve slope reliability

Besides using reinforcing force T, at least two other methods can be used to enhance the reliability of the Hong Kong slope shown in Figure 9.9, namely slope reprofiling (i.e., decreasing the slope angle Ψ_f), and drainage to reduce water pressure in the tension crack and along the discontinuity plane (which is the failure surface inclined at Ψ_p in Figure 9.9). In practice, it is quite common to combine several stabilization methods to enhance the stability of slopes. The target reliability index in RBD can be higher or lower depending on the consequence of failure.

A series of RBD-via-FORM was conducted for the Hong Kong slope, for different combinations of slope angle Ψ_f, mean z_w/z ratio ($\mu_{zw/z}$, which affects water pressure), reinforcing force T and target reliability index β. The results are plotted in Figure 9.10. It is evident that the required reinforcing force T decreases with decreasing values in mean z_w/z ratio, β, and slope angle Ψ_f. For the Hong Kong slope of Figure 9.9, the reliability index is $\beta = 1.89$ if unreinforced (i.e., T = 0). A target β of 2.5 can be achieved with a T force of 146 tonne/m. The same target β of 2.5 (or 3.0 if desired) can be achieved by various combinations of T, Ψ_f and mean z_w/z ratio if these stabilization methods are implemented in combination, as shown in the plots.

9.5 RELIABILITY ANALYSIS OF A FAILED SLOPE IN A LIMESTONE QUARRY

A failed slope of height 30.5 m (Figure 9.11) in a limestone quarry was back-analyzed in Wyllie (2018, p. 137), to obtain the values of shear strength parameters of the discontinuity plane that dips at an angle $\Psi_p = 20°$, followed by probabilistic analysis using Monte Carlo simulation. The desirability of using probability distributions with bounded lower and upper limits was aptly suggested. The random variables were the shear strength

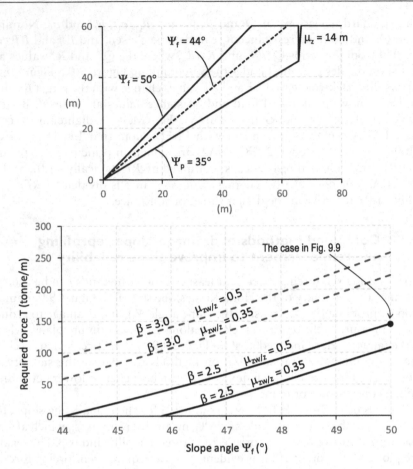

Figure 9.10 Effect of slope angle Ψ_f, reinforcing force T and mean value of z_w/z ratio ($\mu_{zw/z}$) on the reliability of the Sau Mau Ping slope of Hong Kong.

parameters ϕ and c of the discontinuity plane, and the ratio z_w/z where z is the height of the vertical tension crack and z_w is the height of water in the tension crack. Each of the three variables was estimated to have a minimum and a maximum value, and either a mean value or a most likely value, as described below (Wyllie 2018, p. 211):

(1) *Friction angle* ϕ: maximum and minimum values of 15° and 25°, respectively; the mean value was 19° and the standard deviation was 2.3°. This parameter was modelled by a skewed beta distribution with the most likely value of 18°.

(2) *Cohesion c*: most likely value of 90 kPa, and minimum and maximum values of 80 and 125 kPa, respectively; the mean value was 98 kPa. This parameter was modelled by a skewed triangular distribution.

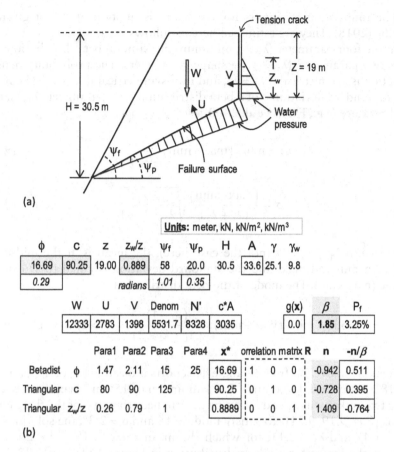

(a)

Units: meter, kN, kN/m², kN/m³

φ	c	z	zw/z	ψf	ψp	H	A	γ	γw
16.69	90.25	19.00	0.889	58	20.0	30.5	33.6	25.1	9.8
0.29			radians	1.01	0.35				

W	U	V	Denom	N'	c*A		g(x)	β	Pf
12333	2783	1398	5531.7	8328	3035		0.0	1.85	3.25%

		Para1	Para2	Para3	Para4	x*	orrelation matrix R			n	-n/β
Betadist	φ	1.47	2.11	15	25	16.69	1	0	0	-0.942	0.511
Triangular	c	80	90	125		90.25	0	1	0	-0.728	0.395
Triangular	zw/z	0.26	0.79	1		0.8889	0	0	1	1.409	-0.764

(b)

Figure 9.11 (a) A slope that failed in a limestone quarry (after Wyllie 2018), (b) reliability analysis indicates failure probability on the high side.

(3) *Water pressure*: expressed as percent filling of the 19 m deep tension crack, z_w/z, ranging from 5 m ($z_w/z = 26\%$) to full ($z_w/z = 100\%$), with the most likely value being 15 m (80%); the mean depth was 13 m (68%). This parameter was modelled by a skewed triangular distribution.

Probabilistic analysis will be done here using the Low and Tang (2007) FORM procedure for comparison with the Monte Carlo simulation result from Wyllie (2018). The three parameters defining the triangular distribution are minimum, mode and maximum. For cohesion c, as described in item (2) above, the inputs are '80, 90, 125.' For the water pressure coefficient z_w/z, as described in item (3) above, the inputs are '0.26, 0.79, 1.0,' where 0.79 is based on 15 m/19 m.

The four-parameter inputs for the beta distribution were not given in Wyllie (2018). They are estimated here as follows.

In the four-parameter (λ_1, λ_2, minimum, maximum) beta distribution, the first two parameters (λ_1, λ_2) are shape parameters. The probability density function is symmetrical if $\lambda_1 = \lambda_2$, and non-symmetrical if $\lambda_1 \neq \lambda_2$. The mean μ and standard deviation σ of a beta distribution with parameters λ_1, λ_2, min and max are (e.g., Evans et al. 2000):

$$\mu = \min + (\max - \min) \times \frac{\lambda_1}{\lambda_1 + \lambda_2} \tag{9.19}$$

$$\sigma = \frac{(\max - \min)}{(\lambda_1 + \lambda_2)} \sqrt{\frac{\lambda_1 \lambda_2}{\lambda_1 + \lambda_2 + 1}} \tag{9.20}$$

Hence, if $\lambda_1 = \lambda_2 = 4$ (not the case here), the mean is at the mid-point between min and max, and the standard deviation is equal to 1/6 of the range (max–min). The mode of the beta distribution is:

$$\text{mode} = \min + (\max - \min) \times \frac{(\lambda_1 - 1)}{(\lambda_1 + \lambda_2 - 2)}; \quad \lambda_1 > 1, \; \lambda_2 > 1 \tag{9.21}$$

Item (1) above, on friction angle ϕ, reports a mean value of 19°, a mode of 18°, a standard deviation of 2.3°, minimum of 15° and maximum of 25°. The two unknowns λ_1 and λ_2 can be determined from two of the three equations (Eqs. 9.19–9.21). To satisfy mode = 18 and $\sigma = 2.3°$, the solutions are $\lambda_1 = 1.47$ and $\lambda_2 = 2.11$, for which the mean value is 19.1° by Eq. 9.19. Hence, the inputs for the beta distribution in Figure 9.11b are '1.47, 2.11, 15°, 25°.'

Figure 9.11b shows the results of FORM analysis, obtaining a reliability index β of 1.85, and a failure probability of $P_f = \Phi(-\beta) = 3.25\%$, compared with a failure probability of 3.4% reported in Wyllie (2018) based on Monte Carlo simulation. The sensitivity indicators (n, or –n/β) in the last two columns of Figure 9.11b indicate that stability is most sensitive to z_w/z, followed by friction angle ϕ and cohesion c of the discontinuity plane.

Instead of triangular distributions, PERT distributions can be used for cohesion c and z_w/z, as shown in Figure 9.12, with the same inputs of 'minimum, most likely, maximum' as the triangular distributions in Figure 9.11b. A negative correlation between friction angle ϕ and cohesion c of the discontinuity plane is modelled. The reliability index is 1.69, and failure probability is 4.56%.

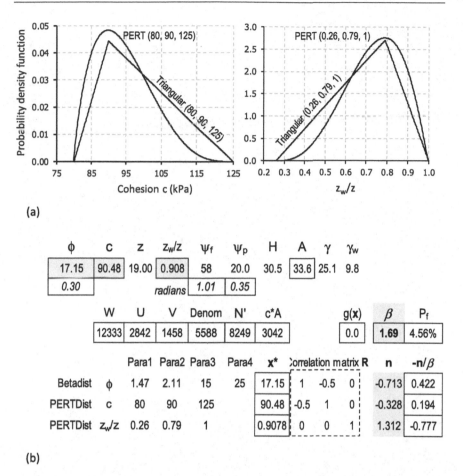

(a)

(b)

Figure 9.12 (a) PERT distribution can be considered for a variable with a most likely value (mode) and lower and upper limits, (b) reliability analysis with PERT distributions for cohesion c and z_w/z, and negatively correlated ϕ and c.

The computed reliability index of β = 1.85 (or 1.69 in Figure 9.12) is lower than the normal target value of β = 2.5 (for P_f = 0.62%), or β = 3.0 (for P_f = 0.13%). The inadequate level of reliability means that failure was not unlikely and did happen.

Chapter 10

Rock slopes with three-dimensional tetrahedral wedges

10.1 INTRODUCTION

The stability analysis of tetrahedral wedges in rock slopes involves resolution of forces in three-dimensional space. The problem has been extensively treated, for example, in Goodman and Taylor (1967), John (1968), Londe et al. (1969), Hendron et al. (1971), Jaeger (1971), Hoek et al. (1973), Goodman (1995), Hoek and Bray (1981) and Wittke (1990). The methods used include stereographic projection technique, engineering graphics and vector analysis. These methods are versatile and able to analyze cases with involved geometry, tension cracks and external forces.

A closed form dimensionless equation for the factor of safety is presented below for rock slopes with 3D wedge mechanisms, based on Low and Einstein (1992) and Low (1997), and verified with the Hoek and Bray (1981) vectorial procedures, prior to probabilistic stability analysis using FORM. It will be shown that the same uncertain entity, for example, friction angle, or water pressure, exhibits different sensitivities on discontinuity planes of different steepness. Also demonstrated is a case with a more critical sliding-on-single-plane mode, even though a safer sliding-on-both-planes mode is indicated by mean values.

The dimensionless factor of safety equation summarized in Figure 10.1 is the culmination of vectorial analysis. It is expressed in terms of intrinsic geometrical angles $\beta_1, \beta_2, \delta_1, \delta_2$. These geometrical angles are easily obtained from dip directions and dips of the two discontinuity planes and the slope face. A merit of the closed form solution is that it enables interpretation of the physical meanings underlying the closed form equation, which may not be obvious in vectorial procedures.

The dip direction of the upper ground surface in Figure 10.1 is assumed to coincide with the dip direction of the slope face, so that the line TX is horizontal. The orientations of the two joint planes B'OD and B'OE are defined by $\beta_1, \delta_1, \beta_2$ and δ_2, where β_1 and β_2 are measured on the horizontal plane STXR, and δ_1 and δ_2 are measured downwards from the horizontal triangle BDE.

DOI: 10.1201/9781003112297-13

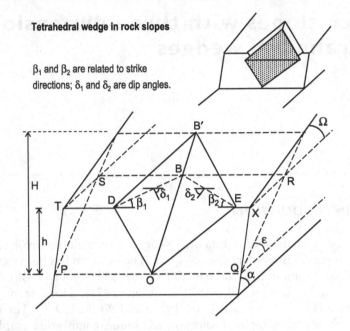

Tetrahedral wedge in rock slopes

β_1 and β_2 are related to strike directions; δ_1 and δ_2 are dip angles.

Sliding on both planes, one of the three modes:

$$F_s = \frac{\text{Resistance}}{\text{Sliding force}} = \frac{(N_1 - U_1)\tan\phi_1 + (N_2 - U_2)\tan\phi_2 + c_1 A_1 + c_2 A_2}{\text{Sliding force along B'O}}$$

Dimensionless equations below embody completed vectorial operations:

$$F_s = \left(a_1 - \frac{b_1 G_{w1}}{s_\gamma}\right) \times \tan\phi_1 + \left(a_2 - \frac{b_2 G_{w2}}{s_\gamma}\right) \times \tan\phi_2 + 3b_1 \frac{c_1}{\gamma h} + 3b_2 \frac{c_2}{\gamma h}$$

in which:

$$a_1 = f(\beta_1 + \beta_2, \delta_1, \delta_2)$$
$$a_2 = f(\beta_1 + \beta_2, \delta_1, \delta_2)$$
$$b_1 = f(\beta_1, \beta_2, \delta_1, \delta_2, \alpha)$$
$$b_2 = f(\beta_1, \beta_2, \delta_1, \delta_2, \alpha)$$
$$G_{w1}, G_{w2} = f(\beta_1, \beta_2, \delta_1, \delta_2, \alpha, \Omega)$$

for pyramidal water pressure distribution along B'O

or $G_{w1}, G_{w2} = \dfrac{3u_1}{\gamma_w h}$ and $\dfrac{3u_2}{\gamma_w h}$

based on average water pressures u_1 and u_2.

Figure 10.1 Closed form dimensionless equation for the factor of safety of tetrahedral wedge in rock slopes. Full equations given in Figures 10.3 and 10.4.

10.2 OBTAINING $\beta_1, \delta_1, \beta_2$ AND δ_2 ANGLES FROM DIP DIRECTIONS AND DIPS

In practice joint orientations are often given in terms of dip direction and dip values. The simple procedure shown in Figure 10.2 can be used to obtain the β and δ angles from given dip and dip direction values, so that the factor of safety can be computed using the equations given here which are in terms of the intrinsic $\beta_1, \delta_1, \beta_2$ and δ_2 angles. The simple transformation procedure in Figure 10.2 involves only sketching three lines (in strike and dip notation) of the two joint planes and the slope face, to form the *horizontal triangle BDE* of Figure 10.1, such that the dip symbol of the slope face (the short line perpendicular to the strike line) is outside the triangle BDE when the slope face is not overhanging its toe. The β_1 and β_2 angles are then obtained easily from the differences between the strike directions. The δ_1 and δ_2 are dip angles if their dip symbols are inside the triangle BDE. If one dip symbol of the two joint planes is outside the triangle BDE, then its δ angle is 180° – dip. The simple steps are explained in Figure 10.2, with an example.

10.3 KINEMATIC REQUIREMENT FOR FORMATION OF WEDGE MECHANISM IN ROCK SLOPES

SPQR in Figure 10.1 is a plane that contains the line of intersection B'O and dips at an angle ε (to be computed below) towards the slope face. PQ is parallel to TX; both are horizontal lines. Angle α is the known dip of the slope face; angle Ω is the known dip of the upper ground surface.

H is the vertical distance between B' and O, and h the vertical distance between the horizontal triangle BDE and O. A simple equation relates H and h, so that if one is known the other can be calculated.

A tetrahedral wedge will be formed if the line of intersection of the two joint planes daylights both on the slope (point O in Figure 10.1) and the upper ground surface (point B'). This means that for an upper ground surface dipping at Ω and a slope face dipping at α, the inclination ε of the plane SPQR has to be greater than Ω and less than α. It can be shown (Low 1997) that:

$$\tan\varepsilon = \frac{\sin(\beta_1 + \beta_2)}{\sin\beta_1 \cot\delta_2 + \sin\beta_2 \cot\delta_1} \tag{10.1}$$

so that the kinematic requirement becomes

$$\tan\Omega < \tan\varepsilon < \tan\alpha \tag{10.2}$$

An alternative method of checking this daylighting requirement is to compare the stereographic projection of the line of intersection with the stereographic projections of the slope face and upper ground surface, Hoek and Bray (1981). The same conclusion will be reached whether Eqs. 10.1 and 10.2 are used, or the stereographic projection method.

10.4 LIMIT EQUILIBRIUM EQUATIONS FOR THE FACTOR OF SAFETY OF WEDGE IN ROCK SLOPE

10.4.1 Sliding along both planes, that is, along the line of intersection

Whenever the kinematic requirement (Eq. 10.2) is satisfied, the factor of safety of the tetrahedral wedge for sliding along both planes can be expressed as follows (Hendron et al. 1971; Hoek and Bray 1981):

$$F_s = \frac{\text{Resistance}}{\text{Driving force}} = \frac{(N_1 - U_1)\tan\phi_1 + (N_2 - U_2)\tan\phi_2 + c_1 A_1 + c_2 A_2}{T_{12}} \quad (10.3)$$

Freehand sketch to obtain the intrinsic β_1, β_2, δ_1 and δ_2 angles:

(1) Draw 3 lines for the two joints and the slope face, using strike and dip symbols. (Strikes are perpendicular to dip directions.)

(2) Form a triangle BDE by shifting parallelly the strike line of the slope face, such that its dip symbol is outside BDE.

(3) Compute angles β_1 and β_2 from the strike directions.

(4) δ = dip, if dip symbol is inside triangle BDE;

   δ = 180 - dip, if dip symbol is outside triangle BDE.

   (At most only one dip symbol can be outside BDE)

Example:

	Dip Dir. / Dip
Plane 1	105 / 45
Plane 2	235 / 70
Slope face	185 / 65

$\beta_1 = 275 - 195 = 80°$

$\beta_2 = 145 - 095 = 50°$

$\delta_1 = 45°$

$\delta_2 = 70°$

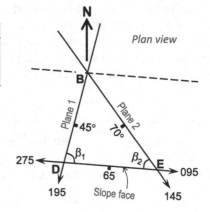

Figure 10.2 Simple freehand sketch to transform dip directions and dip angles to the intrinsic angles of β_1, β_2, δ_1 and δ_2. BDE is the horizontal triangle in Figure 10.1.

where

N = Component of gravitational force, normal to discontinuity plane;

ϕ = Friction angle of discontinuity plane;

c = Cohesion of discontinuity plane;

A = Triangular area of discontinuity plane;

T_{12} = Driving force along the line of intersection B'O;

(Subscripts 1 and 2 identify the two discontinuity planes).

The components in the numerator and denominator of Eq. 10.3 can be obtained from a series of vectorial cròss products and dot products, as explained in Hoek and Bray (1981). Alternatively, Low (1997) presented the following dimensionless F_s expressions which embody completed vectorial operations and trigonometry contractions in terms of the intrinsic angles β_1, δ_1, β_2 and δ_2 angles, which are obtainable from dip directions and dip angles (Figure 10.2), and known inclination angles α and Ω of the slope face and the upper ground surface, respectively:

$$F_s = \left(a_1 - \frac{b_1 G_{w1}}{s_\gamma} \right) \times \tan\phi_1 + \left(a_2 - \frac{b_2 G_{w2}}{s_\gamma} \right) \times \tan\phi_2 + 3b_1 \frac{c_1}{\gamma h} + 3b_2 \frac{c_2}{\gamma h} \qquad (10.4)$$

where γ = density of rock, and c, ϕ and h are as defined previously in connection with Figure 10.1.

In Eq. 10.4, the dimensionless coefficients a_1, a_2, b_1 and b_2 depend only on the orientation angles β_1, δ_1, β_2 and δ_2 of the discontinuity planes and the inclination (α) of the slope face.

Equation 10.4 and its underlying component expressions for sliding along both planes are shown explicitly in Figure 10.3. These equations embody vectorial operations and trigonometry contractions. Vectorial dot products and cross products are no longer required when using the equations.

The factor of safety as given by Eq. 10.4 assumes that the tetrahedral wedge will be in contact with both joint planes B'OD and B'OE (Figure 10.3); the equation is valid only if the parenthesized terms preceding $\tan\phi_1$ and $\tan\phi_2$ are both positive. When these conditions are not satisfied, the following modes of sliding have to be considered: (i) sliding along plane 1 only, (ii) sliding along plane 2 only and (iii) floating failure due to loss of contact on both planes. These three modes are summarized in Figure 10.4, and explained below.

10.4.2 Sliding along plane 1 only

In this case, the net normal force on plane 2 is an uplifting force, which can be resolved into two components, perpendicular and tangential to plane 1, respectively. The perpendicular component is superimposed on the net normal force on plane 1. The tangential component is added vectorially to the driving force along the line of intersection of the two joint planes, thereby

Factor of safety against sliding on both planes:

(a) $F_s = \left(a_1 - \dfrac{b_1 G_{w1}}{s_\gamma}\right) \times \tan\phi_1 + \left(a_2 - \dfrac{b_2 G_{w2}}{s_\gamma}\right) \times \tan\phi_2 + 3b_1\dfrac{c_1}{\gamma h} + 3b_2\dfrac{c_2}{\gamma h}$

(b) $a_1 = \dfrac{\left[\sin\delta_2 \cot\delta_1 - \cos\delta_2 \cos(\beta_1 + \beta_2)\right]}{\sin\psi \sin(\beta_1 + \beta_2)}$

$\dfrac{b_2 G_{w2}}{s_\gamma} = \dfrac{3b_2 u_2}{\gamma h}$ $\dfrac{b_1 G_{w1}}{s_\gamma} = \dfrac{3b_1 u_1}{\gamma h}$

Water pressure input can be G_w or u

(c) $a_2 = \dfrac{\left[\sin\delta_1 \cot\delta_2 - \cos\delta_1 \cos(\beta_1 + \beta_2)\right]}{\sin\psi \sin(\beta_1 + \beta_2)}$

(d) $b_1 = a_0 \sin\beta_2 \sin\delta_2$

(e) $b_2 = a_0 \sin\beta_1 \sin\delta_1$

(f) $s_\gamma = \gamma_{rock}/\gamma_{water}$

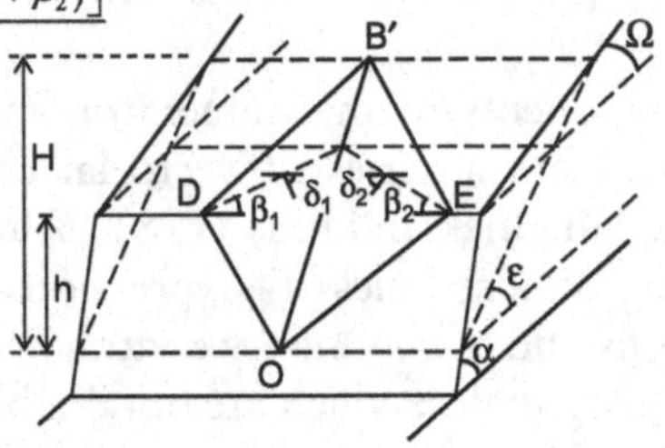

in which:

(g) $\sin\psi = \left|\sqrt{\left\{1 - \left[\sin\delta_1 \sin\delta_2 \cos(\beta_1 + \beta_2) + \cos\delta_1 \cos\delta_2\right]^2\right\}}\right|$

(h) $a_0 = \dfrac{\sin\psi}{\left[\sin(\beta_1 + \beta_2)\sin\delta_1 \sin\delta_2\right]^2 (\cot\varepsilon - \cot\alpha)}$

(i) $\tan\varepsilon = \dfrac{\sin(\beta_1 + \beta_2)}{\sin\beta_1 \cot\delta_2 + \sin\beta_2 \cot\delta_1}$

(j) $G_{w1} = G_{w2} = 0.5\kappa$, for pyramidal water pressure distribution.

(k) $G_{w1} = 3u_1/\gamma_w h$, $G_{w2} = 3u_2/\gamma_w h$ for water pressure u_1 and u_2

(l) $\kappa = \dfrac{H}{h} = \left(1 - \dfrac{\tan\Omega}{\tan\alpha}\right)\bigg/\left(1 - \dfrac{\tan\Omega}{\tan\varepsilon}\right)$

(m) Kinematic requirement for wedge: $\tan\Omega < \tan\varepsilon < \tan\alpha$

(n) Of interest: $h = \dfrac{DE}{(\cot\varepsilon - \cot\alpha) \times (\cot\beta_1 + \cot\beta_2)}$

(o) Of interest: $\dfrac{V}{h^3} = \dfrac{\kappa}{6} \times (\cot\beta_1 + \cot\beta_2) \times (\cot\varepsilon - \cot\alpha)^2$

(p) $A1 = \dfrac{1}{2}\kappa h^2 \dfrac{(\cot\varepsilon - \cot\alpha)}{\sin\beta_1 \sin\delta_1}$ **(q)** $A2 = \dfrac{1}{2}\kappa h^2 \dfrac{(\cot\varepsilon - \cot\alpha)}{\sin\beta_2 \sin\delta_2}$

Figure 10.3 Dimensionless closed-form equation for sliding along the line of intersection B'O, i.e., along both discontinuity planes (Low, 1997).

deviating the resultant driving force from the line of intersection. The following dimensionless equation is obtained:

$$F_s(Plane1) = \dfrac{\left[\left(a_1 - \dfrac{b_1 G_{w1}}{s_\gamma}\right) - \left(\dfrac{b_2 G_{w2}}{s_\gamma} - a_2\right)Z\right]\tan\phi_1 + 3b_1\dfrac{c_1}{\gamma h}}{\sqrt{1 + \left[\left(\dfrac{b_2 G_{w2}}{s_\gamma} - a_2\right)\sin\psi\right]^2}} \qquad (10.5a)$$

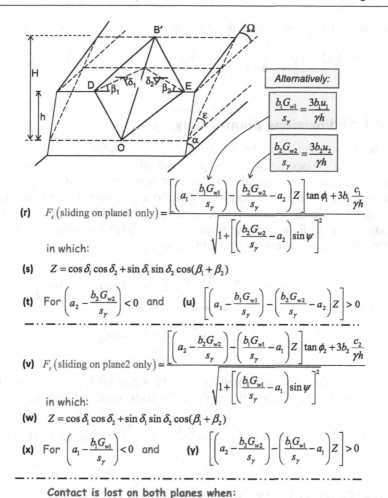

(r) F_s (sliding on plane1 only) $= \dfrac{\left[\left(a_1 - \dfrac{b_1 G_{w1}}{s_\gamma}\right) - \left(\dfrac{b_2 G_{w2}}{s_\gamma} - a_2\right)Z\right]\tan\phi_1 + 3b_1\dfrac{c_1}{\gamma h}}{\sqrt{1 + \left[\left(\dfrac{b_2 G_{w2}}{s_\gamma} - a_2\right)\sin\psi\right]^2}}$

in which:

(s) $Z = \cos\delta_1\cos\delta_2 + \sin\delta_1\sin\delta_2\cos(\beta_1 + \beta_2)$

(t) For $\left(a_2 - \dfrac{b_2 G_{w2}}{s_\gamma}\right) < 0$ and $\quad$ **(u)** $\left[\left(a_1 - \dfrac{b_1 G_{w1}}{s_\gamma}\right) - \left(\dfrac{b_2 G_{w2}}{s_\gamma} - a_2\right)Z\right] > 0$

(v) F_s (sliding on plane2 only) $= \dfrac{\left[\left(a_2 - \dfrac{b_2 G_{w2}}{s_\gamma}\right) - \left(\dfrac{b_1 G_{w1}}{s_\gamma} - a_1\right)Z\right]\tan\phi_2 + 3b_2\dfrac{c_2}{\gamma h}}{\sqrt{1 + \left[\left(\dfrac{b_1 G_{w1}}{s_\gamma} - a_1\right)\sin\psi\right]^2}}$

in which:

(w) $Z = \cos\delta_1\cos\delta_2 + \sin\delta_1\sin\delta_2\cos(\beta_1 + \beta_2)$

(x) For $\left(a_1 - \dfrac{b_1 G_{w1}}{s_\gamma}\right) < 0$ and $\quad$ **(y)** $\left[\left(a_2 - \dfrac{b_2 G_{w2}}{s_\gamma}\right) - \left(\dfrac{b_1 G_{w1}}{s_\gamma} - a_1\right)Z\right] > 0$

Contact is lost on both planes when:

(z1) $\left[\left(a_1 - \dfrac{b_1 G_{w1}}{s_\gamma}\right) - \left(\dfrac{b_2 G_{w2}}{s_\gamma} - a_2\right)Z\right] < 0$ & **(z2)** $\left[\left(a_2 - \dfrac{b_2 G_{w2}}{s_\gamma}\right) - \left(\dfrac{b_1 G_{w1}}{s_\gamma} - a_1\right)Z\right] < 0$

Figure 10.4 Dimensionless closed-form equation for sliding along discontinuity plane 1 or plane 2 only (Low, 1997).

in which:

$$Z = \cos\delta_1\cos\delta_2 + \sin\delta_1\sin\delta_2\cos(\beta_1 + \beta_2) \qquad (10.5b)$$

All other symbols in Eqs. 10.5a and 10.5b are as defined in connection with Eq. 10.4 and Figure 10.3.

Equation 10.5a for sliding along plane 1 is to be used only when:

$$\left(a_2 - \frac{b_2 G_{w2}}{s_\gamma}\right) < 0 \quad \text{and} \quad \left[\left(a_1 - \frac{b_1 G_{w1}}{s_\gamma}\right) - \left(\frac{b_2 G_{w2}}{s_\gamma} - a_2\right) Z\right] > 0 \quad (10.5c)$$

10.4.3 Sliding along plane 2 only

In this case, the net normal force on plane 1 is an uplifting force, which can be resolved into two components, perpendicular and tangential to plane 2, respectively. The perpendicular component is superimposed on the net normal force on plane 2. The tangential component is added vectorially to the driving force along the line of intersection of the two joint planes, thereby deviating the resultant driving force from the line of intersection. The following dimensionless equation is obtained:

$$F_s(Plane2) = \frac{\left[\left(a_2 - \frac{b_2 G_{w2}}{s_\gamma}\right) - \left(\frac{b_1 G_{w1}}{s_\gamma} - a_1\right) Z\right]\tan\phi_2 + 3b_2 \frac{c_2}{\gamma h}}{\sqrt{1 + \left[\left(\frac{b_1 G_{w1}}{s_\gamma} - a_1\right)\sin\psi\right]^2}} \quad (10.6a)$$

where all symbols are as defined for Eqs. 10.4 and 10.5a, and Figure 10.3. Equation 16a for sliding along plane 2 is to be used only when:

$$\left(a_1 - \frac{b_1 G_{w1}}{s_\gamma}\right) < 0 \quad \text{and} \quad \left[\left(a_2 - \frac{b_2 G_{w2}}{s_\gamma}\right) - \left(\frac{b_1 G_{w1}}{s_\gamma} - a_1\right) Z\right] > 0 \quad (10.6b)$$

10.4.4 Contact is lost on both planes

This means that the wedge floats as a result of water pressure acting on planes 1 and 2. In this case, the factor of safety falls to zero. This scenario is realized when:

$$\left[\left(a_1 - \frac{b_1 G_{w1}}{s_\gamma}\right) - \left(\frac{b_2 G_{w2}}{s_\gamma} - a_2\right) Z\right] < 0 \quad (10.7a)$$

and

$$\left[\left(a_2 - \frac{b_2 G_{w2}}{s_\gamma}\right) - \left(\frac{b_1 G_{w1}}{s_\gamma} - a_1\right) Z\right] < 0 \quad (10.7b)$$

In summary, Eqs. 10.4, 10.5, 10.6, 10.7 calculate, respectively, the factor of safety against wedge sliding on both planes (hereafter referred to as the

BiPlane mode), along plane 1 only (*Plane1* mode), along plane 2 only (*Plane2* mode), and failure of wedge by floating on both planes, respectively. Also, when the kinetic requirement of Eqs. (m) and (i), Figure 10.3, is not satisfied, tetrahedral wedge mechanism is not possible and the slope is safe as far as tetrahedral wedge sliding is concerned.

Note that h (not H), as defined in Figure 10.1, appears in Eqs. 10.4–10.6. If H is known, h can be obtained from H/κ, where κ is given by Eqs. (l) and (i) in Figure 10.3. If only the width DE of Figure 10.1 is known, h can be obtained from Eq. (n) in Figure 10.3. Equations (o), (p) and (q), for volume and areas of planes 1 and 2, are not required for computation of factors of safety using Eqs. (a)–(l) for *BiPlane* mode, Eqs. (r)–(t) for *Plane1* mode, Eqs. (v)–(x) for *Plane2* mode, or Eqs. (z1)–(z2) for *floating* mode, but they are useful for assessing the consequence of failure based on volume, and for understanding the sensitivities of cohesion parameters c_1 and c_2, and water pressures u_1 and u_2, which could be more sensitive when acting on larger areas.

10.4.5 Water pressure coefficients G_{w1} and G_{w2}, and relationship with water pressures u_1 and u_2

Equation (j) in Figure 10.3 for the dimensionless parameters G_{w1} and G_{w2} is based on pyramidal water pressure conditions:

$$G_{w1} = G_{w2} = 0.5\kappa, \quad \text{for pyramidal water pressure distribution. (10.8)}$$

in which κ is given by Eqs. (*l*) and (i) of Figure 10.3.

This means that within the triangular joint planes B′OD and B′OE (Figure 10.1), the water pressure distribution is assumed to be represented by pyramids with B′OD and B′OE as bases and with apices at a height of $0.5\gamma_wH$ above the mid-point of B′O; water pressure is zero along the edges B′D, DO, OE, EB′.

Alternatively, one may wish to assign average water pressure u_1 and u_2 acting on planes B′OD and B′OE, respectively. The factor of safety is in this case given by Eqs. 10.4–10.7 with the following substitutions (Eq. k in Figure 10.3):

$$G_{w1} = \frac{3u_1}{\gamma_w h} \tag{10.9a}$$

$$G_{w2} = \frac{3u_2}{\gamma_w h} \tag{10.9b}$$

This will change Eq. 10.4 (for *BiPlane* mode) to the following:

$$F_s = \left(a_1 - \frac{3b_1 u_1}{\gamma h} \right) \times \tan\phi_1 + \left(a_2 - \frac{3b_2 u_2}{\gamma h} \right) \times \tan\phi_2 + 3b_1 \frac{c_1}{\gamma h} + 3b_2 \frac{c_2}{\gamma h} \tag{10.10}$$

in which γ is the unit weight of rock, and similar replacement of the $b_i G_w/s_\gamma$ terms for the other three failure modes with the terms $3b_i u_i/\gamma h$.

For the water condition represented by Eq. 10.8, the equivalent average water pressures are:

$$u_1 = u_2 = \frac{1}{6}\kappa\gamma_w h, \quad \text{for pyramidal water pressure distribution.} \quad (10.11)$$

where κ is given by Eqs. (l) and (i) of Figure 10.3.

The four failure modes of Eqs. 10.4–10.7 are the same as those considered in the vectorial procedure of Hoek and Bray (1981, Appendix 2), except that Eqs. 10.4–10.7 (and component equations Figures 10.3 and 10.4) are based on intrinsic angles β_1, δ_1, β_2 and δ_2, whereas the vectorial procedure in Hoek and Bray (1981) are in terms of dip directions (α) and dip angles (ψ). As illustrated in Figure 10.2, the intrinsic angles β_1, δ_1, β_2 and δ_2 can be obtained easily from dip directions and dip angles using a simple freehand sketch.

10.5 COMPARISON WITH STEREOGRAPHIC METHOD

Hoek and Bray (1981) presented a procedure for tetrahedral wedge analysis which can account for the effects of cohesion, water pressure and obliquely inclined upper ground surface. The factor of safety was calculated from the following equation:

$$F = \frac{3}{\gamma H}\left(c_A.X + c_B.Y\right) + \left(A - \frac{\gamma_w}{2\gamma}.X\right)\tan\phi_A + \left(B - \frac{\gamma_w}{2\gamma}.Y\right)\tan\phi_B \quad (10.12)$$

in which X, Y, A and B are given by in Hoek and Bray (1981, Eqs. 92–95). They are dimensionless factors that depend on the geometry of the wedge and the slope. The eight angles (Ψ_5, $\theta_{na,nb}$, θ_{24}, θ_{45}, $\theta_{2,na}$, θ_{13}, θ_{35} and $\theta_{1,nb}$) required for calculating the coefficients X, Y, A and B were measured on a stereoplot. An example is shown in Figure 10.5a, from Wyllie (2018), which presented the Hoek-Bray solution in S.I. units.

The solution to the above problem, using the Low (1997) closed form equations shown in Figure 10.3, is shown in Figure 10.5b. It is first necessary to convert the dip direction and dip values to the intrinsic β and δ angles, using the simple freehand sketch procedure shown in Figure 10.2. The values (Figure 10.2) are $\beta_1 = 80°$, $\beta_2 = 50°$, $\delta_1 = 45°$ and $\delta_2 = 70°$, shown at the top left of Figure 10.5b. The value $h = 30.55$ m was obtained from $h = H/\kappa$, where $H = 40$ m, and $\kappa = 1.309$ from Eq. (l) in Figure 10.3. The values of G_{w1} and G_{w2} were calculated as 0.5κ, for pyramidal water pressure

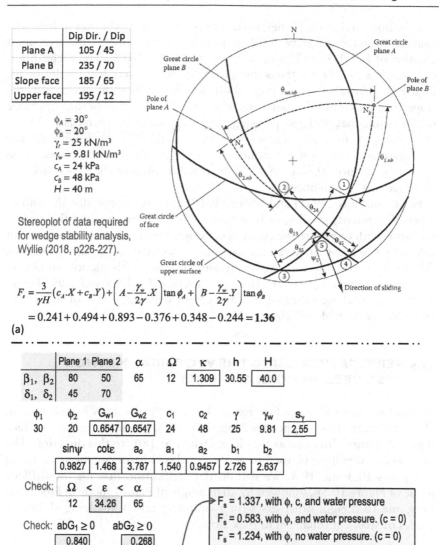

	Dip Dir. / Dip
Plane A	105 / 45
Plane B	235 / 70
Slope face	185 / 65
Upper face	195 / 12

$\phi_A = 30°$
$\phi_B = 20°$
$\gamma_r = 25$ kN/m³
$\gamma_w = 9.81$ kN/m³
$c_A = 24$ kPa
$c_B = 48$ kPa
$H = 40$ m

Stereoplot of data required
for wedge stability analysis,
Wyllie (2018, p226-227).

$$F_s = \frac{3}{\gamma H}(c_A.X + c_B.Y) + \left(A - \frac{\gamma_w}{2\gamma}.X\right)\tan\phi_A + \left(B - \frac{\gamma_w}{2\gamma}.Y\right)\tan\phi_B$$

$$= 0.241 + 0.494 + 0.893 - 0.376 + 0.348 - 0.244 = \mathbf{1.36}$$

(a)

	Plane 1	Plane 2	α	Ω	κ	h	H
β_1, β_2	80	50	65	12	1.309	30.55	40.0
δ_1, δ_2	45	70					

ϕ_1	ϕ_2	G_{w1}	G_{w2}	c_1	c_2	γ	γ_w	s_γ
30	20	0.6547	0.6547	24	48	25	9.81	2.55

sinψ	cotε	a_0	a_1	a_2	b_1	b_2
0.9827	1.468	3.787	1.540	0.9457	2.726	2.637

Check:
Ω	<	ε	<	α
12		34.26		65

Check: abG₁ ≥ 0   abG₂ ≥ 0

0.840	0.268

F_s = 1.337, with ϕ, c, and water pressure
F_s = 0.583, with ϕ, and water pressure. (c = 0)
F_s = 1.234, with ϕ, no water pressure. (c = 0)
F_s = 1.988, with ϕ, c, no water pressure.

abG₁

F_s = 1.337

$$F_s = \left(a_1 - \frac{b_1 G_{w1}}{s_\gamma}\right) \times \tan\phi_1 + \left(a_2 - \frac{b_2 G_{w2}}{s_\gamma}\right) \times \tan\phi_2 + 3b_1\frac{c_1}{\gamma h} + 3b_2\frac{c_2}{\gamma h}$$

(b)

Figure 10.5 (a) An example from Hoek and Bray (1981) and Wyllie (2018), (b)
solution using closed form equation based on intrinsic angles β_1, β_2,
δ_1 and δ_2 from Figure 10.2.

distribution on the two planes, with maximum at the mid-point of B'O. The calculated factor of safety (1.34) is practically the same as the stereoplot solution of $F_s = 1.36$ in Figure 10.5a, from Wyllie (2018). The slight difference is due mainly to the assumption, in the closed form solution of Figure 10.5b, that the upper ground surface dips in the same direction as the slope face. This is not the case here, where the dip direction of the upper face is 195° while that of the slope face is 185°. The eight angles measured from the stereoplot may also cause some very slight difference between the stereoplot-and-equation procedure of Figure 10.5a and the closed form solution of Figure 10.5b even when the dip directions of the upper face and the slope face are identical.

Hoek and Bray (1981) and Wyllie (2018) aptly recommended the importance of sensitivity studies, and for the case in hand investigated three other scenarios with respect to the effect of cohesion and drainage on the factor of safety. Figure 10.5b analyzed the same three scenarios of (i) zero cohesion, (ii) zero cohesion and zero water pressure and (iii) with friction and cohesion, but zero water pressure. The F_s values of 0.58, 1.23 and 1.99 (Figure 10.5b, text box) are in reasonable agreement with the values of 0.62, 1.24 and 1.98 mentioned in Wyllie (2018).

10.6 VERIFICATION USING THE VECTORIAL METHOD DESCRIBED IN HOEK AND BRAY

A versatile vectorial procedure for wedge stability analysis is described in Hoek and Bray (1981, Appendix 2, pp. 337–341). The upper ground surface dips in the same direction as the slope face or at 180° to this direction. The method can therefore be used to verify the closed form equations (a) to (z) of Figures 10.3 and 10.4. We will do the verification first for the *BiPlane* mode of Figure 10.5b, followed by an example of *Plane1* mode in Hoek and Bray, and then eight other cases (in a summary table) involving all failure modes of wedge mechanism in rock slopes.

10.6.1 Verification of an example with *BiPlane* failure mode

The symbol H in Hoek & Bray's vectorial method denotes the height to the crest of the slope face; it has the same meaning as the symbol h in Figure 10.1. Therefore, to verify the closed form equations of the case in Figure 10.5b, the input for Hoek and Bray's vectorial method is as given by the table at the top left of Figure 10.5a, except that H is 30.55 m, and the dip dir./dip of upper ground surface are 185/12, and water pressures u_1 and u_2 are both equal to 65.4 kPa (from Eq. 10.9). The computed factor of safety is 1.337, virtually identical to that in Figure 10.5b which is based on closed form solution.

10.6.2 Verification of an example with sliding failure along a single plane

Figure 10.6a shows an example from Hoek and Bray (1981, pp. 340–341), involving wedge failure by sliding along plane 1. The vectorial procedure computed a factor of safety $F_s = 0.626$, by sliding along plane 1 (and moving away from plane 2). If there is no water pressure, the factor of safety is 1.154 by vectorial procedure.

The solution using the closed form solution based on intrinsic angles β_1, δ_1, β_2 and δ_2 is shown in Figure 10.6b, including a sketch of the three strike lines of the slope face and the two planes for determining the β_1 and β_2 angles, same in principle to that in Figure 10.2. The G_{w1} and G_{w2} values of 0.4587 are calculated from the u_1 and u_2 values of 30 kN/m^2 using Eq. 10.9. The spreadsheet solution template automatically determines that the *Plane1* mode of Eq. 10.5 (i.e., Eqs. (r), (s) and (t) in Figure 10.4) should be used and computes a factor of safety for *Plane1* mode (FsP1) of 0.627, virtually identical to the value of 0.626 in Figure 10.6a.

If there is no water pressure ($G_{w1} = G_{w2} = 0$), Figure 10.6b will compute a factor of safety of 1.154, still a *Plane1* mode, identical in value to that mentioned in the vectorial procedure of Hoek and Bray (Figure 10.6a).

The closed form solution can show that when there is no water pressure, potential sliding mode of the wedge is *Plane1* when angle DBO of Figure 10.1 is greater than 90°, and *Plane2* when angle EBO is greater than 90°. Even when both angles DBO and EBO are acute, variations in joint orientations and in the water pressure parameter (G_{w1} and G_{w2}) can induce single-plane sliding mode based on the criteria stated in Eqs. (t) and (x) of Figure 10.4.

10.6.3 Verification for all four failure modes of wedge mechanism in rock slope

Figure 10.7 provides a summary of verification of the closed form solutions of Figure 10.3 and 10.4, using eight hypothetical cases. The original orientations of the two planes are given in dip directions and dip values in the first rows, for analysis using vectorial procedure, and corresponding intrinsic angles β_1, δ_1, β_2 and δ_2 in the second rows (obtained using freehand sketch as illustrated in Figure 10.2), for analysis using closed form equations. For all eight cases, the dip direction and dip are 145/65 for the slope face, and 145/10 for the upper ground surface. Other common properties for the eight cases are the height to the crest of the slope face (h) equal to 20 m, rock unit weight (γ) equal to 25 kN/m^3, water unit weight (γ_w) equal to 10 kN/m^3, fiction angle of plane 1 (ϕ_1) equal to 30°, fiction angle of plane 2 (ϕ_2) equal to 35°, cohesion of plane 1 (c_1) equal to 25 kN/m^2, cohesion of plane 2 (c_2) equal to 0. The average water pressure u is assumed to be the same on both planes, but different for the eight cases. The value of G_w in each of the

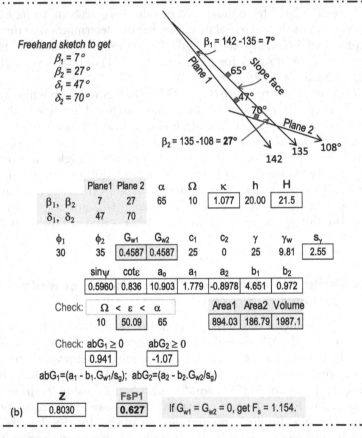

Hoek and Bray (1981, p340-341 example)

Calculate the factor of safety against wedge failure of a slope for which the following data applies

Plane	1	2	3	4
Dip dir.	052	018	045	045
Dip	47	70	10	65

$$\gamma = 25 \text{ kN/m}^3, \quad H = 20 \text{ m}, \quad c_1 = 25 \text{ kN/m}^2, \quad c_2 = 0$$
$$\phi_1 = 30°, \quad \phi_2 = 35°, \quad u_1 = u_2 = 30 \text{ kN/m}^2, \quad \eta = +1$$

Note: H in Hoek and Bray (1981) denotes height to the crest of slope face, not total height

Solution by Hoek and Bray, using a series of vector operations:
There is contact on plane 1 only. $F_s = 0.626$. Hence the slope is unstable.

(a)   If there is no water pressure, $F_s = 1.154$.

Freehand sketch to get
$$\beta_1 = 7°$$
$$\beta_2 = 27°$$
$$\delta_1 = 47°$$
$$\delta_2 = 70°$$

$\beta_1 = 142 - 135 = 7°$

$\beta_2 = 135 - 108 = 27°$

	Plane1	Plane 2	α	Ω	κ	h	H
β_1, β_2	7	27	65	10	1.077	20.00	21.5
δ_1, δ_2	47	70					

ϕ_1	ϕ_2	G_{w1}	G_{w2}	c_1	c_2	γ	γ_w	s_γ
30	35	0.4587	0.4587	25	0	25	9.81	2.55

$\sin\psi$	$\cot\varepsilon$	a_0	a_1	a_2	b_1	b_2
0.5960	0.836	10.903	1.779	-0.8978	4.651	0.972

Check:

$\Omega < \varepsilon < \alpha$			Area1	Area2	Volume
10	50.09	65	894.03	186.79	1987.1

Check: $abG_1 \geq 0$ $abG_2 \geq 0$

0.941	-1.07

$abG_1 = (a_1 - b_1 \cdot G_{w1}/s_g); \quad abG_2 = (a_2 - b_2 \cdot G_{w2}/s_g)$

z	FsP1
0.8030	0.627

(b)   If $G_{w1} = G_{w2} = 0$, get $F_s = 1.154$.

Figure 10.6 (a) An example from Hoek and Bray (1981) with sliding on plane 1 only, (b) solution using closed form equation based on intrinsic angles $\beta_1, \beta_2, \delta_1$ and δ_2.

For all cases:				(*In S.I. units*)	
α	Ω	h	γ	γ_w	
65	10	20	25	10	DipDir/Dip
ϕ_1	ϕ_2	c_1	c_2	Slope face	145/65
30	35	25	0	Upper Ground	145/10

Notes on Hoek and Bray's methods:
- For vectorial method, H = height to crest = h in Fig. 10.1
- For stereographic method (Fig. 10.5a), H = total height = H in Fig. 10.1

Vectorial	DipDir1	Dip1	DipDir2	Dip2	u	Mode	F_s
Equation	β_1	δ_1	β_2	δ_2	G_w	Mode	F_s
Case 1	152	47	118	70	30	Plane 1	0.6267
	7	47	27	70	0.450	Plane 1	0.6267
Case 2	034	23	357	80	53	Plane 1	2.171
	111	23	32	100	0.795	Plane 1	2.171
Case 3	042	33	174	72	33	BiPlane	1.603
	103	33	29	72	0.495	BiPlane	1.603
Case 4	069	37	186	50	58	BiPlane	0.9403
	76	37	41	50	0.870	BiPlane	0.9403
Case 5	240	70	163	20	48	Plane 2	1.172
	85	110	18	20	0.720	Plane 2	1.172
Case 6	266	79	185	28	35	Plane 2	0.7883
	59	101	40	28	0.525	Plane 2	0.7883
Case 7	094	58	201	76	47	Floats	0
	51	58	56	76	0.705	Floats	0
Case 8	115	88	202	78	49		
	30	88	57	78	0.735	NoWedge	

Figure 10.7 Verification of closed form equations using the vectorial procedure described in Hoek and Bray (1981), *involving sliding along plane 1 only, along plane 2 only, along both planes ('BiPlane'),* and loss of contact on both planes ('floats').

second rows is computed from the average water pressure u using Eq. 10.9. The values mentioned above are arbitrary for purpose of verification; other values can be used if desired.

The closed form equations (a) to (z) of Figures 10.3 and 10.4 are verified in Figure 10.7 by the virtual identity between the values of F_s calculated from closed form equations (in the second rows) and those from the vectorial procedure (in the first rows). In the Monte Carlo simulation results mentioned in Section 10.9, the virtual identical results between the failure probability based on closed form solution and that based on vectorial procedure can be taken as another verification of the closed form equations.

Although more succinct and convenient than sequential vectorial operations, the closed form equations of Figures 10.3 and 10.4 assume the same dip direction for the upper ground and the slope face; also, tension crack and reinforcing force are not dealt with. The closed form equations are not as comprehensive as the vectorial procedure described in Hoek and Bray (1981) and Wyllie (2018), which can be used for cases involving tension cracks, oblique upper ground surface and reinforcing force.

We next show that the above deterministic approach, whether based on closed form equations or the implicit vectorial procedure, can be extended easily to probability-based design and analysis to account for uncertainty in the values of parameter inputs. Among the merits of the probability-based analysis and design are the insights provided at the design point (or most probable failure point) of the first-order reliability method (FORM), and thought-inspiring implications and potential guidance for prevailing partial factor design methods (e.g., Eurocode 7) which are evolving.

10.7 RELIABILITY ANALYSIS AND RBD OF A TETRAHEDRAL ROCK WEDGE WITH A DOMINANT FAILURE MODE

In this chapter, β without subscript refers to reliability index, whereas β_1 and β_2 are angles (Figures 10.1 and 10.2) related to the difference between the strike directions of the discontinuity planes 1 and 2 and the slope face.

An example evaluation of reliability index using the Low and Tang (2007) spreadsheet FORM method for a tetrahedral wedge in rock slope is shown in Figure 10.8. The random variables are the orientation parameters β_1, β_2, δ_1, δ_2 of the two planes, waters pressures u_1 and u_2 and friction angles ϕ_1 and ϕ_2, on planes 1 and 2, respectively, with mean values (μ) and standard deviations (σ) as shown in Figure 10.8b. These variables are assumed to be uncorrelated and normally distributed. The transition from the deterministic template of Figure 10.8a to the first-order reliability method (FORM) set-up of Figure 10.8b is straightforward: one merely enters cell addresses in the cells of β_1, β_2, δ_1, δ_2, u_1, u_2, ϕ_1 and ϕ_2 in Figure 10.8a, to refer to the $\mathbf{x}^*$ column values in Figure 10.8b. Initially the $\mathbf{n}$ values in Figure 10.8b were zeros.

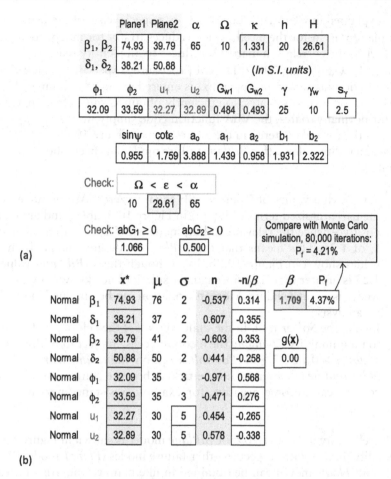

(a)

(b)

Figure 10.8 Reliability analysis of tetrahedral wedge in rock slope: (a) deterministic template, (b) reliability analysis with eight random variables.

Since the variables are uncorrelated, the equation for the eight-dimensional
ellipsoid is an extension of the canonical form of the equation for two-
dimensional ellipse. The problem is therefore formulated as follows:

$$\text{Minimize:} \quad \beta = \sqrt{\sum_{i=1}^{i=8} n_i^2} \quad \left(\text{``= sqrt}\left(\text{sumsq}(n)\right)\text{,'' in spreadsheet}\right) \quad (10.13a)$$

$$\text{Subject to:} \quad F_s\left(\beta_1, \beta_2, \delta_1, \delta_2, \phi_1, \phi_2, u_1, u_2, \ldots\right) \leq 1 \quad (10.13b)$$

$$\text{Or:} \quad g(\mathbf{x}) = F_s - 1 \leq 0 \quad (10.13c)$$

As explained in Chapter 2, Eq. 10.13a is a special case of the usual array formula $\sqrt{\mathbf{n}^T \mathbf{R}^{-1} \mathbf{n}}$ for the β cell, which reduces to $\sqrt{n_i^2}$ for independent variables when the correlation matrix $\mathbf{R}$ reduces to an identity matrix.

In the Low and Tang (2007) FORM procedure, a specially created simple VBA function (as shown in Appendix) is entered in the $\mathbf{x}^*$ column of Figure 10.8b for nonnormal variates, to efficiently obtain x_i values from n_i values. But for normal variates, the VBA function code simply means $x_i = \mu_i + n_i \sigma_i$, which is the formula entered in the $\mathbf{x}^*$ column of Figure 10.8b.

The reliability index β (for *BiPlane* failure mode) has been obtained easily as follows:

1. Start with n values of Figure 10.8b equal to zeros. At this stage, both the parenthesized terms of Eq. (a) in Figure 10.3, abG1 and abG2, are positive, hence the relevant sliding mode is *BiPlane*, with a calculated F_s of 1.25. This means that the *BiPlane* mode governs at the mean values shown in Figure 10.8b. Even though the F_s(*BiPlane*) value of 1.25 is greater than 1.0, the engineer still does not know how safe the wedge is against sliding. More information can be obtained in reliability analysis.
2. Invoke the *Solver* tool. In the dialog box, *Set* the cell of β value equal to minimum, *By Changing* the n column cells, *Subject to* F_s(*BiPlane*) ≤ 1, $abG1 \geq 0$, $abG2 \geq 0$. (The *Solver* option '*Make unconstrained variables non-negative*' must not be ticked, because some values in the n column can become negative, for example, resistance parameters ϕ_1 and ϕ_2.)

The reliability index is 1.709 for *BiPlane* mode, as shown in Figure 10.8b. Reliability indices with respect to other failure modes (*Plane1* mode, *Plane2* mode and *Floats* mode) can be obtained in like manner and are all greater than 5.0. This means that for the case in Figure 10.8, *BiPlane* mode is the dominant and governing mode. The β index of 1.709 is smaller than the target β values for ultimate limit states, which are typically in the range from 2.5 to 3.5, depending on the consequence of failure. We will next investigate probability-based design of drainage, to reduce water pressures on the two planes, so that higher reliability against wedge sliding can be achieved.

10.8 RELIABILITY-BASED DESIGN OF DRAINAGE TO ENHANCE STABILITY OF WEDGE IN ROCK SLOPE

It is assumed that drainage will be able to reduce the water pressure on the two planes on which the wedge sits, but that there is uncertainty on the magnitude of the reduced water pressure, Hence, the design water pressure (assumed equal on both planes) is treated as a variable, with a mean value μ and a standard deviation equal to 1/6 of its mean value. The objective now is to find the mean water pressure that will increase the reliability index from 1.709 to 2.50. Reliability analysis using different trial mean values of water pressures show that this higher target β of 2.5 will be achieved when the water pressure is reduced from 30 kPa (Figure 10.8, last two rows) to 24.07 kPa (Figure 10.9, last two rows).

The following may be noted:

(1) The case in Figure 10.8 is FORM analysis of a situation, not reliability-based design (RBD). The x^* values constitute the most probable point (MPP) of failure, not the design point, strictly speaking. This MPP of failure is where the expanding 8D dispersion ellipsoid first touches the limit state surface (LSS). In contrast, the case in Figure 10.9 is RBD for a target β of 2.50. The x^* values constitute the design point. It is also

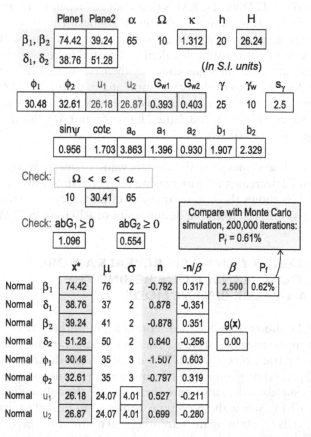

Figure 10.9 Reliability-based design of water pressure reduction on discontinuity planes for target β value 2.5. The c.o.v. of u_1 and u_2 remains at 1/6.

the MPP of failure, that is, the first contact point between the LSS and an expanding dispersion ellipsoid.

(2) Although the mean values of friction angle on planes 1 and 2 are the same (35°) in Figures 10.8 and 10.9, the sensitivity indicator values (n), or their direction cosine values (−n/β), are different: that of ϕ_1 about twice that of ϕ_2. This means that, for the case in hand, despite identical mean values of the resistance parameters ϕ_1 and ϕ_2, the friction angle ϕ_1 on the less steep plane 1 (where dip δ is 37°) is a more sensitive parameter than the friction angle ϕ_2 on the steeper plane 2 (where dip δ is 50°)

(3) In contrast to item (2) above, the different n values of the load parameters u_1 and u_2 mean that the water pressure on plane 2 is a more sensitive parameter than the water pressure on plane 1, despite their identical mean values of 30 kPa in Figure 10.8 and 24.07 kPa in Figure 10.9. The orientation of planes 1 and 2 and their areas are two factors affecting the sensitivity of water pressures acting on those surfaces.

(4) A merit of RBD-via-FORM is that context-dependent sensitivities will be reflected at its design point (the x^* values). Parameters of the same physical nature, like ϕ_1 and ϕ_2, or u_1 and u_2, can have different sensitivities (and different back-calculated partial factors) if the underlying factors affecting the significance of their roles are different.

(5) The design values of the orientation angles β_1, β_2, δ_1 and δ_2 of the two planes indicate that a 'wider' wedge (i.e., larger 180° − β_1 − β_2 value) and steeper planes 1 and 2 (i.e., larger δ_1 and δ_2 values) are the more dangerous scenarios.

Figure 10.10a shows that the design value of mean water pressure on planes 1 and 2 decreases with increasing target reliability index, as expected. Figure 10.10b shows the non-uniqueness of partial factors (with respect to mean values) inferred from the design point of RBD-via-FORM.

10.9 RELIABILITY ANALYSIS REVEALS A MORE CRITICAL *PLANE I* MODE BEHIND THE MEAN-VALUE *BIPLANE* MODE

Figure 10.11 shows a wedge of orientations and other properties different from that in the previous section. The 10 random variables consist of four geometrical parameters (β_1, β_2, δ_1, δ_2) of the wedge, shear strength parameters (ϕ_1, ϕ_2 c_1, c_2) of planes 1 and 2, and water pressures (u_1, u_2). The mean values and standard deviations are shown under the column labeled μ and σ in Figure 10.11. When the n column values were initially zeros, the *abG1* and *abG2* cells (containing the parenthesized terms in Eq. (a) of Figure 10.3) are positive, indicating that the sliding mode at mean values of input is a *BiPlane* mode, for which the reliability index is 2.34, the first value in the

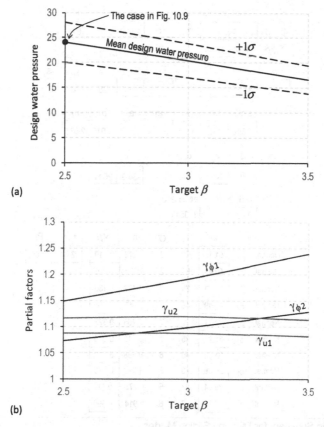

Figure 10.10 (a) Reliability-based design of water pressure on discontinuity planes for different target β values, (b) context-dependent partial factors for ϕ_1 and ϕ_2, and for u_1 and u_2, based on mean values.

row labelled β in the table at the bottom of Figure 10.11. Reliability index values for the other three failure modes are also calculated, obtaining β values equal to 1.33 for the *Plane1* mode, 5.88 for the *Plane2* mode and 5.08 for contact lost on both planes ('floats' mode).

The following may be noted:

(1) Although the governing failure mode at mean values of the 10 variables is *BiPlane*, the reliability index β for *Plane1* mode (1.325) is more critical than that for *BiPlane* mode (2.34). This information would not be revealed in a deterministic analysis using mean values, or in a reliability analysis that considers only one failure mode. The *BiPlane* reliability index of 2.34, taken by itself, would suggest a probability of failure ($\Phi(-2.34)$), or 0.96%, which falls far short of the Monte Carlo total failure probability of 9.5% in (3) below.

	Plane1	Plane2	α	Ω	κ	h	H
β_1, β_2	61.16	18.73	70	0	1.000	16	16.00
δ_1, δ_2	50.48	48.49					

(In S.I. units)

ϕ_1	ϕ_2	u_1	u_2	G_{w1}	G_{w2}	γ	γ_w	s_γ
29.71	34.21	28.64	32.49	0.537	0.609	25	10	2.5

$\sin\psi$	$\cot\varepsilon$	a_0	a_1	a_2	b_1	b_2
0.852	1.057	3.805	0.598	0.680	0.915	2.571

Check: $\Omega < \varepsilon < \alpha$ c_1 c_2

0	43.43	70		34.9	25.66

Check: $abG_1 \geq 0$ $abG_2 \geq 0$

0.401	0.054

(BiPlane mode)

		x^*	μ	σ	n	$-n/\beta$	β	P_f
Normal	β_1	61.16	62	3	-0.279	0.119	2.34	0.96%
Normal	δ_1	50.48	50	2	0.239	-0.102		
Normal	β_2	18.73	20	3	-0.422	0.180		
Normal	δ_2	48.49	48	2	0.247	-0.105		
Normal	ϕ_1	29.71	35	7	-0.755	0.322		
Normal	ϕ_2	34.21	35	7	-0.112	0.048		
Normal	c_1	34.90	40	8	-0.638	0.272		
Normal	c_2	25.66	40	8	-1.793	0.765		
Normal	u_1	28.64	27	6	0.273	-0.117		
Normal	u_2	32.49	27	6	0.914	-0.390		

Scenario Summary for Different Sliding Modes				
Failure mode	Biplane	Plane1	Plane2	Floats
Constraints	$abG1 \geq 0$ $abG2 \geq 0$ Fs(Biplane) $\leq$ 1	$abG2 \leq 0$ $abGZ1 \geq 0$ Fs(Plane1) $\leq$ 1	$abG1 \leq 0$ $abGZ2 \geq 0$ Fs(Plane2) $\leq$ 1	$abGZ1 \leq 0$ $abGZ2 \leq 0$
β	2.34	1.325	5.88	5.08
Failure probability = $\Phi(-\beta)$	0.96%	9.26%	0.00%	0.00%
Monte Carlo (equations)*	0.54%	8.98%	0.00%	0.00%
Monte Carlo (vectorial)**	0.54%	8.97%	0.00%	0.00%

* Closed form equationsof Figs.10.3 and 10.4, 100,000 iterations
** Hoek and Bray (1981) vectorial procedure, 100,000 iterations
- Expressions for *abG1* and *abG2* are given in the parenthesized terms in Eq.(a) of Fig. 10.3 and expressions for *abGZ1* and *abGZ2* are given in Eqs. (u) and (y) of Fig. 10.4

Figure 10.11 Reliability analyses reveal that *Plane1* failure mode (β = 1.325) is more critical than *BiPlane* mode (β = 2.34), even though the mean-value point is a *BiPlane* mode.

(2) The failure domain for *Plane1* mode is bounded by three surfaces: $abG2 = 0$, $abGZ1 = 0$, and $F_s(Plane1) = 1$, where $abG2$ and $abGZ1$ denote the two expressions in Eqs. (t) and (u) of Figure 10.4. An ellipsoid expanding from the *BiPlane* mode at mean values encountered the most probable failure point for *Plane1* mode at $F_s < 1.0$, not $F_s = 1.0$. This point of the *Plane1* mode satisfies $abG2 = 0$ and hence is on the $abG2 = 0$ failure surface. The correct β index for *Plane1* mode (1.325) was obtained with the constraint $F_s(Plane1) \leq 1$ (bottom table of Figure 10.11), which allows *Solver* to consider all surfaces bounding the failure region of the *Plane1* mode. Had one imposed the constraint $F_s(Plane1) = 1$, a misleadingly higher value of β index would have been obtained by the Solver tool.

(3) Monte Carlo simulation was conducted using the @RISK software, on the closed form equations of Figures 10.3 and 10.4, and on the vectorial procedure as presented in Hoek and Bray (1981). As shown in the last two rows of Figure 10.11, the Monte Carlo simulations obtain *BiPlane* and *Plane1* failure probabilities of 0.54% and 8.98%, based on the equations in Figures 10.3 and 10.4, virtually identical to the values of 0.54% and 8.97% based on vectorial procedure. That *Plane1* failures outnumber *BiPlane* failures is consistent with the findings noted in (1) and (2). Since the four failure modes are mutually exclusive, the total failure probability is just the sum of the failure probabilities for all the modes, equal to 10.2% based on FORM, 9.52% based on MC simulation on closed form equations, and 9.51% based on Monte Carlo simulation on vectorial procedures.

(4) In the transition from the mean-value *BiPlane* mode to the *Plane1* mode, there is a gradual loss of the frictional resistance on plane 2, but an abrupt loss of cohesion on plane 2 as soon as the *Plane1* mode is reached. This accounts for the *Plane1* mode being far more critical than the mean-value *BiPlane* mode, for the case in hand. The small value of $abG2$ (0.054) in Figure 10.11 suggests that the mean-value *BiPlane* mode is close to the *Plane1* mode boundary. The closed-form equations of Figures 10.3 and 10.4, and the Hoek-Bray vectorial procedure, assume (logically) zero friction and zero cohesion resistance on the plane (plane 2 in this case) that the wedge loses contact with.

(5) The '*failure domain*' constraint (Fs ≤ 1) should be used when dealing with multiple failure modes, or when the '*average point*' (where all random variables assume their mean values) is already inside the unsafe domain. In the latter case, a reliability index equal to zero would be correctly obtained, meaning that the most probable failure point is the unsafe '*average point*' itself. In contrast, the '*failure surface*' constraint ($F_s = 1$) would, if the '*average point*' is unsafe, lead to

a reliability index $\beta > 0$, which is still meaningful if properly interpreted as a negative reliability index: it indicates 'how far' the (unsafe) *average point* is from the safe region. The possibility of negative reliability index was noted in Tichy (1993, pp. 150–151, 160–161).

In conducting Monte Carlo simulation for the case in Figure 10.11 using the Hoek and Bray vectorial procedure, it is necessary to use dip directions and dips instead of the intrinsic angles β_1, β_2, δ_1 and δ_2. Again, a freehand sketch as shown in Figure 10.12, in like manner to Figure 10.2, enables the conversion (in reverse).

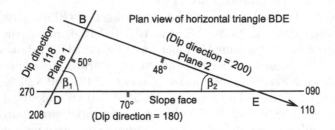

In this case, $\Omega = 0$; H = h = 16 m; $\gamma = 25$ kN/m³.
The intrinsic β_1 and β_2 geometrical angles of the wedge are not changed if the dip directions of the slope face and Planes 1 and 2 rotate by the same angle. For example, the dip directions could have been 270, 208 and 290, instead of 180, 118, 200 shown above.

(a)

			Plane1	Plane2	Upper face	Slope face
	Dip	ψ	50.00	48.00	0	70
(In S.I. units)	Dip dir.	α	118.00	200.00	180	180

		x^*	μ	σ
Normal	DipDir α_1	118.00	118	3
Normal	Dip ψ_1	50.00	50	2
Normal	DipDir α_2	200.00	200	3
Normal	Dip ψ_2	48.00	48	2
Normal	ϕ_1	35.00	35	7
Normal	ϕ_2	35.00	35	7
Normal	c_1	40.00	40	8
Normal	c_2	40.00	40	8
Normal	u_1	27.00	27	6
Normal	u_2	27.00	27	6

Monte Carlo simulations based on **Hoek and Bray** (1981) vectorial procedure, 100,000 iterations:

Failure probability of BiPlane mode: 0.54%
Failure probability of Plane1 mode: 8.97%
Total failure probability = **9.51%**

Monte Carlo simulations based on **closed form** equations (Figs. 10.3 & 10.4), 100,000 iterations:

Failure probability of BiPlane mode: 0.54%
Failure probability of Plane1 mode: 8.98%
Total failure probability = **9.52%**

(b)

Figure 10.12 (a) Simple freehand sketch to convert intrinsic β_1 and β_2 angles to dip directions and dip angles of the slope face and the two planes, (b) statistical inputs for Monte Carlo simulation, for verification using the Hoek and Bray vectorial procedure.

10.10 SENSITIVITY COMPUTATION IN THE ELLIPSOID APPROACH

The sensitivity factors α_i are an integral part of the classical reliability computation procedure (e.g., Rackwitz 1976; Ang and Tang 1984). These factors convey the relative importance of the random variables in affecting reliability. They could be useful in cost analysis and design planning.

In the Low and Tang (2007) spreadsheet FORM procedure, the sensitivity factors α_i can be computed once the reliability index β and the point ($\mathbf{x}^*$) are known:

$$\alpha_i = -\left(\frac{x_i^* - \mu_i}{\beta\sigma_i}\right) = -\frac{n_i}{\beta} \qquad (10.14)$$

where x_i^* is the value of x_i at the most probable failure point on the failure surface (i.e., the point where the dispersion ellipsoid is tangent to the failure surface), and μ_i and σ_i are the mean and standard deviations of the x_i's. Both n and $-n/\beta$ are shown in Figures 10.8, 10.9 and 10.11. Both are sensitivity indicators for their respective purposes. The n values are preferred when showing their variation with increasing target reliability indices; the values of $-n/\beta$ tend to vary little when plotted against target reliability indices. The partial factors implied at the design point of RBD-via-FORM are also related to the n values.

Another way of obtaining sensitivity information is to perform numerical differentiation of the reliability index β with respect to the mean values or the standard deviations. The mean or standard deviation of the random variable concerned is given a small perturbation ($\Delta\mu_i$ or $\Delta\sigma_i$); the β index is recalculated using the Solver tool, and finally $\Delta\beta/\Delta\mu_i$ or $\Delta\beta/\Delta\sigma_i$ are evaluated. A user-defined function can be written to perform this on the spreadsheet. For the tetrahedral wedge problem of Figure 10.11, numerical differentiation of the reliability index β with respect to 1% perturbation on the mean value and the standard deviation of δ_1 resulted in $\Delta\beta/\Delta\mu_{\delta1} = -0.0507$ and $\Delta\beta/\Delta\sigma_{\delta1} = -0.0122$. These values are virtually identical to those (-0.0510 and -0.0122) computed from the following established relationships (Madsen et al. 1986, p. 122), assumsing the variables are independent:

$$\frac{\partial\beta}{\partial m_i} = \frac{\alpha_i}{\sigma_i} \qquad (10.15a)$$

$$\frac{\partial\beta}{\partial\sigma_i} = -\frac{\beta}{\sigma_i}\left(\alpha_i\right)^2 \qquad (10.15b)$$

in which α_i is computed from Eq. 10.14.

For independent variables, therefore, sensitivity information can be obtained using Eqs. 10.15a and 10.15b. When the x_i's are not independent, however, the α_i's have no direct physical meaning (Melchers 1987, p. 110). In such cases, sensitivity information can still be obtained using the standard method of numerical differentiation, as described above.

10.11 CHAPTER SUMMARY

Closed form equations based on limit equilibrium have been presented for stability analysis of tetrahedral wedges with multiple failure modes. The equations allow calculation of the factor of safety for the four failure modes: (i) sliding along two bounding joint planes (Figure 10.3), (ii) sliding on plane1, (iii) sliding on plane 2 and (iv) floating failure of wedge, Figure 10.4. Simple expressions are also provided, Eqs. (m) and (i), to check whether wedge formation is possible.

The water pressure effect can be input as G_{w1} and G_{w2}, Eq. (j) in Figure 10.3, or average water pressures u_1 and u_2, as indicated below Eq. (a) in Figure 10.3.

The simple freehand sketch procedure in Figure 10.2 allows exact conversion from dip direction and dip values to the intrinsic joint orientation angles (β_1, β_2, δ_1 and δ_2, Figure 10.1) used in the closed form equations of Figures 10.3 and 10.4. The reverse conversion, from intrinsic wedge orientation angles (β_1, β_2, δ_1 and δ_2) to dip direction and dip values, is as easy, as shown in Figure 10.12a. The closed form equations have been verified (Figures 10.5, 10.6 and 10.7) for different failure modes by comparison with the vectorial procedure of Hoek and Bray (1981).

The deterministic factor-of-safety approach was extended probabilistically, including FORM analysis (Figure 10.8), RBD-via-FORM (Figure 10.9), connections with partial factor design (Figure 10.10), and detection of a more dangerous mode behind a seemly safe mean-value mode (Figure 10.11). Reliability analysis can reveal subtleties of practical significance that would not have been revealed in a deterministic approach or in a single-mode reliability analysis. Sensitivity information at the most probable failure point of FORM was discussed.

Chapter 11

Underground excavations in rock

11.1 INTRODUCTION

This chapter presents deterministic solutions followed by probability-based analysis and design of both stress-controlled and structurally controlled instability mechanisms in underground rock excavations. The chapter opens with a procedure for deterministic ground–support interaction analysis using automatic Newton–Raphson method in spreadsheet, for a shaft excavated in sandstone and with internal support pressure. Reliability analysis is then conducted to identify *in situ* stress and Young's modulus of the rock mass as two parameters of high impact on the behaviour of the excavation. The effects of mean values and standard deviations of these two parameters on the limit state of support system are investigated. This is followed by reliability analysis of a symmetric roof wedge in a circular tunnel, where the ambiguity caused by two different definitions of factor of safety is resolved. The valuable context-dependent sensitivity information offered by FORM is discussed. A final example applies FORM to a circular tunnel supported with elastic rockbolts in a homogeneous and isotropic elasto-plastic ground obeying the Coulomb failure criterion. The similarities and differences between the ratios of design-point values to mean values, on the one hand, and the partial factors of limit state design, on the other hand, are discussed. The reliability-based design of the length and spacing of rock bolts for a target reliability index is illustrated. Comparisons are made with second-order reliability method, and Monte Carlo simulations with and without importance sampling.

11.2 DETERMINISTIC GROUND–SUPPORT INTERACTION ANALYSIS OF A SHAFT EXCAVATED IN SANDSTONE

Hoek et al. (2000, pp. 107–108) illustrates the ground–support interaction analysis of a 6 m diameter reinforced shaft excavated in a fair quality, blocky sandstone, Figure 11.1. The estimated cohesion c and friction angle ϕ of the rock mass are 2.6 MPa and 30°, respectively. The *in situ* axisymmetric stress

DOI: 10.1201/9781003112297-14

Ground-Support interaction example from Hoek, Kaiser and Bawden (2000, p107-108)

A 6 metre diameter shaft is excavated in a fair quality, blocky sandstone. The estimated strength characteristics of the rock mass are a cohesion c = 2.6 MPa and friction angle ϕ = 30°. The in situ stress is p_o = 10 MPa. The support consists of mechanically anchored rockbolts. The estimated maximum support pressure *Psm* is 0.34 MPa. The maximum elastic displacement which can be withstood by these bolts is u_{sm} = 21 mm. An initial radial displacement of u_{so} = 25 mm is assumed prior to installation of the support system.

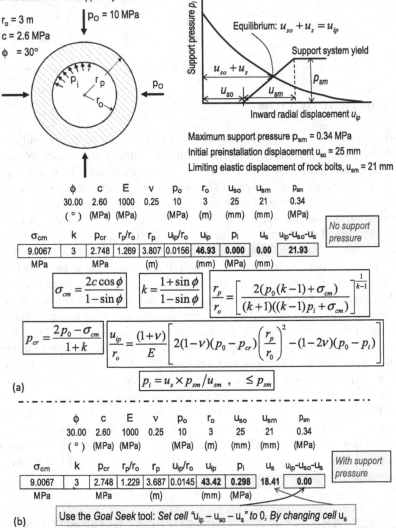

r_o = 3 m
c = 2.6 MPa
ϕ = 30°
p_o = 10 MPa

Equilibrium: $u_{so} + u_s = u_{ip}$

Support system yield

$u_{so} + u_s$

u_{so} u_{sm} p_{sm}

Inward radial displacement u_{ip}

Maximum support pressure p_{sm} = 0.34 MPa
Initial preinstallation displacement u_{so} = 25 mm
Limiting elastic displacement of rock bolts, u_{sm} = 21 mm

ϕ	c	E	v	p_o	r_o	u_{so}	u_{sm}	p_{sm}
30.00	2.60	1000	0.25	10	3	25	21	0.34
(°)	(MPa)	(MPa)		(MPa)	(m)	(mm)	(mm)	(MPa)

σ_{cm}	k	p_{cr}	r_p/r_o	r_p	u_{ip}/r_o	u_{ip}	p_i	u_s	$u_{ip}-u_{so}-u_s$
9.0067	3	2.748	1.269	3.807	0.0156	46.93	0.000	0.00	21.93
MPa		MPa		(m)		(mm)	(MPa)	(mm)	

No support pressure

$$\sigma_{cm} = \frac{2c\cos\phi}{1-\sin\phi} \qquad k = \frac{1+\sin\phi}{1-\sin\phi} \qquad \frac{r_p}{r_o} = \left[\frac{2(p_0(k-1)+\sigma_{cm})}{(k+1)((k-1)p_i+\sigma_{cm})}\right]^{\frac{1}{k-1}}$$

$$p_{cr} = \frac{2p_0 - \sigma_{cm}}{1+k} \qquad \frac{u_{ip}}{r_o} = \frac{(1+v)}{E}\left[2(1-v)(p_0-p_{cr})\left(\frac{r_p}{r_o}\right)^2 - (1-2v)(p_0-p_i)\right]$$

(a)

$$p_i = u_s \times p_{sm}/u_{sm}, \quad \leq p_{sm}$$

- -

ϕ	c	E	v	p_o	r_o	u_{so}	u_{sm}	p_{sm}
30.00	2.60	1000	0.25	10	3	25	21	0.34
(°)	(MPa)	(MPa)		(MPa)	(m)	(mm)	(mm)	(MPa)

σ_{cm}	k	p_{cr}	r_p/r_o	r_p	u_{ip}/r_o	u_{ip}	p_i	u_s	$u_{ip}-u_{so}-u_s$
9.0067	3	2.748	1.229	3.687	0.0145	43.42	0.298	18.41	0.00
MPa		MPa		(m)		(mm)	(MPa)		

With support pressure

(b)

Use the *Goal Seek* tool: Set cell "$u_{ip} - u_{so} - u_s$" to 0, By changing cell u_s

Figure 11.1 Deterministic ground–support interaction analysis of a 6 metre diameter shaft excavated in a fair quality, blocky sandstone: (a) no support pressure, (b) with support pressure (after Hoek et al., 2000).

(p_o) is 10 MPa. The estimated maximum available support pressure p_{sm} from anchored rock bolts is 0.34 MPa, and the maximum elastic displacement (u_{sm}) which can be withstood by these bolts is 21 mm. An initial radial displacement (u_{so}) of 25 mm is assumed prior to installation of the support system.

As explained in Hoek et al. (2000), the internal support pressure p_i is in practice often smaller than the critical pressure p_{cr} required for elastic behaviour of the rock mass. Hence, a plastic zone will be formed near the excavated circular wall, extending to a radius r_p, accompanied by a total inward radial displacement u_{ip} of the walls of the excavation. For a continuous, homogeneous and isotropic rock mass under hydrostatic *in situ* stress and obeying the Mohr-Coulomb failure criterion, the analytical solution of the plastic zone extent (r_p/r_o) and inward radial displacement ratio around the excavated perimeter (u_{ip}/r_o) can be calculated using equations in Hoek et al. (2000), which are shown in Figure 11.1. The equations for the plastic zone radius and the inward radial displacement of the excavation perimeter are:

$$\frac{r_p}{r_o} = \left[\frac{2\left(p_0\left(k-1\right)+\sigma_{cm}\right)}{\left(k+1\right)\left(\left(k-1\right)p_i+\sigma_{cm}\right)} \right]^{\frac{1}{k-1}} \tag{11.1}$$

$$\frac{u_{ip}}{r_o} = \frac{\left(1+v\right)}{E}\left[2\left(1-v\right)\left(p_0-p_{cr}\right)\left(\frac{r_p}{r_o}\right)^2 - \left(1-2v\right)\left(p_0-p_i\right) \right] \tag{11.2}$$

in which k, σ_{cm} and p_{cr} are given by equations shown in Figure 11.1, and r_o is the original radius of the shaft, r_p the radius of the plastic zone, p_o the *in situ* hydrostatic stress, k the slope of the σ_1 versus σ_3 line, σ_{cm} the uniaxial compressive strength of the rock mass, p_i the uniform internal support pressure, p_{cr} the critical internal support pressure below which plastic zone will appear, u_{ip} the inward radial plastic displacement of the excavation perimeter, v the Poisson's ratio and E the Young's modulus of the rock mass, and (in the equation of σ_{cm}) c the cohesive strength and ϕ the angle of friction of the rock mass.

The equations at the bottom of Figure 11.1a have been entered in the labeled boxed cells of Figure 11.1a, for σ_{cm}, k, p_{cr}, r_p/r_o, r_p ($= r_p/r_o \times r_o$), u_{ip}/r_o, u_{ip} ($= u_{ip}/r_o \times r_o$), p_i ($= \min(u_s{}^*p_{sm}/u_{sm}, p_{sm})$). For the input values in the top row of Figure 11.1a, the required threshold support pressure to prevent any plastic zone arising due to the cylindrical excavation is 2.748 MPa, as shown by the p_{cr} cell. The support elongation u_s is equal to zero for this case with zero support pressure p_i. The computed plastic zone radius r_p is 3.8 m, and the inward radial displacement (u_{ip}) at the perimeter of the shaft wall is 46.9 mm. These computed values agree with those mentioned in Hoek et al. (2000), where $p_{cr} = 2.75$ MPa, $r_p = 3.8$ m and $u_{ip} = 47$ mm.

Installed anchored rock bolts around the perimeter of the shaft will be stressed into tension as they are lengthened by the ground deformation, thereby providing an internal support pressure p_i of magnitude shown by the following equation (for rock bolts with linearly elastic and perfectly plastic behaviour):

$$p_i = u_s \times \frac{p_{sm}}{u_{sm}}, \quad \leq p_{sm} \tag{11.3a}$$

that is,

$$p_i = \frac{\left(u_{ip} - u_{so}\right)}{u_{sm}} \times p_{sm}, \quad \leq p_{sm} \tag{11.3b}$$

in which u_{so} is the initial radial displacement prior to installation of the support system, p_{sm} is the maximum available support pressure from the rock bolts and u_{sm} is the elastic displacement at which p_{sm} is realized, as shown in the ground–support interaction curves in Figure 11.1a.

In theory, Eq. 11.3b can substitute for the p_i term in Eqs. 11.2 and 11.1, resulting in an expression similar to Eq. 11.2 that contains only one unknown on both sides of the nonlinear expression, namely the inward radial displacement u_{ip} at the perimeter of the excavated shaft. However, iteration is needed to solve for u_{ip} because of the way p_i (and hence u_{ip}) appears in Eq. 11.1 which is contained in Eq. 11.2.

A more convenient solution procedure is to use the *Goal Seek* tool of the spreadsheet to solve for the equilibrium value of u_{ip} and u_s, as shown in Figure 11.1b. Initially, the u_s cell in Figure 11.1b is zero. The *Goal Seek* tool performs Newton–Raphson method iteratively to obtain the equilibrium value of u_{ip} and hence u_s and p_i. The solution values of u_s and p_i (18.41 mm and 0.298 MPa) are below the given threshold values of 21 mm and 0.34 MPa, hence are deemed acceptable. The solution value of u_{ip} (43.4 mm) is only slightly lower than that (46.9 mm, Figure 11.1a) of the case without support pressure. The above values of p_i and u_{ip} agree with the remarks in Hoek et al. (2000) that the load–displacement curve of the support system intersects the ground response curve at a support pressure value of about 0.3 MPa, and at a displacement of approximately 43 mm.

As Hoek et al. (2000) and Hoek (2007) aptly emphasized, it is important to conduct sensitivity studies to determine which parameter/s have a significant influence on the behaviour of underground openings. When different values of the parameters in the top row of Figure 11.1b are tried, it will be found that Young's modulus E and *in situ* stress p_o are important parameters affecting the ground–support interaction behaviour of the excavation in this case, with E exerting a stabilizing effect, and p_o exerting a destabilizing effect. We may refer to E as a resistance parameter, and p_o as a load parameter for the case in hand. (As we will see later, *in situ* stress p_o can exert a stabilizing effect in some cases of structurally-controlled instability mechanisms.)

One important merit of FORM analysis in the next section is that the sensitivity measure of a parameter is indicated by the normalized difference between its most probable failure value and its mean value. The extension of the deterministic analysis in Figure 11.1b to the probabilistic FORM analysis of Figure 11.2a is to illustrate valuable sensitivity information obtainable from reliability analysis. The statistical inputs in the next section are illustrative, not based on actual data; the effects of different statistical inputs can be investigated without affecting the conclusion on sensitivities drawn for the case in hand.

11.3 RELIABILITY ANALYSIS INVOLVING GROUND–SUPPORT INTERACTION OF A SHAFT EXCAVATED IN SANDSTONE

The random variables for our reliability analysis of the preceding ground–support interaction example will be the shear strength parameters c and ϕ of the rock mass, Young's modulus E of the rock mass and the hydrostatic *in situ* stress p_o around the cylindrical excavation. Mean values and standard deviations of the parameters are needed for reliability analysis.

The values of $\phi = 30°$, $c = 2.6$ MPa, E $= 1000$ MPa and $p_o = 10$ MPa used in the deterministic analysis of Figure 11.1 result in an equilibrium support pressure p_i of 0.3 MPa and u_s of 18.4 mm, not far from the support system yield stress of 0.34 MPa or the threshold bolt elongation (u_s) of 21 mm. The implication seems to be that the design is acceptable even if equilibrium support pressure reaches the maximum yield value of 0.34 MPa, and the bolt elongation reaches the threshold value of 21 mm. This is in line with EC7 and LRFD principle, where a design is deemed to be acceptable even when factored resistance is equal to factored actions/loads.

For reliability analysis of the deterministic example from the preceding section, we will assume that the mean values of resistance parameters (ϕ, c and E) are somewhat higher than the values of 30°, 2.6 MPa and 1000 MPa used in the deterministic analysis of Figure 11.1, and the mean value of the load parameter p_o to be lower (by about 12%) than its value of 10 MPa used in the deterministic analysis. The rationale is similar to that when comparing reliability analysis with partial factor design methods like the Eurocode 7 (EC7). As explained in Chapter 4, in EC7, partial factors are applied to conservative characteristic values to arrive at design values. For resistance parameters (like ϕ, c and E in this case), the mean values in reliability analysis should be higher than the characteristic values in EC7. This is because the conservative characteristic strength values in EC7 are lower than mean strength values. For example, the characteristic strength values could be taken at the 20 percentiles of their probability distributions.

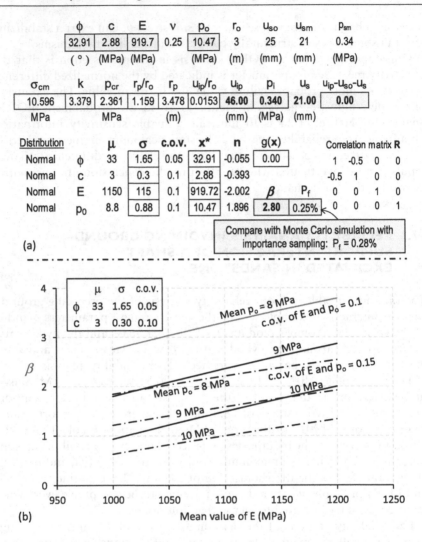

φ	c	E	ν	p_o	r_o	u_{so}	u_{sm}	p_{sm}
32.91	2.88	919.7	0.25	10.47	3	25	21	0.34
(°)	(MPa)	(MPa)		(MPa)	(m)	(mm)	(mm)	(MPa)

σ_{cm}	k	p_{cr}	r_p/r_o	r_p	u_{ip}/r_o	u_{ip}	p_i	u_s	$u_{ip}-u_{so}-u_s$
10.596	3.379	2.361	1.159	3.478	0.0153	46.00	0.340	21.00	0.00
MPa		MPa		(m)		(mm)	(MPa)	(mm)	

Distribution		μ	σ	c.o.v.	x*	n	g(x)	Correlation matrix R			
Normal	φ	33	1.65	0.05	32.91	-0.055	0.00	1	-0.5	0	0
Normal	c	3	0.3	0.1	2.88	-0.393		-0.5	1	0	0
Normal	E	1150	115	0.1	919.72	-2.002	β P_f	0	0	1	0
Normal	p_o	8.8	0.88	0.1	10.47	1.896	2.80 0.25%	0	0	0	1

Compare with Monte Carlo simulation with importance sampling: $P_f = 0.28\%$

(a)

(b)

Figure 11.2 (a) Probabilistic ground–support interaction analysis using first-order reliability method, (b) reliability index β changes with the mean values of Young's modulus E of the rock mass and the *in situ* stress p_o and their coefficients of variation.

Specifically, the mean values of φ, c, E and p_o for reliability analysis in Figure 11.2a are 33°, 3 MPa, 1150 MPa and 8.8 MPa, with coefficients of variation (c.o.v.) equal to 0.05 0.1, 0.1 and 0.1, respectively. They are assumed to be normally distributed. The 20-percentile values of φ, c and E, and the 80-percentile value of p_o, are about equal to the deterministic values of 30°, 2.6 MPa, 1000 MPa and 10 MPa in Figure 11.1. (The inputs of φ, c, E and p_o in Figure 11.1 are not based on characteristic values of EC7, but may also not be mean values. The discussion here is merely to emphasize that FORM needs mean values.)

A correlation coefficient of -0.5 in the correlation matrix R of Figure 11.2a is used to model negatively correlated ϕ and c. Standard deviation σ is the product of c.o.v. and mean value, as shown in the column labeled σ. These statistical inputs of mean values and standard deviations are for illustrative purpose of reliability analysis. The effects of different mean values and standard deviations on reliability of the ground–support interaction will be investigated later.

The top-row cells of ϕ, c, E and p_o in Figure 11.2a contain cell addresses which read values from the x^* column. The equation for the performance function is:

$$g(\mathbf{x}) = p_{sm} - p_i \quad \text{(limit state of support system)} \qquad (11.4)$$

The x^* column contains the equations $x_i = \mu_i + n_i\sigma_i$ when dealing with normal variates, which is the case here. As explained in Eq. 2.9 of Chapter 2, the array formula for the β cell at the bottom of Figure 11.2a is $\sqrt{\mathbf{n}^T \mathbf{R}^{-1} \mathbf{n}}$, which implements nested matrix multiplications involving the R matrix and the n column cells.

Initially the n column cells were zeros, and the cell u_s (for rock bolt elongation) was also zero. The *Solver* tool was then invoked, *to set* the β cell to minimum, *by changing* the n column cells and the u_s cell, subject to the constraints that the $g(\mathbf{x})$ cell and the '$u_{ip} - u_{so} - u_s$' cell be zeros. The solutions of reliability index β, the most probable failure values (x^*) of variables, and the sensitivity indicators n, are as shown in the figure.

The $g(\mathbf{x})$ cell value of zero means that the support system is at its limit state, with internal support pressure p_i equal to maximum available support pressure p_{sm}. This also means that the rockbolt elongation is also at its limit state of $u_s = u_{sm} = 21$ mm. The distance from the safe mean-value point (represented by the μ column values) to the most probable failure point (the x^* column values) on the limit state surface, in units of directional standard deviations, is the reliability index β, found to 2.80 by the Solver tool. The corresponding failure probability, based on $P_f = \Phi(-\beta)$, is 0.25%. Five Monte Carlo simulation with importance sampling was also conducted, each with 8000 iterations, obtaining failure probabilities of 0.279%, 0.274%, 0.285%, 0.283%, 0.279%, or an average of 0.28%, practically the same as the P_f of 0.25% based on the reliability index value of $\beta = 2.80$.

FORM analysis in Figure 11.2a results in sensitivity indicators (under the n column) of values -2.002 for E, 1.896 for p_o, and much smaller values for c and ϕ. This means that the parameters E and p_o are the two most important ones of the four random variables modeled, with respect to their influence on the ultimate limit state of support yield stress (p_{sm}) of 0.34 MPa, Eq. 11.4. The most probable failure value of E (919.72, under the x^* column) is $2.002 \times \sigma_E$ *smaller* than the mean value (1150) of E. The most probable failure value of the *in situ* stress p_o (10.47, under the $\mathbf{x^*}$ column) is $1.896 \times \sigma_{po}$ *larger* than the mean value (8.8 MPa) of p_o. For uncorrelated E and p_o, the negative sign of the sensitivity indicator of E means that the failure direction is in decreasing E, and the positive sign of the sensitivity indicator of p_o means that the failure direction is in increasing p_o. This also means that E is a stabilizing parameter, while p_o is a destabilizing parameter for this case. (The phenomenon of correlated sensitivities is discussed in Section 2.5, and in Low 2020b.)

The effects of the mean values of E and p_o and their standard deviations on the ultimate limit state of support yielding are shown in Figure 11.2b. The solid lines are for c.o.v. of E and p_o equal to 0.1, while the dashed lines are for c.o.v. of E and p_o equal to 0.15. The plots show that the reliability against support yielding increases with increasing values of E, and decreasing values of p_o and coefficients of variation of E and p_o.

Hudson and Harrison (1997) explained the difference between stress-controlled instability mechanisms and structurally controlled instability mechanisms. The case of ground–support interaction in Figures 11.1 and 11.2 may be regarded as a stress-controlled instability mechanism. We next investigate a case of structurally controlled instability mechanism, where the sensitivities of friction angle and *in situ* stress can vary under different conditions.

11.4 ROOF WEDGE IN TUNNEL, A TALE OF TWO FACTORS OF SAFETY

A symmetric roof wedge of central height h and apical angle 2α in a circular tunnel of radius R is shown in Figure 11.3. An analytical approach for assessing the stability is based on Bray's 1977 (Unpublished note, Imperial College, London), two-stage relaxation procedure, as described, for example, in Sofianos et al. (1999) and Brady and Brown (2006). The first stage computes the confining lateral force H_0 on the wedge from the stress field and the geometries of the wedge and tunnel, for an assumed homogeneous, isotropic, linearly elastic, weightless medium. The second stage then assumes deformable joints and a rigid rock mass, to arrive at the normal force N acting on each joint surface.

Two different definitions of the factor of safety against wedge falling have been reported in the literature, each with its own rationale. The first appeared in Sofianos et al. (1999) and Brady and Brown (2006), for example. It is the ratio of the pull-out resistance of the wedge to the weight of the wedge and was expressed as follows:

$$FS_1 = \frac{2MH_0}{W} \tag{11.5}$$

where
$M = f(\phi, i, \alpha, k_s/k_n)$, and
$H_o = f(p, R, K_0, h/R)$

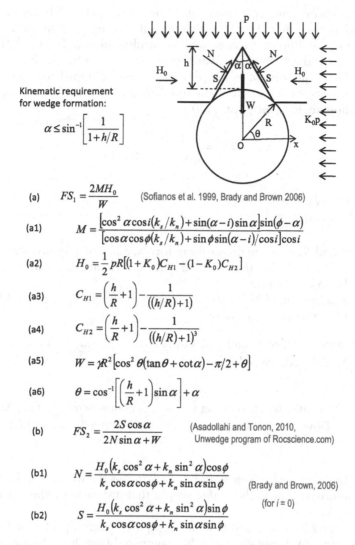

Kinematic requirement
for wedge formation:

$$\alpha \le \sin^{-1}\left[\frac{1}{1+h/R}\right]$$

(a) $$FS_1 = \frac{2MH_0}{W}$$ (Sofianos et al. 1999, Brady and Brown 2006)

(a1) $$M = \frac{\left[\cos^2\alpha\cos i(k_s/k_n)+\sin(\alpha-i)\sin\alpha\right]\sin(\phi-\alpha)}{\left[\cos\alpha\cos\phi(k_s/k_n)+\sin\phi\sin(\alpha-i)/\cos i\right]\cos i}$$

(a2) $$H_0 = \frac{1}{2}pR[(1+K_0)C_{H1}-(1-K_0)C_{H2}]$$

(a3) $$C_{H1} = \left(\frac{h}{R}+1\right)-\frac{1}{((h/R)+1)}$$

(a4) $$C_{H2} = \left(\frac{h}{R}+1\right)-\frac{1}{((h/R)+1)^3}$$

(a5) $$W = \gamma R^2\left[\cos^2\theta(\tan\theta+\cot\alpha)-\pi/2+\theta\right]$$

(a6) $$\theta = \cos^{-1}\left[\left(\frac{h}{R}+1\right)\sin\alpha\right]+\alpha$$

(b) $$FS_2 = \frac{2S\cos\alpha}{2N\sin\alpha+W}$$ (Asadollahi and Tonon, 2010,
Unwedge program of Rocscience.com)

(b1) $$N = \frac{H_0\left(k_s\cos^2\alpha+k_n\sin^2\alpha\right)\cos\phi}{k_s\cos\alpha\cos\phi+k_n\sin\alpha\sin\phi}$$ (Brady and Brown, 2006)

(for $i = 0$)

(b2) $$S = \frac{H_0\left(k_s\cos^2\alpha+k_n\sin^2\alpha\right)\sin\phi}{k_s\cos\alpha\cos\phi+k_n\sin\alpha\sin\phi}$$

Figure 11.3 Notations, and two equations for the factor of safety of a symmetric roof wedge in a circular tunnel.

In the above equations, W is the weight of the wedge, α the semi-apical angle of the wedge, ϕ and i the effective friction and dilation angles of the joints, k_s and k_n the shear stiffness and normal stiffness of the joints, R the radius of the tunnel, p and K_0 the vertical *in situ* stress and the coefficient of horizontal *in situ* stress, h the clear height of the wedge (measured from the tunnel crown) and θ the angle denoted in Figure 11.3. The equations for M, H_0 and W are shown in Figure 11.3 as Eqs. a1, a2 and a5. The terms C_{H1} and C_{H2} in Eq. a2 are calculated by Eqs. a3 and a4; the term θ in Eq. a5 is calculated by Eq. a6.

The second definition is similar in principle to that which has been long and widely used in soil and rock slope stability analysis, in the *Unwedge* programme of *Rocscience.com*, and in Asadollahi and Tonon (2010), for example. It is the ratio of the available shear strength to the shear strength required for equilibrium. In the present context of tunnel roof wedge, this definition was given in Asadollahi and Tonon (2010) as follows (assuming the dilation angle of the joints $i = 0$):

$$FS_2 = \frac{2S\cos\alpha}{2N\sin\alpha + W} \tag{11.6}$$

where
$N = f(H_0, \phi, \alpha, k_s, k_n)$, and
$S = f(H_0, \phi, \alpha, k_s, k_n)$

where N and S are normal and shear forces, and other symbols as defined for FS_1 in Eq. 11.5. The terms N and S in Eq. 11.6 are calculated from Eqs. b1 and b2 in Figure 11.3.

Low and Einstein (2013) recast the two definitions, Eqs. 11.5 and 11.6, in terms of N, W, α and ϕ, as follows:

$$FS_1 = \frac{\text{Limiting wedge weight}}{\text{Actual wedge weight}} = \frac{2N\tan\phi\cos\alpha - 2N\sin\alpha}{W} = \frac{\tan\phi / \tan\alpha - 1}{W / (2N\sin\alpha)} \tag{11.7}$$

$$FS_2 = \frac{\text{Maximum available resisting forces}}{\text{Downward driving forces}} = \frac{2N\tan\phi\cos\alpha}{2N\sin\alpha + W} = \frac{\tan\phi/\tan\alpha}{1 + W/(2N\sin\alpha)} \tag{11.8}$$

The 'Limiting wedge weight' in Eq. 11.7 means the wedge weight at limiting equilibrium, that is, the wedge weight that just causes failure. It is negative if $\phi < \alpha$.

The same FS_1 is obtained whether computed from Eq. 11.5 or Eq. 11.7, and the same FS_2 is obtained whether computed from Eq. 11.6 or Eq. 11.8. Nevertheless, the rationales, similarities and differences between FS_1 and FS_2 are rendered much more transparent in Eqs. 11.7 and 11.8 than in Eqs. 11.5 and 11.6. That FS_1 can be negative when $\phi < \alpha$ is also readily appreciated from Eq. 11.7. One may note that FS_1 by Eq. 11.7—which is mathematically equivalent to Eq. 11.5—can be very large and positive if W/N is small and $\phi > \alpha$, and negative if $\phi < \alpha$.

The arbitrary and unsatisfactory nature of safety factor definitions has been discussed in the context of rock slopes in Einstein and Baecher (1982), who noted that FS is non-invariant to mechanically equivalent definitions of failure, except at FS = 1.0. The tunnel roof wedge example in the present study provides further corroborations to Einstein and Baecher (1982), and also affords reconciliations for the different *FS* definitions via the first-order reliability method (FORM).

The two definitions as given by Eqs. 11.7 and 11.8, and hence Eqs. 11.5 and 11.6, are mathematically equivalent when $FS_1 = FS_2 = 1$. This is shown in Figure 11.4a, where the FS_1 contours (solid lines) of 1, 10, 20 and 30 are shown together with the FS_2 contours (dashed lines) of 1.0, 1.15, 1.30, 1.48

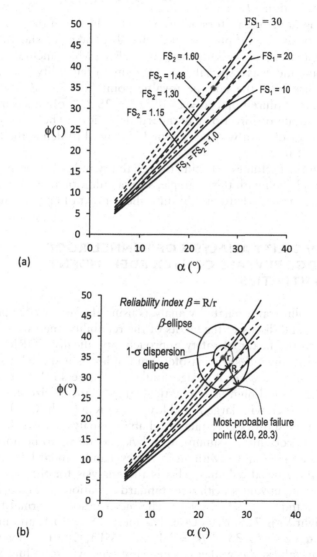

(a)

(b)

Figure 11.4 (a) Contours of FS_1 and FS_2 and the mean-value point where $FS_1 = 30.1$ and $FS_2 = 1.48$, (b) dispersion ellipses in the space of α and ϕ. The plots are for $i = 0$, $k_s/k_n = 0.1$, $p = 1$ MPa, $K_0 = 1.5$, $R = 2$ m, $h/R = 0.85$ and $\gamma = 0.027$ MN/m³ (Low and Einstein 2013).

and 1.60. The $FS_1 = 1.0$ and $FS_2 = 1.0$ contours coincide perfectly. For the input values given in the figure caption, the factors of safety at the mean-value point $\alpha = 25°$ and $\phi = 35°$ are $FS_1 = 30.1$ and $FS_2 = 1.48$, but Figure 11.4a clearly shows that the mean FS_1 is only as safe as the mean FS_2, despite its being 20 times higher than FS_2, because the same (α, ϕ) point for both is at the same distance from the limit sate surface (LSS, where $FS_1 = FS_2 = 1$) which separates the safe combinations $(FS > 1.0)$ of values of (α, ϕ) from the unsafe combinations $(FS < 1.0)$ of (α, ϕ).

The ambiguity in the factors of safety (Figure 11.4a) of the roof wedge in tunnel can be avoided if one uses the reliability index, as shown in Figure 11.4b. The first-order reliability method (FORM, which includes the earlier Hasofer–Lind index as a special case) computes a reliability index, β, which is the distance from the safe mean-value point $(\alpha = 25°, \phi = 35°)$ to the most-probable failure point $(\alpha = 28.01°, \phi = 28.33°,$ obtained in the next section), in units of directional standard deviations. The same β value is computed regardless of whether FS_1 or FS_2 is used to define the limit state surface $FS = 1.0$.

Although the usefulness of using the reliability index has been quite convincingly explained with this example, several additional comments appear to be useful, particularly to put this into the context of engineering design.

11.5 RELIABILITY ANALYSIS OF TUNNEL ROOF WEDGE REVEALS CONTEXT-DEPENDENT SENSITIVITIES

This section illustrates reliability analysis using the roof wedge problem as an example, and discusses the merits of the reliability analysis approach as compared to the factor of safety approach. Specifically, FORM reliability analysis (which includes the Hasofer–Lind index as a special case) will be conducted on a tunnel roof wedge similar to Figure 11.3, first with small tunnel radius R and high *in situ* vertical stress p and horizontal stress $K_0 p$ (with $K_0 > 1$), then with large R and low p and $K_0 p$ (with $K_0 < 1$).

Figure 11.5a shows the results of reliability analysis using the Low and Tang (2007) procedure, assuming α, ϕ, k_s/k_n, p and K_0 to be normally distributed random variables, with mean values μ and standard deviations σ as shown in the labeled columns. This is the analysis for the case in Figure 11.4b. Other input values (with zero standard deviations) are $i = 0$, R = 2.0 m, h/R = 0.85, and γ = 0.027 MN/m³. The mean ratio of normal force N to wedge weight W is 71 for this case. The mean $FS_1 = 30.1$, and mean $FS_2 = 1.48$, with $\alpha = 25°$, $\phi = 35°$, $k_s/k_n = 0.1$, p = 1 MPa, $K_0 = 1.5$. Hence, at these mean input values, the wedge is safe. However, wedge failure will occur when the mean values descend/ascend to the most probable failure point values shown in the x^* column in Figure 11.5a, analogous to Figure 11.4b but in five-dimensional space. The most probable point values are $\alpha^* = 28.01°$,

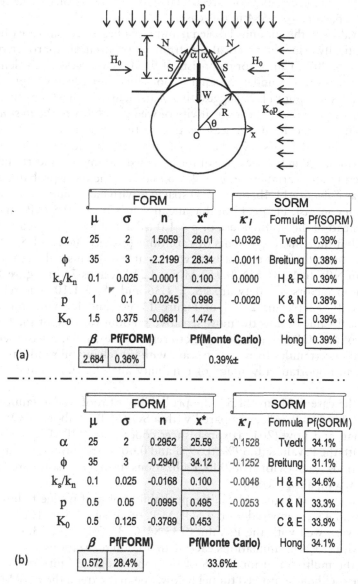

	FORM				SORM		
	μ	σ	n	x*	κ_I	Formula	Pf(SORM)
α	25	2	1.5059	28.01	-0.0326	Tvedt	0.39%
φ	35	3	-2.2199	28.34	-0.0011	Breitung	0.38%
k_s/k_n	0.1	0.025	-0.0001	0.100	0.0000	H & R	0.39%
p	1	0.1	-0.0245	0.998	-0.0020	K & N	0.38%
K_0	1.5	0.375	-0.0681	1.474		C & E	0.39%
	β	Pf(FORM)		Pf(Monte Carlo)		Hong	0.39%
(a)	2.684	0.36%		0.39%±			

	FORM				SORM		
	μ	σ	n	x*	κ_I	Formula	Pf(SORM)
α	25	2	0.2952	25.59	-0.1528	Tvedt	34.1%
φ	35	3	-0.2940	34.12	-0.1252	Breitung	31.1%
k_s/k_n	0.1	0.025	-0.0168	0.100	-0.0048	H & R	34.6%
p	0.5	0.05	-0.0995	0.495	-0.0253	K & N	33.3%
K_0	0.5	0.125	-0.3789	0.453		C & E	33.9%
	β	Pf(FORM)		Pf(Monte Carlo)		Hong	34.1%
(b)	0.572	28.4%		33.6%±			

Figure 11.5 FORM and SORM reliability analysis of tunnel roof wedges: (a) R = 2 m, high p and K_0, (b) R = 6 m, low p and K_0.

$\phi^* = 28.34°$, $(k_s/k_n)^* = 0.1$, $p^* = 0.998$ MPa, $K_0^* = 1.474$. The wedge is on the verge of failure at these values, due mainly to a greater semi-apical angle ($\alpha^* = 28.01°$ > mean α value of 25°) and a lower friction angle of the joints ($\phi^* = 28.34°$ is smaller than the mean φ value of 35°). The distance from the safe mean-value point to this most probable failure combination of parametric values, in units of directional standard deviations, is the reliability index β, equal to 2.684 in this case.

The values in the columns labeled n and x^* in Figure 11.5a were obtained automatically using the practical constrained optimization approach of Low and Tang (2007). The vector n consists of the dimensionless equivalent standard normal variates under the square root sign of Eq. 2.9 of Chapter 2, and their values in Figure 11.5a suggest that ϕ and α are the most sensitive of the five random variables. The insensitivity of wedge stability to the *in situ* stress parameters p and K_0 in this case is due to the large mean N/W value of 2.779 MN/0.0391 MN = 71. Given the large gripping force arising from *in situ* stresses, and in the presence of the assumed parametric uncertainties for the five random variables (α, ϕ, k_s/k_n, p and K_0), the most probable failure point (as in Figure 11.4b, but in 5-D space) is with increasing α and decreasing ϕ. In contrast, Figure 11.5b has mean values of p = 0.5 MPa and K_0 = 0.5, instead of the mean values of p = 1.0 MPa and K_0 = 1.5 used in Figure 11.5a, but the same coefficients of variations (σ/μ) as Figure 11.5a. Also, R = 6.0 m instead of 2.0 m (but h/R remains at 0.85). This reduces the value of mean N/W to 1.131 MN/0.352 MN = 3.214 (versus 71 of Figure 11.5a). The mean factors of safety are FS_1 = 1.36 and FS_2 = 1.10. The reliability index is 0.572 (versus 2.684 of Figure 11.5a), and K_0, with the largest numerical value of n, is the most sensitive random variable of the five. The ability of the FORM reliability index to reflect different parametric sensitivities and uncertainties from case to case without relying on rigid partial factors is an important advantage of reliability analysis and reliability-based design.

For the case in Figure 11.5a, the probability of roof wedge failure based on $P_f = \Phi(-\beta)$ is 0.36%, compared with a SORM P_f of about 0.39% using the Chan and Low (2012a) spreadsheet codes. The SORM P_f agrees very well with the P_f values (0.39%, 0.39% and 0.40%) from three Monte Carlo simulations each with 100,000 realizations using the software @RISK (http://www.palisade.com).

For the case in Figure 11.5b, the probability of roof wedge failure based on β is 28.4%, compared with a SORM P_f of about 33.5%. The SORM P_f agrees very well with the P_f values (33.84%, 33.73% and 33.37%) from three Monte Carlo simulations each with 50,000 realizations.

For the multi-dimensional LSS of the roof wedge stability model above and the rockbolt reinforced tunnel below, one may extend the FORM analysis into the SORM analysis. Should the curvatures of the LSS turn out to be negligible, the curvature values (κ_i) in Figure 11.5a and 11.5b will be practically zeros, and all the SORM formulas will reduce to $\Phi(-\beta)$, with the result that the computed SORM probability of failure will be the same as FORM probability of failure. As a practical alternative, one may also note that reliability-based design typically requires a target β index of 2.5 or 3.0, corresponding to FORM P_f of 0.62% and 0.13%, respectively, and hence some inaccuracy in the FORM P_f based on $\Phi(-\beta)$ is of no practical concern because the correct P_f will still be below 1%. For the case in Figure 11.5a, the P_f of 0.36%, estimated based on FORM β of 2.68, is lower than the more accurate 0.39%, but this discrepancy may not justify much practical concern.

The next example deals with reliability-based design of the length and spacing of rock bolts supporting a circular tunnel. The deterministic procedure implemented in spreadsheet is verified first by comparing with the plots in the source, before extending into probability-based design involving implicit formulations.

11.6 DETERMINISTIC VERIFICATION USING BOBET AND EINSTEIN (2011) FORMULATION OF REINFORCED TUNNEL

Bobet and Einstein (2011) presented closed-form solutions for a circular tunnel supported with elastic rockbolts in homogeneous and isotropic elasto-plastic ground with brittle failure governed by the Coulomb criterion and non-associated flow rule. The formulation assumed axisymmetric *in situ* stress and approximated the three-dimensional effects due to excavation with the β_σ method, where the unreinforced opening just prior to rockbolt installation is subjected to an internal pressure $\sigma_i = \beta_\sigma\sigma_o$ and far-field stress σ_o. (In this section, the β of the β-method is shown with a subscript σ, to distinguish it from the reliability index β.)

As explained in Bobet and Einstein (2011), the β_σ method consists of two steps. In the first step, the tunnel is excavated and a pressure equal to $\beta_\sigma\sigma_o$ is applied to the perimeter of the opening, where $0 < \beta_\sigma < 1$ is the stress reduction factor. In the second step, the reinforcement is placed and the stress $\beta_\sigma\sigma_o$ is shared by the ground and the reinforcement. The magnitude of β_σ is related to the 'delay' in placing the bolts. A large unreinforced length of the tunnel is associated with a small β_σ, and vice versa. A large factor β_σ results in small displacements of the ground, small radial displacements at the perimeter of the excavation and potentially large loads of the reinforcement, and vice versa. The values of β_σ strongly depend on the construction method, ground stiffness, plasticity model, K_o, unreinforced tunnel length, etc. (Möller 2006). A typical value of β_σ is 0.5, but it can range from 0.2 to 0.8. In the reliability-based design of the next section, β_σ is treated as a random variable to reflect its uncertainty.

The rockbolts were modeled in Bobet and Einstein (2011) in one of three ways: (i) discretely mechanically or frictionally coupled (DMFC), (ii) continuously mechanically coupled (CMC) or (iii) continuously frictionally coupled (CFC). Equations were derived for radial stress σ_r, tangential stress σ_θ, radius r_p defining the boundary between the plastic and elastic zones, and radial displacements U_r prior to and after rockbolt reinforcement. The analytical solutions were compared with results obtained using the FEM software ABAQUS. The agreement was good.

The solution of a tunnel with untensioned DMFC bolts was obtained with the assumption that the contribution of the bolt can be approximated as a uniformly distributed pressure (smeared approach). At the tunnel excavation boundary, the support pressure is $p_o = T/(S_\theta S_z)$, where T is the bolt tension, S_θ is the circumferential spacing of the bolts at the tunnel perimeter (Figure 11.6) and S_z is the spacing in the axial direction. The closed-form equation for p_o of untensioned DMFC bolts, given in the Appendix of Bobet and Einstein (2011), is linked to a system of nested equations such that an iterative solution procedure for p_o is required.

Low and Einstein (2013) coded the Bobet and Einstein (2011) system of equations in the VBA programming environment of the Microsoft Excel platform and used the Excel Solver routine to obtain the solution for the

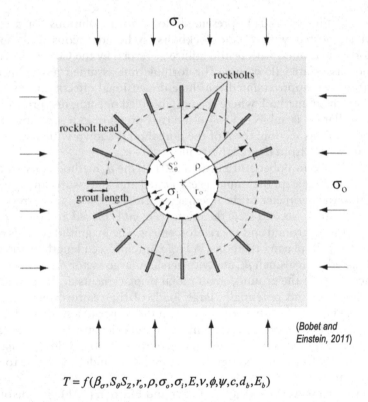

$$T = f(\beta_\sigma, S_\theta S_z, r_o, \rho, \sigma_o, \sigma_i, E, \nu, \phi, \psi, c, d_b, E_b)$$

Mobilized bolt force T is a function of tunnel radius r_o, tunnel internal support pressure σ_i, ground properties E, ν, ϕ, c and ψ (dilation angle), in situ stress σ_o, bolt diameter d_b and bolt modulus E_b, bolt spacings S_θ and S_z, effective bolt length (ρ - r_o), and the parameter β_σ for the 3-D excavation/bolt installation effect

Figure 11.6 Tunnel with DMFC rockbolts (Bobet and Einstein 2011).

support pressure p_o automatically. Excel's GoalSeek tool, which performs Newton-Ralphson iterations automatically, can also be used. Another alternative is mathematical software MatLab.

The Bobet and Einstein (2011) model will be used in reliability-based design of the next section. Hence, it is first necessary to verify that the systems of equations in Bobet and Einstein (2011) have been correctly coded. The verification is done in Figure 11.7a, which shows that the deterministic plots of normalized radial stress σ_r, tangential stress σ_θ and radial displacements U_r are all in agreement with those in Bobet and Einstein (2011). The plots are based on bolt spacing in the axial direction $s_z = 1$ m, tunnel radius $r_0 = 3$ m, rock bolt length plus tunnel radius $= \rho = 6$ m, far-field stress $\sigma_o = 1$ MPa, internal pressure $\sigma_i = 0$, Young's modulus and Poisson's ratio of the ground are $E = 500$ MPa and $\nu = 0.2$, peak and residual internal friction angles $\phi_p = \phi_r = 30°$, dilation angle $\psi = 15°$, peak and residual cohesion $c_p = c_r = 0.1$ MPa, diameter of bolt $d_b = 25$ mm, Young's modulus and Poisson's ratio of bolt are $E_b = 210$ GPa and $\nu_b = 0.3$, and three-dimensional excavation effect parameter $\beta_\sigma = 0.3$.

11.7 RELIABILITY-BASED DESIGN OF THE LENGTH AND SPACING OF ROCKBOLTS FOR A TARGET β VALUE

Low and Einstein (2013) investigated RBD-via-FORM of the length and spacings $S_z S_\theta$ of rockbolts for a reliability index β of 2.5 against bolt tension exceeding the bolt rupture strength of 160 kN. The implicit mechanical model for mobilized bolt force as formulated in Bobet and Einstein (2011) was used, for a circular tunnel supported with elastic rockbolts in a homogeneous and isotropic elasto-plastic ground obeying the Coulomb failure criterion. Four lognormally distributed random variables were modeled, namely the 3-D effect parameter β_σ, and the shear strength parameters ϕ and c and modulus E of the rock mass, with a negative correlation coefficient ρ of –0.5 between ϕ and c. The mean values of β_σ, ϕ, c and E are 0.3, 30°, 0.1 MN/m² and 500 MN/m²; their standard deviations are 0.05, 4°, 0.01 MN/m² and 100 MN/m². The four parameters β_σ, ϕ, c and E are treated as lognormally distributed random variables, and dilation angle ψ is assumed to be equal to 0.5ϕ.

The LSS is described by the following performance function when it evaluates to zero during constrained optimization:

$$g(\mathbf{x}) = T_{Limiting} - T \tag{11.9}$$

where $T_{Limiting} = 0.160$ MN and T is the mobilized bolt force which depends on the 14 parameters shown in the function of T in Figure 11.6. The

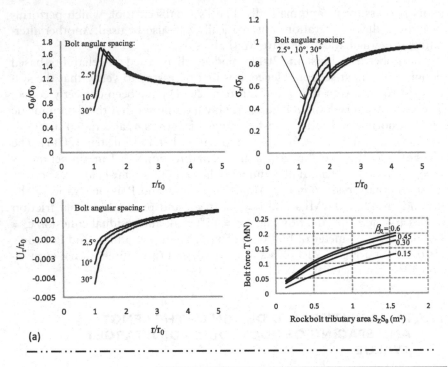

		Probability of failure			
Rock bolt length $(\rho - r_0)$	Required $S_z S_\theta$ for $\beta = 2.5$	FORM $P_f = \Phi(-\beta)$	SORM P_f	Six simulations with Importance Sampling each 3,000 random sets	Six Monte Carlo simulations each 3,000 random sets
2 m	0.59 m²	0.62%	0.44%	0.43%, 0.47%, 0.45%, 0.46%, 0.45%, 0.41%	0.50%, 0.33%, 0.40%, 0.40%, 0.57%, 0.47%
3 m	0.70 m²	0.62%	0.45%	0.46%, 0.48%, 0.44%, 0.48%, 0.47%, 0.45%	0.57%, 0.53%, 0.60%, 0.40%, 0.43%, 0.50%
4 m	0.81 m²	0.62%	0.47%	0.47%, 0.47%, 0.48%, 0.47%, 0.46%, 0.46%	0.37%, 0.60%, 0.73%, 0.53%, 0.30%, 0.47%

(b)

Figure 11.7 (a) Deterministic verification by comparing plots of normalized rock stresses, inward radial displacements U_r and bolt force T with those in Bobet and Einstein (2011), (b) reliability-based design of the length and spacing of rockbolts for a target reliability index β of 2.5, and comparisons of probability of failure based on FORM, SORM, importance sampling and ordinary Monte Carlo simulations. (Low and Einstein, 2013)

mobilized bolt force T is computed using the following expression in Eq. (A4) of Bobet and Einstein (2011):

$$T = K\left[\left(U_r\big|_{r=\rho}^{final} - U_r\big|_{r=\rho}^{initial}\right) - \left(U_r\big|_{r=r_0}^{final} - U_r\big|_{r=r_0}^{initial}\right)\right]$$ (11.10)

where K is the bolt's spring constant, $K = \dfrac{E_b A_b}{\rho - r_0}$, in which E_b is the bolt modulus, A_b the bolt cross-sectional area, and $\rho - r_0$ the effective bolt length. The bracketed term ([.]) computes the net elongation of the rockbolt as the difference in radial displacement U_r between the head (where r = r_0) and tail (where r = ρ) of the rock bolt *after* rockbolt installation. The radial displacement ($U_r|^{initial}$) prior to rockbolt installation is subtracted from the total displacement $U_r|^{final}$ to obtain the displacement after rockbolt installation. The radial displacements $U_r|^{initial}$ and $U_r|^{final}$ are computed from a system of nested equations.

Figure 11.7b shows that the target reliability index β of 2.5 will be achieved when $S_z S_\theta = 0.70$ m² (i.e., $S_z = S_\theta \approx 0.84$ m) for rockbolts of length ($\rho - r_0$) = 3 m. The design point values (x*) are β_σ* = 0.363, ϕ* = 23.7°, c* = 0.103 MPa and E* = 383 MPa, compared with their mean values μ of (0.3, 30°, 0.1 MPa, 500 MPa). The sensitivity indicators (n) values of β_σ, ϕ, c and E are 1.226, –1.703, 0.372, –1.245, respectively, indicating that ϕ is the most sensitive parameter, followed by E and β_σ. The ability of the reliability analysis to seek the most probable failure x* values (also called the design point) without presuming any partial factors is an important and desirable feature. A β of 2.5 implies a failure probability of 0.62%. Higher β values (e.g., 3.0 or 3.5) can be specified if the consequence of failure is severe. In contrast, a target design factor of safety cannot reflect the uncertainties of the input parameters and may even be ambiguous as illustrated in Figure 11.4 where FS_1 = 30 is as safe as FS_2 = 1.48.

In a limit state design based on partial factors, the design value of each parameter is obtained by applying partial factors to the characteristic values of loads and to the characteristic values of strength parameters, such that *Amplified destabilizing effect = Reduced resistance*. In contrast, in a reliability-based design, one does not specify the partial factors. The design point values (x*) are determined automatically and reflect sensitivities, standard deviations, correlation structure, and probability distributions in a way that prescribed partial factors cannot reflect. For example, for the case with rock bolt length of 3 m and $S_z S_\theta$ of 0.7 m² in Figure 11.7b, the ratio of μ_ϕ/ϕ* = 30/23.73 ≈ 1.26 (or, in terms of $\mu_{tan\phi}/tan\phi$*, tan(30°)/tan(23.73°) = 1.31), is similar in nature to the partial factor on the characteristic value of tanφ in limit state design (e.g., EC7). These ratios are informative by-products (corollaries) of a reliability-based design; they incorporate the two-tier safety levels of characteristic values and partial factors but, unlike partial factors, may vary from case to case with uncertainties and sensitivities. In other words, a partial factor in EC7 is ≥ 1.0 and denotes *Design load/Characteristic* *load*, or *Characteristic strength/Design strength*, where the *Characteristic load* (if unfavourable) is a conservative upper-percentile estimate, that is, higher than mean load, and the *Characteristic strength* is a conservative lower-percentile estimate, that is, lower than mean strength. The *Design load* is then obtained as partial factor × *Characteristic load*, and the *Design strength* is obtained as *Characteristic strength*/partial factor. In contrast, the ratio of *Mean strength/Design strength* or *Design Load/Mean load* in the outcome of RBD-via-FORM, while analogous to a code-specified partial factor, reflects uncertainty, correlation and context-dependent sensitivity directly, not through the two-tier safety levels of characteristic values and partial factors.

If rockbolts of length $(\rho-r_0) = 4$ m are used, the target reliability index β of 2.5 will be achieved when $S_z S_\theta = 0.81$ m^2 (i.e., $S_z = S_\theta \approx 0.90$ m). On the other hand, if rockbolts of length $(\rho-r_0) = 2$ m are used, the target reliability index β of 2.5 will be achieved when $S_z S_\theta = 0.59$ m^2 (i.e., $S_z = S_\theta \approx 0.77$ m). These are summarized in the first two columns in Figure 11.7b.

The FORM probability of failure in column 3 of Figure 11.7b is computed from $\Phi(-\beta)$, which is approximate if the limit state surface is curved or the variables follow nonnormal distributions. Comparisons with other methods are shown in columns 4–6, namely SORM, Importance Sampling method, and the ordinary Monte Carlo simulation method. SORM P_f is based on estimated curvatures near the FORM design point (as discussed in Section 2.9 and Figure 11.5). The six Monte Carlo simulations in column 6 were done using the commercial software @RISK with Latin hypercube sampling concentrating near the mean values of the lognormal random variables β_o, ϕ, c and E. The six importance samplings in column 5 were done with @RISK sampling near the FORM design-point values of β_o, ϕ, c and E. The equations for the direct method of importance sampling as proposed in Melchers (1984) based on Shinozuka (1983) and Rubinstein (1981) were used. For the same sample size of 3000 per simulation, Monte Carlo simulations with importance sampling achieve more precise values (much narrower range in column 5) than the ordinary Monte Carlo simulations of column 6. (The spreadsheet-based Monte Carlo simulations incorporating importance sampling was used for laterally loaded piles in Chan and Low 2012b, with similar efficiency.)

The results of SORM and the averages of importance sampling simulations in Figure 11.7b are in very good agreement and indicate a probability of failure of about 0.45%, compared with a FORM P_f of about 0.6% based on $\beta = 2.5$.

In terms of reliability level (equal to $1 - P_f$), it is 99.4% reliability level against rock bolt failure (Eq. 11.9) based on FORM β, versus 99.55% reliability level based on SORM.

Engineers have to decide whether for practical purposes (i) it is adequate (using a target FORM β of 2.5, for example) that the implied *small* probability of failure 0.6%, based on $\Phi(-\beta)$, is sometimes approximate but similarly small as the theoretically correct probability of failure (about 0.45% for the cases in Figure 11.7b), or (ii) engage in more precise reliability-based design via SORM or importance sampling for a target probability of failure. If the latter is the desired action, a practical approach is as follows. In the light of the comparison of FORM P_f of 0.62% vs. actual P_f of 0.45% in Figure 11.7b, suppose a probability of failure of 0.2% is the target P_f for the case in hand, one may do a FORM design of the rock bolt length and spacing for a reliability index corresponding to a probability of failure $P_f = \Phi(-\beta)$ of $(0.62\%/0.45\%)^*0.2\% = 0.28\%$, i.e., $\beta = \Phi^{-1}(1 - 0.0028) =$ Excel function NormSInv(0.9972) = 2.77. For the case with rock bolt length of 3 m, the required $S_z S_\theta$ is found by the Solver tool to be about 0.67 m^2 for a target $\beta = 2.77$. To verify, with input $S_z S_\theta = 0.67$ m^2 and rock bolt length of 3 m, three Monte Carlo simulations incorporating importance sampling each of 10,000 random sets yielded probabilities of failure of 0.206%, 0.199% and 0.203%, respectively; the average is practically the target P_f of 0.2%.

The above reliability-based design procedure may be summarized as follows:

(i) Conduct reliability-based design based on initial target reliability index β_i of 2.5 or 3.0. The implied FORM P_f is $\Phi(-\beta_i)$, equal to 0.62% and 0.13%, respectively.

(ii) For the design solution of (i), obtain more accurate P_f using SORM or importance sampling on the foundation of FORM design point. This P_f is denoted as $P_{f,SORM}$.

(iii) Decide on a target P_f (< 1%), and repeat the reliability-based design based on an updated target reliability index (β_{rev}) as follows:

$$\beta_{rev} = \Phi^{-1}\left(1 - \frac{\Phi(-\beta_i)}{P_{f,SORM}} \times \text{target} P_f\right) \qquad (11.11)$$

The final design solution is obtained via repeated FORM analyses using different designs until the reliability index is equal to β_{rev}. Linear interpolation between two designs and their respective reliability indices can also be done. The final design based on β_{rev} will have failure probability close to the target P_f.

If the design $S_z S_\theta$ is to be obtained by direct trial and error of different designs each checked against Monte Carlo simulations (with or without importance sampling), the computation time is likely to be longer than the FORM-and-SORM approach involving Eq. 11.11.

Appendix

An efficient spreadsheet algorithm for **FORM**

The matrix formulation (Veneziano, 1974; Ditlevsen, 1979) of the Hasofer and Lind (1974) index β is:

$$\beta = \min_{\mathbf{x} \in F} \sqrt{(\mathbf{x} - \boldsymbol{\mu})^T \mathbf{C}^{-1} (\mathbf{x} - \boldsymbol{\mu})} \tag{A1a}$$

or, equivalently:

$$\beta = \min_{\mathbf{x} \in F} \sqrt{\left[\frac{x_i - \mu_i}{\sigma_i} \right]^T \mathbf{R}^{-1} \left[\frac{x_i - \mu_i}{\sigma_i} \right]} \tag{A1b}$$

where $\mathbf{x}$ is a vector representing the set of random variables x_i, $\boldsymbol{\mu}$ the vector of mean values μ_i, $\mathbf{C}$ the covariance matrix, $\mathbf{R}$ the correlation matrix, σ_i the standard deviations and F the failure domain. The point denoted by the x_i values, which minimize Eq. (A1) and satisfy $\mathbf{x} \in F$, is the most probable failure point, also referred to as the design point in RBD with a target reliability index. This is the point of tangency of an expanding dispersion ellipsoid with the limit state surface (LSS), which separates safe combinations of parametric values from unsafe combinations, as shown in Figure 2.3 of Chapter 2. The one-standard-deviation (1σ) dispersion ellipse and the β-ellipse in that figure are tilted by virtue of cohesion c and friction angle ϕ being negatively correlated. The quadratic form in Eq. (A1) appears also in the negative exponent of the established probability density function of the multivariate normal distribution. As a multivariate normal dispersion ellipsoid expands from the mean-value point, its expanding surfaces are contours of decreasing probability values. Hence, to obtain β by Eq. (A1) means maximizing the value of the multivariate normal probability density function and is graphically equivalent to finding the smallest ellipsoid tangent to the LSS at the most probable failure point (the *design point*). This intuitive and visual understanding of the *design point* is consistent with the more mathematical approach in Shinozuka (1983), in which all variables were standardized and the limit state equation was written in terms of standardized variables.

In FORM, for correlated nonnormal variates, one can rewrite Eq. (A1b) as follows (Low and Tang, 2004) and regard the computation of β as that of finding the smallest equivalent hyperellipsoid (centred at the equivalent normal mean-value point μ^N and with equivalent normal standard deviations σ^N) that is tangent to the limit state surface (LSS):

$$\beta = \min_{x \in F} \sqrt{\left[\frac{x_i - \mu_i^{\,N}}{\sigma_i^{\,N}}\right]^T \mathbf{R}^{-1} \left[\frac{x_i - \mu_i^{\,N}}{\sigma_i^{\,N}}\right]} \qquad (A2)$$

where μ_i^N and σ_i^N can be calculated by the Rackwitz and Fiessler (1978) equivalent normal transformation:

Equivalent normal standard deviation: $\quad \sigma^N = \dfrac{\varphi\left\{\Phi^{-1}\left[F(x)\right]\right\}}{f(x)} \qquad (A3)$

Equivalent normal mean: $\quad \mu^N = x - \sigma^N \times \Phi^{-1}\left[F(x)\right] \qquad (A4)$

where x is the original nonnormal variate, $\Phi^{-1}[.]$ is the inverse of the cumulative probability (CDF) of a standard normal distribution, $F(x)$ is the original nonnormal CDF evaluated at x, $\phi\{.\}$ is the probability density function (PDF) of the standard normal distribution and $f(x)$ is the original nonnormal probability density ordinate at x.

Hence, for correlated nonnormals, the ellipsoid perspective still applies in the original coordinate system, except that the nonnormal distributions are replaced by an equivalent normal ellipsoid, centred not at the original mean values of the nonnormal distributions, but at the equivalent normal mean μ^N.

Eq. (A2) and the Rackwitz–Fiessler equations of A3 and A4 were used in the spreadsheet-automated constrained optimization computational approach of FORM in Low and Tang (2004). An alternative to the 2004 FORM procedure is given in Low and Tang (2007), which uses the following equation for the reliability index β:

$$\beta = \min_{x \in F} \sqrt{\mathbf{n}^T \mathbf{R}^{-1} \mathbf{n}} \qquad \left(\text{No need to compute } \mu_i^N \text{ and } \sigma_i^N\right) \qquad (A5)$$

The computational approaches of Eqs. (A1b), (A2) and (A5) and associated ellipsoidal perspective are complementary alternatives to the classical u-space computational approach and may help overcome the conceptual and language barriers.

The two spreadsheet-based computational approaches of FORM are compared in Figures 2.4 and 2.6 of Chapter 2. Either method can be used as an alternative to the classical u-space FORM procedure. The vectors **n** and u can be obtained from one another, $n = Lu$ and $u = L^{-1}n$, as follows (e.g., Low et al., 2011):

$$\beta = \min_{x \in F} \sqrt{n^T R^{-1} n} = \min_{x \in F} \sqrt{n^T (LU)^{-1} n} = \min_{x \in F} \sqrt{(L^{-1}n)^T (L^{-1}n)} \quad \text{(A6a)}$$

i.e. $\quad \beta = \min_{x \in F} \sqrt{u^T u}, \quad$ where $\quad u = L^{-1}n, \quad$ and $\quad n = Lu \quad$ (A6b)

in which L is the lower triangular matrix of R. When the random variables are uncorrelated, $u = n$ by Eq. (A6a), because then $L^{-1} = L = I$ (the identity matrix).

The expressions which enable x_i to be computed automatically as a function of n_i are shown in Figure A1. The derivations of the closed form expressions $x_i = h(n_i)$ are straightforward for normal, lognormal, Gumbel, exponential, uniform, triangular and Weibull distributions, as shown below. For normal variates, $x_i = \mu_i + n_i \sigma_i$.

Type of F(x)	x = F⁻¹[Φ(n)]
Normal: mean μ_x, std. dev. σ_x	$x = \mu_x + n\sigma_x$
Lognormal: mean μ_x, std. dev. σ_x	$x = \exp[\lambda + \zeta n], \quad \zeta = \sqrt{\ln(1 + (\sigma_x/\mu_x)^2)}, \quad \lambda = \ln\mu_x - 0.5\zeta^2$
Extreme Value (Gumbel): mean μ_x, std. dev. σ_x	$x = u - \dfrac{\ln[-\ln(\Phi(n))]}{\alpha}, \quad \alpha = \dfrac{1}{\sqrt{6}}\left(\dfrac{\pi}{\sigma_x}\right), \quad u = \mu_x - \dfrac{0.5772}{\alpha}$
Exponential: mean = μ_x	$x = -\mu_x \ln(1 - \Phi(n))$
Uniform: x_{min} x_{max}	$x = x_{min} + (x_{max} - x_{min}) \times \Phi(n)$
Triangular: a m c	$x = a + \sqrt{\Phi(n) \times (m-a) \times (c-a)}, \quad \text{if } \Phi(n) \le (m-a)/(c-a).$ $x = c - \sqrt{(1-\Phi(n)) \times (c-a) \times (c-m)}, \text{ otherwise.}$
Weibull CDF: $1 - \exp[-(x/\lambda)^\alpha]$	$x = \lambda[-\ln(1-\Phi(n))]^{1/\alpha}$
Gamma with PDF: $f(x) = \dfrac{x^{\alpha-1}\exp(-x/\alpha)}{\lambda^\alpha \Gamma(\alpha)}$	Para1 = α, Para2 = λ Mean: $\mu_x = \alpha \times \lambda$ Newton method used to obtain x from n, starting from μ_x.
Beta distribution	Para 1 = α, Para2 = λ, Para3 = min, Para4 = max Mean: μ_x = min + (max − min)$\alpha/(\alpha + \lambda)$ Newton method used to obtain x from n, starting from μ_x.
PERT distribution	Para1 = min, Para2 = mode, Para3 = max Mean: μ_x = (min + 4*mode +max)/6 Newton method used to obtain x from n, starting from μ_x.

Figure A1 Obtaining x from n, based on F(x) = Φ(n). (Low and Tang, 2007)

For a lognormal variate x with mean μ_x and standard deviation σ_x, the CDF is known to be:

$$CDF = P[X \le x] = \Phi\left(\frac{\ln x - \lambda}{\zeta}\right) \tag{A7}$$

where $\zeta = \sqrt{\ln\left(1 + (\sigma_x/\mu_x)^2\right)}$, $\lambda = \ln \mu_x - 0.5\zeta^2$ and $\Phi(.)$ is the standard normal cumulative distribution function. Equating the above CDF to $\Phi(n)$ and, after rearrangement, one obtains:

$$x = \exp[\lambda + \zeta n]$$

For Type 1 extreme value distribution (Gumbel), equating its CDF with the CDF of standard normal n:

$$Exp\left[-Exp(-\alpha(x - u))\right] = \Phi(n)$$

from which one obtains: $$x = u - \frac{\ln\left[-\ln(\Phi(n))\right]}{\alpha}$$

where the parameters α and u are known functions of the mean and standard deviation of the extreme value variable x as:

$$\alpha = \frac{1}{\sqrt{6}}\left(\frac{\pi}{\sigma_x}\right), \text{ and } u = \mu_x - \frac{0.5772}{\alpha}$$

For exponential distribution, equating its CDF to the CDF of n:

$$1 - \exp\left(-\frac{x}{\mu_x}\right) = \Phi(n)$$

hence $$x = -\mu_x \ln(1 - \Phi(n)) = -\mu_x \ln(1 - normsdist(n))$$

where $normsdist(.)$ is Microsoft Excel's built-in function for evaluating $\Phi(.)$.

For uniform and triangular distributions, the derivations of x as a function of n follow the same principle.

For the two-parameter Weibull distribution, equating its CDF to the standard normal CDF of n:

$$1 - \exp\left[-\left(\frac{x}{\lambda}\right)^\alpha\right] = \Phi(n)$$

from which one obtains: $$x = \lambda\left[-\ln(1 - \Phi(n))\right]^{1/\alpha}$$

```
Function x_i(DistributionName, paralist, ni) As Double
  para1 = paralist(1):  para2 = paralist(2):  para3 = paralist(3):  para4 = paralist(4)
  With Application.WorksheetFunction
  Select Case UCase(Trim(DistributionName))        'Trim leading/trailing spaces and convert to uppercase
    Case "NORMAL":    x_i = para1 + ni * para2
    Case "LOGNORMAL":   lamda = Log(para1) - 0.5 * Log(1 + (para2 / para1) ^ 2)
      zeta = Sqr(Log(1 + (para2 / para1) ^ 2)):  x_i = Exp(lamda + zeta * ni)
    Case "EXTVALUE1":  alfa = 1.28255 / para2:  u = para1 - 0.5772 / alfa
      x_i = u - Log(-Log(.NormSDist(ni))) / alfa
    Case "EXPONENTIAL":  mean = para1:  x_i = -mean * Log(1 - .NormSDist(ni))
    Case "UNIFORM":    min = para1:  max = para2:   x_i = min + (max - min) * .NormSDist(ni)
    Case "TRIANGULAR":  a = para1:  m = para2:  c = para3:  tem = .NormSDist(ni):  maca = (m - a) / (c - a)
      If tem <= maca Then x_i = a + Sqr(tem * (m - a) * (c - a)) Else x_i = c - Sqr((1 - tem) * (c - a) * (c - m))
    Case "WEIBULL":    x_i = para2 * (-Log(1 - .NormSDist(ni))) ^ (1 / para1)
    Case "GAMMA":   xprev = para1 * para2
      For i = 1 To 100
        CDF = .GammaDist(xprev, para1, para2, True): pdf = .GammaDist(xprev, para1, para2, False)
        xnew = xprev - (CDF - .NormSDist(ni)) / pdf
        If Abs((xnew - xprev) / xprev) < 0.000001 Then Exit For
        If xnew <= 0 Then xnew = 0.5 * xprev
        xprev = xnew
      Next i
      x_i = xnew
    Case "BETADIST":   a1 = para1: a2 = para2: min = para3: max = para4:  xprev = min + (max - min) * a1 / (a1 + a2)
8:    For i = 1 To 100
        CDF = .BetaDist(xprev, a1, a2, min, max)
        BetaFunc = Exp(.GammaLn(a1) + .GammaLn(a2) - .GammaLn(a1 + a2))
        pdf = 1 / BetaFunc * (xprev - min) ^ (a1 - 1) * (max - xprev) ^ (a2 - 1) / (max - min) ^ (a1 + a2 - 1)
        xnew = xprev - (CDF - .NormSDist(ni)) / pdf
        If Abs((xnew - xprev) / xprev) < 0.000001 Then Exit For
        If xnew <= min Then xnew = 0.5 * (min + xprev):  If xnew >= max Then xnew = 0.5 * (max + xprev)
        xprev = xnew
      Next i
      x_i = xnew
    Case "PERTDIST":   min = para1:  Mode = para2:  max = para3:  mean = (min + 4 * Mode + max) / 6:  xprev = mean
      If Mode = mean Then f = 6 Else f = (2 * Mode - min - max) / (Mode - mean)
      a1 = (mean - min) * f / (max - min):  a2 = a1 * (max - mean) / (mean - min)
      GoTo 8
  End Select
  End With
End Function
```

Figure A2. VBA Function x_i for obtaining x_i from n_i, coded inside Microsoft Excel, based on Figure A1.

The above results are summarized in Figure A1, and coded in the VBA programming environment of Microsoft Excel in Figure A2.

In general, the objective is to find the value x such that the nonnormal cumulative probability distribution $F(x)$ at x is equal to the standard normal cumulative distribution $\Phi(n)$, as illustrated above. If closed form expression cannot be obtained, one can use a refined Newton method (coded in Figure A2 under the cases Gamma, BetaDist and PERTDist) to obtain the solution by iteration via Eq. (A8) below.

For gamma, beta and PERT distributions, the Newton-Raphson (or Newton) iteration method is used to determine x such that $F(x) = \Phi(n)$:

$$x_{k+1} = x_k - \frac{F(x_k) - \Phi(n)}{\frac{d}{dx}\left[F(x_k) - \Phi(n)\right]} = x_k - \frac{F(x_k) - \Phi(n)}{f(x_k)} \tag{A8}$$

where $F(x)$ is cumulative probability and $f(x)$ is probability density function. The initial value prior to iterative solution by the Newton method is the mean value of the nonnormal distribution, shown in the last three rows of

Figure A1 and embodied under cases Gamma, BetaDist and PERTDist in Figure A2. The Newton method is coded within the iteration loops of cases Gamma, BetaDist and PERTDist. A simple refinement has been incorporated in the Newton method to ensure that the iterations stay within the permissible range of the Gamma, beta and the PERT distributions. The program Function x_i(…) returns the solution of x_i based on the equations in Figure A1. Its arguments are *DistributionName*, *Paralist* and n_i. In the fourth line of the program in Figure A2, the *Select Case* control structure alters the flow of execution to one of several code segments, depending on the input *DistributionName*. In the third line of the VBA program, '*With Application. WorksheetFunction*' enables liberal calls to Excel's built-in functions, using the syntax '.NormSDist(.),' '.GammaDist(…),' '.BetaDist(…),' '.GammaLn,' where 'Application' stands for Microsoft Excel. The use of these Excel objects (each a container of program codes) results in much simplicity, clarity and brevity of the program codes and structure.

References

@RISK. (n.d.). Version 7.6 [Computer software], Palisade, Ithaca, NY.

AASHTO. (2020). *AASHTO LRFD Bridge Design Specifications*, 9th edn, American Association of State Highway and Transportation Officials (AASHTO), Washington, DC.

Ang, H.S., and Tang, W.H. (1976). *Probability Concepts in Engineering Planning and Design, Vol. I: Basic Principles*, John Wiley & Sons, New York, 409 p.

Ang, H.S., and Tang, W.H. (1984). *Probability Concepts in Engineering Planning and Design, Vol. II: Decision, Risk, and Reliability*, John Wiley & Sons, New York, 562 pp.

Ang, H.S., and Tang, W.H. (2006). *Probability Concepts in Engineering: Emphasis on Applications in Civil & Environmental Engineering*, 2nd edn, John Wiley, New York.

Asadollahi, P., and Tonon, F. (2010). "Definition of factor of safety for rock blocks", *International Journal of Rock Mechanics and Mining Sciences*, 47, 1384–1390.

Baecher, G.B., and Christian, J.T. (2003). *Reliability and Statistics in Geotechnical Engineering*, John Wiley, Chichester, West Sussex, England; Hoboken, NJ.

Bathurst, R.J., and Javankhoshdel, S. (2017). "Influence of model type, bias and input parameter variability on reliability analysis for simple limit states in soil–structure interaction problems", *Georisk: Assessment and Management of Risk for Engineered Systems and Geohazards*, Taylor & Francis, 11(1):42–54.

Benjamin, J.R., and Cornell, C.A. (1970). *Probability, Statistics, and Decisions for Civil Engineers*, McGraw-Hill, Inc., New York.

Bobet, A., and Einstein, H.H. (2011). "Tunnel reinforcement with rockbolts", *Tunnelling and Underground Space Technology*, 26, 100–123.

Bond, A.J., Schuppener, B., Scarpelli, G., and Orr, T.L.L. (2013). *"Eurocode 7: Geotechnical design worked examples"*, Worked Examples presented at the *Workshop "Eurocode 7: Geotechnical Design"*, Dublin, 1–14 June 2013, https://eurocodes.jrc.ec.europa.eu/doc/2013_06_WS_GEO/report/2013_06_WS_GEO.pdf

Bond, A.J., Burlon, S., Van Seters, A., and Simpson, B. (2015). *"Planned changes in Eurocode 7 for the second generation of Eurocodes"*, Geotechnical Engineering for Infrastructure and Development – Proceedings of the XVI European Conference on Soil Mechanics and Geotechnical Engineering (ECSMGE 2015), Vol. 7, pp. 4217–4222.

Bowles, J.E. (1996). *Foundation Analysis and Design*, 5th edn, McGraw-Hill, New York.

Brady, B.H.G., and Brown, E.T. (2006). *Rock Mechanics for Underground Mining*, 3rd edn, Kluwer, Dordrecht.

BS 8002. (1994). *Code of Practice for Earth Retaining Structures*, British Standards Institution, London.

Burland, J.B., and Burbidge, M.C. (1985). "Settlements of foundations on sand and gravel", *Proceedings of the Institution of Civil Engineers*, 78(1), 1325–1381.

Chan, C.L., and Low, B.K. (2012a). "Practical second-order reliability analysis applied to foundation engineering", *International Journal for Numerical and Analytical Methods in Geomechanics*, 36(11), 1387–1409.

Chan, C.L., and Low, B.K. (2012b). "Probabilistic analysis of laterally loaded piles using response surface and neural network approaches", *Computers and Geotechnics*, 43, 101–110.

Chan, C.L., Low, B.K., and Ng, M.W.M. (2012). *"Probabilistic sensitivity analysis illustrated for an underground rock excavation"*, *Proceedings of the 13th World Conference of the Associated Research Centers for the Urban Underground Space (ACUUS 2012)*, 7–9 November 2012, Society for Rock Mechanics and Engineering Geology Singapore, Singapore, pp. 1227–1238.

Chen, Z.Y., and Morgenstern, N.R. (1983). "Extensions to the generalized method of slices for stability analysis", *Canadian Geotechnical Journal*, 20(1), 104–119. doi:10.1139/t83-010

Ching, J.Y., Phoon, K.-K., and Hu, Y.G. (2009). "Efficient evaluation of reliability for slopes with circular slip surfaces using importance sampling", *Journal of Geotechnical and Geoenvironmental Engineering*, ASCE, 135(6), 768–777.

Choa, V., Wong, K.S., and Low, B.K. (1990). *New Airport at Chek Lap Kok, Geotechnical Review and Assessment*, Consulting Report to Maunsell Pvt Ltd., Singapore.

Clayton, C.R.I., Woods, R.I., Bond, A.J., and Milititsky, J. (2013). *Earth Pressure and Earth-Retaining Structures*, 3rd edn, CRC Press, Taylor & Francis Group, Boca Raton, FL, 588 pp.

Coates, R.C., Coutie, M.G., and Kong, F.K. (1994). *Structural Analysis*, 3rd edn, Chapman & Hall, London.

Cornell, C.A. (1967). "Bounds on the reliability of structural systems", *Journal of the Structural Division*, ASCE, 93(1), 171–200.

CSA (Canadian Standards Association). (2019). *Canadian Highway Bridge Design Code*. CAN/CSA-S6-19. CSA, Mississauga, ON, Canada.

Der Kiureghian, A. (2005). "First- and second-order reliability methods" (Chapter 14), *Engineering Design Reliability Handbook*, edited by E. Nikolaidis, D.M. Ghiocel, and S. Singhal, CRC Press, Taylor & Francis Group, Boca Raton, FL.

Der Kiureghian, A., and Liu, P.L. (1986). "Structural reliability under incomplete probability information", *Journal of Engineering Mechanics*, ASCE, 112(1), 85–104.

Der Kiureghian, A., Lin, H.Z., and Hwang, S.J. (1987). "Second-order reliability approximations", *Journal of Engineering Mechanics*, ASCE, 113(8), 1208–1225.

Ditlevsen, O. (1979). "Generalized second moment reliability index", *Journal of Structural Mechanics*, 7(4), 435–451.

Ditlevsen, O. (1981). *Uncertainty Modeling: With Applications to Multidimensional Civil Engineering Systems*, McGraw-Hill, New York.

Duncan, J.M. (2000). "Factors of safety and reliability in geotechnical engineering", *Journal of Geotechnical and Geoenvironmental Engineering*, 126(4), 307–316.

Duncan, J.M., and Buchignani, A.L. (1973). "Failure of underwater slope in San Francisco Bay", *Joubal of the Soil Mechanics and Foundation Division*, ASCE, 99(9), 687–703.

Duncan, J.M., and Wright, S.G. (2005). *Soil Strength and Slope Stability*, Wiley & Sons, 297 pp.

Duncan, J.M., Wright, S.G., and Brandon, T.L. (2014). *Soil Strength and Slope Stability*, 2nd edn, Wiley, Hoboke, NJ.

Einstein, H.H., and Baecher, G.B. (1982). "Probabilistic and statistical methods in engineering geology, I. Problem statement and introduction to solution", *Rock Mechanics*, Springer-Verlag, 12(Suppl.), 47–61.

EN7-1. (2004). *Eurocode 7: Geotechnical Design – Part 1: General Rules*, BS EN 1997-1:2004, British Standards Institution, London.

Eurocode 7 (EC7). (2020). *Geotechnical Design – Part 3: Geotechnical Structures*, prEN 1997-3-Working Document, 296 pp.

Evans, M., Hastings, N., and Peacock, B. (2000). *Statistical Distributions*, 3rd edn, Wiley, New York.

Everitt, B.S., and Skrondal, A. (2010). *The Cambridge Dictionary of Statistics*, 4th edn, Cambridge University Press, Cambridge.

Foott, R., Koutsoftas, D.C., and Handfelt, L.D. (1987). "Test gill at Chek Lap Kok, Hong Kong", *Journal of Geotechnical Engineering*, ASCE, 113(2), 106–126.

Frank, R., Bauduin, C., Driscoll, R., Kavvadas, M., Krebs Ovesen, N., Orr, T.L.L., and Schuppener, B. (2005). *Designers' Guide to EN 1997-1 Eurocode 7: Geotechnical Design – General Rules*, Thomas Telford, London.

Geotechnical Engineering Office of Hong Kong. (2006). *Geo Publication No. 1/2006, Foundation Design and Construction*, Civil Engineering and Development Department.

Goodman, R.E. (1989). *Introduction to Rock Mechanics*, 2nd edn, John Wiley, New York, 562 pp.

Goodman, R.E. (1995). "Thirty-fifth Rankine lecture: 'Block theory and its application'", *Geotechnique*, 45(3), 383–423.

Goodman, R.E., and Taylor, R.L. (1967). "Methods of analysis for rock slopes and abutments: A review of recent developments", *Failure and Breakage of Rocks*, edited by C. Fairhurst, AIME, pp. 303–320.

Haldar, A., and Mahadevan, S. (2000). *Probability, Reliability and Statistical Methods in Engineering Design*, John Wiley, New York.

Harr, M.E. (1987). *Reliability-based Design in Civil Engineering*, McGraw-Hill, New York.

Hasofer, A.M., and Lind, N.C. (1974). "Exact and invariant second-moment code format", *Journal of Engineering Mechanics*, ASCE, 100, 111–121.

Hendron, A.J., Cording, E.J., and Aiyer, A.K. (1971). "Analytical and graphical methods for the analysis of slopes in rock masses", *U.S. Army Engrg. Nuclear Cratering Group Technical Rep. No. 36*.

Hetenyi, M. (1946). *Beams on Elastic Foundations*, University of Michigan Press, Ann Arbor, MI.

Hoek, E. (2007). "Practical rock engineering", https://www.rocscience.com/learning/hoeks-corner/course-notes-books

Hoek, E., and Bray, J. (1981a). *Rock Slope Engineering*, revised 3rd edn, Institute of Mining and Metallurgy, London, 358 pp.

Hoek, E., and Bray, J. (1981b). *Rock Slope Engineering*, 3rd edn, Institute of Mining and Metallurgy, London, UK.

Hoek, E., Bray, J.W., and Boyd, J.M. (1973). "The stability of a rock slope containing a wedge resting on two intersecting discontinuities", *Quarterly Journal of Engineering Geology & Hydrogeology*, 6(1), 1–55.

Hoek, E., Kaiser, P.K., and Bawden, W.F. (2000). *Support of Underground Excavations in Hard Rock*, Balkema, Rotterdam, The Netherlands, CRC Press, Taylor & Francis Group, Boca Raton, FL.

Hudson, J.A., and Harrison, J.P. (1997). *Engineering Rock Mechanics: An Introduction to the Principles*, 2nd impression 2000, Elsevier, Pergamon.

Jaeger, J.C. (1971). "Friction of rocks and the stability of rock slopes", *Geotechnique*, 21(2), 97–134.

Jaksa, M.B., Kaggwa, W.S., and Brooker, P.I. (1999). *"Experimental evaluation of the scale of fluctuation of a stiff clay"*, *Proceedings of the 8th International Conference on the Application of Statistics and Probability*, Sydney, Australia, Vol. 1, pp. 415–422, AA Balkema, Rotterdam.

Jeffrey, J.R., Brown, M.J., Knappett, J.A., Ball, J.D., and Caucis, K. (2016). "Continuous helical displacement pile performance, Part I: Physical modelling", *Proceedings of the Institution of Civil Engineers: Geotechnical Engineering*, 169(GE5), 421–435.

John, K.W. (1968). "Graphical stability analysis of slopes in jointed rock", *Journal of the Soil Mechanics and Foundations Division*, ASCE, 94(2), 497–526.

Karlsrud, K., Hansen, S.B., Dyvik, R., and Kalsnes, B. (1993). "NGI's pile tests at Tilbrook and Pentre—review of testing procedures and results", *Large-Scale Pile Tests in Clay*, edited by Clarke, J., Thomas Telford, London, pp. 549–583.

Knappett, J., and Craig, R.F. (2019). *Craig's Soil Mechanics*, 9th edn, CRC Press, Taylor & Francis Group, Boca Raton, FL.

Kolk, H.J., and van der Velde, E. (1996). *"A reliable method to determine friction capacity of piles driven into clays"*, *Proceedings of the Offshore Technology Conference*, Houston, TX, Paper OTC 7993.

Kottegoda, N.T., and Rosso, R. (2008). *Applied Statistics for Civil and Environmental Engineers*, 2nd edn, Wiley-Blackwell, New York, 736 pp.

Koutsoftas, D.C., Foott, R., and Handfelt, L.D. (1987). "Geo-technical investigations offshore Hong Kong", *Journal of Geotechnical Engineering*, ASCE, 113(2), 87–105.

Kreyszig, E. (1988). *Advanced Engineering Mathematics*, 6th edn, Wiley, New York, pp. 972–973.

Lambe, T.W., and Whitman, R.V. (1979). *Soil Mechanics* [SI version], John Wiley & Sons, 553 pp.

Li, G. (2000). "Soft clay consolidation under reclamation fill and reliability analysis", *Ph.D. thesis*, Nanyang Technological University, School of Civil & Environmental Engineering, Singapore.

Londe, P., Vigier, G., and Vormeringer, R. (1969). "The stability of rock slopes, a three-dimensional study", *Journal of the Soil Mechanics and Foundations Division*, ASCE, 95(1), 235–262.

Low, B.K. (1997). "Reliability analysis of rock wedges", *Journal of Geotechnical and Geoenvironmental Engineering*, ASCE, 123(6), 498–505.

Low, B.K. (2003a). *"Practical probabilistic slope stability analysis"*, Proceedings, Soil and Rock America 2003, 12th Panamerican Conference on Soil Mechanics and Geotechnical Engineering and 39th U.S. Rock Mechanics Symposium, MIT, Cambridge, MA, 22–26 June 2003, Verlag Glückauf GmbH Essen, Vol. 2, pp. 2777–2784.

Low, B.K. (2003b). "Theories, computations, and design procedures involving vertical drains" (Chapter 2), *Soil Improvement: Prefabricated Vertical Drain Techniques*, edited by M.W. Bo, J. Chu, B.K. Low, and V. Choa, Thomson Learning, Thomson Asia Pvt Ltd., pp. 5–56.

Low, B.K. (2005). "Reliability-based design applied to retaining walls", *Geotechnique*, 55(1), 63–75.

Low, B.K. (2007). "Reliability analysis of rock slopes involving correlated nonnormals", *International Journal of Rock Mechanics and Mining Sciences*, Elsevier, 44(6), 922–935.

Low, B.K. (2008a). "Practical reliability approach using spreadsheet" (Chapter 3), *Reliability-Based Design in Geotechnical Engineering: Computations and Applications*, edited by K.-K. Phoon, Taylor & Francis, pp. 134–168.

Low, B.K. (2008b). *"Settlement analysis of Chek Lap Kok trial embankments with probabilistic extensions"*, Proceedings of the Sixth International Conference on Case Histories in Geotechnical Engineering and Symposium in Honor of Professor James K. Mitchell, Arlington, Virginia, USA, 11–16 August 2008, 12 pp.

Low, B.K. (2010). *"Slope reliability analysis: Some insights and guidance for practitioners"*, Proceedings of the 17th Southeast Asian Geotechnical Conference, Taipei, Taiwan, 10–13 May 2010, Vol. 2, pp. 231–234.

Low, B.K. (2015). "Reliability-based design: Practical procedures, geotechnical examples, and insights" (Chapter 9), *Risk and Reliability in Geotechnical Engineering*, edited by K.-K. Phoon and J. Ching, CRC Press, Taylor & Francis Group, Boca Raton, FL, pp. 355–393.

Low, B.K. (2017). "Insights from reliability-based design to complement load and resistance factor design approach", *Journal of Geotechnical and Geoenvironmental Engineering*, ASCE, 143(11). (2019 ASCE Thomas A. Middlebrooks award).

Low, B.K. (2019). "Probabilistic insights on a soil slope in San Francisco and a rock slope in Hong Kong", *Georisk: Assessment and Management of Risk for Engineered Systems and Geohazards*, 13(4), 326–332. doi:10.1080/17499518.2019.1606923

Low, B.K. (2020a). *"Geotechnical insights from reliability-based design to improve partial factor design methods"*, ASCE Geo-Congress 2020 (GSP 316), Minneapolis, Minnesota, 25–28 February 2020, pp. 686–695.

Low, B.K. (2020b). "Correlated sensitivities in reliability analysis and probabilistic Burland-and-Burbidge method", *Georisk: Assessment and Management of Risk for Engineered Systems and Geohazards*, Taylor & Francis. doi:10.1080/1749951 8.2020.1861636

Low, B.K., and Duncan, J.M. (2013). *"Testing bias and parametric uncertainty in analyses of a slope failure in San Francisco Bay mud"*, Proceedings of Geo-Congress 2013, San Diego, CA, 3–6 March 2013, ASCE, pp. 937–951.

Low, B.K., and Einstein, H.H. (1992). *"Simplified reliability analysis for wedge mechanisms in rock slopes"*, Proceedings, Sixth International Symposium on Landslides, New Zealand, Vol. 1, pp. 499–507, A.A. Balkema Publishers.

Low, B.K., and Einstein, H.H. (2013). "Reliability analysis of roof wedges and rock-bolt forces in tunnels", *Tunnelling and Underground Space Technology*, 38, 1–10.

Low, B.K., and Phoon, K.-K. (2015a). *"Geotechnical reliability-based designs and links with LRFD"*, *12th International Conference on Applications of Statistics and Probability in Civil Engineering*, Vancouver, BC, Canada, 12–15 July 2015.

Low, B.K., and Phoon, K.-K. (2015b). "Reliability-based design and its complementary role to Eurocode 7 design approach", *Computers and Geotechnics*, Elsevier, 65, 30–44.

Low, B.K., and Tang, W.H. (1997). "Reliability analysis of reinforced embankments on soft ground", *Canadian Geotechnical Journal*, 34(5), 672–685.

Low, B.K., and Tang, W.H. (2004). "Reliability analysis using object-oriented constrained optimization", *Structural Safety*, Elsevier, 26(1), 69–89.

Low, B.K., and Tang, W.H. (2007). "Efficient spreadsheet algorithm for first-order reliability method", *Journal of Engineering Mechanics*, 1378–1387. doi:10.1061/(ASCE)0733-9399(2007)133:12(1378)

Low, B.K., Gilbert, R.B., and Wright, S.G. (1998). "Slope reliability analysis using generalized method of slices", *Journal of Geotechnical and Geoenvironmental Engineering*, ASCE, 124(4), 350–362.

Low, B.K., Lacasse, S., and Nadim, F. (2007). "Slope reliability analysis accounting for spatial variation", *Georisk: Assessment and Management of Risk for Engineered Systems and Geohazards*, 1(4), 177–189.

Low, B. K., Teh, C. I., and Tang, W.ilson H. (2001). *"Stochastic nonlinear p-y analysis of laterally loaded piles"*., *Proceedings of the Eight International Conference on Structural Safety and Reliability, ICOSSAR '01*, Newport Beach, CA, 17–22 June 2001, A.A. Balkema Publishers, 8 pp.

Low, B.K., Zhang, J., and Tang, W.H. (2011). "Efficient system reliability analysis illustrated for a retaining wall and a soil slope", *Computers and Geotechnics*, Elsevier, 38(2), 196–204.

Low, B.K., et al. (2017). "Lead discusser of Chapter 4: EXCEL-based direct reliability analysis and its potential role to complement Eurocodes", *Final Report of the Joint TC205/TC304 Working Group on "Discussion of Statistical/Reliability Methods for Eurocodes"*, pp. 79–101, https://www.icsmge2017.org/download/19th%20ICSMGE_Workshop_TC205&304.pdf

Lü, Q., and Low, B.K. (2011). "Probabilistic analysis of underground rock excavations using response surface method and SORM", *Computers and Geotechnics*, Elsevier, 38, 1008–1021.

Madsen, H.O., Krenk, S., and Lind, N.C. (1986). *Methods of Structural Safety*, Prentice Hall, Englewood Cliffs, NJ, 403 pp.

Melchers, R.E. (1984). "Efficient Monte-Carlo probability integration", *Research Report No. 7*, Dept of Civil Engineering, Monash University, Clayton, Australia.

Melchers, R.E. (1987). *Structural Reliability: Analysis and Prediction*, Ellis Horwood Ltd., Chichester, West Sussex, England.

Melchers, R.E. (1999). *Structural Reliability: Analysis and Prediction*, 2nd edn, John Wiley, New York.

Melchers, R.E., and Beck, A.T. (2018). *Structural Reliability: Analysis and Prediction*, 3rd edn, John Wiley, New York.

Möller, S. (2006). "Tunnel induced settlements and structural forces in linings", *Ph.D. thesis*, Universität Stuttgart, Germany.

Nash, D. (1987). "A comparative review of limit equilibrium methods of stability analysis", *Slope Stability*, edited by M.G. Anderson and K.S. Richards, John Wiley & Sons Ltd, pp. 11–75.

Orr, T.L.L. (2017). "Defining and selecting characteristic values of geotechnical parameters for designs to Eurocode 7", *Georisk: Assessment and Management of Risk for Engineered Systems and Geohazards*, 11(1), 103–115.

Phoon, K.-K. (2017). "Role of reliability calculations in geotechnical design", *Georisk: Assessment and Management of Risk for Engineered Systems and Geohazards*, 11(1), 4–21.

Phoon, K.-K., and Tang, C. (2019). "Characterisation of geotechnical model uncertainty", *Georisk: Assessment and Management of Risk for Engineered Systems and Geohazards*, 13(2), 101–130. doi:10.1080/17499518.2019.1585545

Poulos, H.G., and Davis, E.H. (1980). *Pile Foundation Analysis and Design*, John Wiley, New York.

Rackwitz, R. (1976). "Practical probabilistic approach to design", *Bulletin 112*, Comite European du Beton, Paris, France.

Rackwitz, R. (2001). "Reliability analysis: A review and some perspectives", *Structural Safety*, Elsevier, 23(4), 365–395.

Rackwitz, R., and Fiessler, B. (1978). "Structural reliability under combined random load sequences", *Computers & Structures*, Elsevier, 9, 484–494.

Randolph, M.F., and Murphy, B.S. (1985). "*Shaft capacity of driven piles in clay*", *Proceedings of the 17th Annual Offshore Technology Conference*, Houston, TX, Vol. 1, pp. 371–378.

Randolph, M.F., and Wroth, C.P. (1978). "Analysis of deformation of vertically loaded piles", *Journal of the Geotechnical Engineering Division*, American Society of Civil Engineers, 104, 1465–1488.

Rubinstein, R.Y. (1981). *Simulation and the Monte Carlo Method*, John Wiley & Sons, New York.

Sahai, H., and Khurshid A. (2001). *Pocket Dictionary of Statistics*, McGraw-Hill, Boston.

Salgado, R. (2008). *The Engineering of Foundations*, McGraw-Hill, Boston.

Salgado, R., and Kim, D. (2014). "Reliability analysis of load and resistance factor design of slopes", *Journal of Geotechnical and Geoenvironmental Engineering*, 57–73. doi:10.1061/(ASCE)GT.1943-5606.0000978

Schneider, H.R., and Schneider, M.A. (2013). "Dealing with uncertainties in EC7 with emphasis on determination of characteristic soil properties", *Modern Geotechnical Design Codes of Practice*, edited by P. Arnold, G.A. Fenton, M.A. Hicks, T. Schweckendiek, and B. Simpson, IOS Press, Amsterdam, pp. 87–101.

Semple, R.M., and Rigden, W.J. (1984). "*Shaft capacity of driven piles in clay*", *Proceedings of the Symposium on Analysis and Design of Pile Foundations*, San Francisco, CA, pp. 59–79.

Shahin, M.A., Maier, H.R., and Jaksa, M.B. (2002). "Predicting settlement of shallow foundations using neural networks", *Journal of Geotechnical and Geoenvironmental Engineering*, ASCE, 128 (9), 779–785.

Shinozuka, M. (1983). "Basic analysis of structural safety", *Journal of the Structural Engineering*, ASCE, 109(3), 721–740.

Simpson, B. (2017). "*Eurocode 7 and robustness*", *Geo-Risk 2017: Keynote Lectures*, Denver, CO, 4–7 June 2017, Geotechnical Special Publication (GSP 282), pp. 52–68.

Skempton, A.W., and LaRochelle, P. (1965). The Bradwell slip: A short-term failure in London clay, *Geotechnique*, 15(3), 221–242.

Smith, I. (2014). *Smith's Elements of Soil Mechanics*, 9th edn, Wiley-Blackwell, Chichester, West Sussex, 471 pp.

Sofianos, A.I., Nomikos, P., and Tsoutrelis, C.E. (1999). "Stability of symmetric wedge formed in the roof of a circular tunnel: Nonhydrostatic natural stress field", (Technical Note), *International Journal of Rock Mechanics and Mining Sciences*, 36, 687–691.

Spencer, E. (1973). "Thrust line criterion in embankment stability analysis", *Geotechnique*, 23:85–100.

Tandjiria, V., Teh, C.I., and Low, B.K. (2000). "Reliability analysis of laterally loaded piles using response surface methods", *Structural Safety*, Elsevier, 22(4), 335–355.

Terzaghi, K., Peck, R.B., and Mesri, G. (1996). *Soil Mechanics in Engineering Practice*, 3rd edn, Wiley, New York. ISBN: 978-0-471-08658-1.

Tichy, M. (1993). *Applied Methods of Structural Reliability*, Kluwer Academic, Dordrecht, Boston, 402 pp.

Tomlinson, M.J. (1994). *Pile Design and Construction Practice*, 4th edn, E & FN Spon, London.

Tomlinson, M.J. (2001). *Foundation Design and Construction*, 7th edn, Longman Scientific, Harlow.

Tomlinson, M.J., and Woodward, J. (2015). *Pile Design and Construction Practice*, 6th edn, CRC Press, Taylor & Francis Group, Boca Raton, FL.

Tung, Y.K., Yen, B.C., and Melching, C.S. (2006). *Hydrosystems Engineering Reliability Assessment and Risk Analysis*, McGraw-Hill, New York.

Upton, G., and Cook, I. (2014). *The Oxford Dictionary of Statistics*, 3rd edn, Oxford Univrsity Press, Oxford.

Vanmarcke, E.H. (1980). "Probabilistic stability analysis of earth slopes", *Engineering Geology*, 16, 29–50.

Veneziano, D. (1974). "Contributions to second moment reliability", *Research Report No. R74-33*, Department of Civil Engrg., MIT, Cambridge, Massachusetts.

Wittke, W. (1990). *Rock Mechanics: Theory and Applications with Case Histories* (English translation), Springer-Verlag, New York.

Wu, T.H. (2008). "Reliability analysis of slopes" (Chapter 11), *Reliability-Based Design in Geotechnical Engineering: Computations and Applications*, edited by K.-K. Phoon, Taylor & Francis, London, pp. 413–447.

Wu, T.H., and Kraft, L.M. (1970). "Safety analysis of slopes", *Journal of the Soil Mechanics and Foundations Division*, 96(2), 609–630.

Wyllie, D.C. (2018). *Rock Slope Engineering: Civil Applications*, 5th edn, Based on the third edition by Hoek and Bray, Taylor & Francis, CRC Press, Boca Raton, FL.

Xu, B., and Low, B.K. (2006). "Probabilistic stability analyses of embankments based on finite-element method", *Journal of Geotechnical and Geoenvironmental Engineering*, ASCE, 132(11), 1444–1454.

Young, W.C., and Budynas, R.G. (2002). *Roark's Formulas of Stress and Strain*, 7th edn, McGraw-Hill, New York.

Further reading

Baecher, G.B., and Christian, J.T. (2010). "Risk modeling issues and appropriate technology", *Geotechnical Special Publication*, 199, 1944–1951.

Bathurst, R.J. (2015). "LRFD calibration of simple limit state functions in geotechnical soil-structure design" (Chapter 8), *Risk and Reliability in Geotechnical Engineering*, edited by K.-K. Phoon and J. Ching, CRC Press, Boca Raton, FL, pp. 339–354.

Bjerager, P., and Krenk, S. (1989). "Parametric sensitivity in first order reliability theory", *Journal of Engineering Mechanics*, ASCE, 115(7), 1577–1582.

Bolton, M.D. (1993). "What are partial factors for?" *The International Symposium on Limit State Design in Geotechnical Engineering*, 1993-5 to –, Copenhagen, Denmark, pp. 565–584.

Breitung, K. (1984). "Asymptotic approximations for multinormal integrals", *Journal of Engineering Mechanics*, ASCE, 110(3), 357–366.

BSI (British Standards Institution). (2004). "Eurocode 7: Geotechnical design. Part 1: General rules", EN 1997-1, London.

Cai, G.Q., and Elishakoff, I. (1994). "Refined second-order reliability analysis", *Structural Safety*, Elsevier, 14(4), 267–276.

Chan, C.L., and Low, B.K. (2009). "Reliability analysis of laterally loaded piles involving nonlinear soil and pile behavior", *Journal of Geotechnical and Geoenvironmental Engineering*, ASCE, 135(3), 431–443.

Chan, C.L., and Low, B.K. (2012). "Sensitivity analysis of laterally loaded pile involving correlated non-normal variables", *International Journal of Geotechnical Engineering*, 6(2), 163–169.

Chen, X., and Lind, N.C. (1983). "Fast probability integration by three-parameter normal tail approximations", *Structural Safety*, 1(4), 269–276.

Ching, J.Y., Li, D.Q., and Phoon, K.-K. (2016). "Statistical characterization of multivariate geotechnical data" (Chapter 4), *Reliability of Geotechnical Structures in ISO2394*, edited by K.-K. Phoon and J.V. Retief, CRC Press/Balkema, London, pp. 89–126.

Chowdhury, R.N., and Xu, D.W. (1995). "Geotechnical system reliability of slopes", *Reliability Engineering and System Safety*, Elsevier, 47, 141–151.

Christian, J.T., Ladd, C.C., and Baecher, G.B. (1994). "Reliability applied to slope stability analysis", *Journal of Geotechnical Engineering*, ASCE, 120(12), 2180–2207.

Dai, S.H., and Wang, M.O. (1992). *Reliability Analysis in Engineering Applications*, Van Nostrand Reinhold, New York, 433 pp.

Duncan, J.M., and Sleep, M. (2016). "The need for judgement in geotechnical reliability studies", *Georisk: Assessment and Management of Risk for Engineered Systems and Geohazards*, 11(1), 42–54.

Einstein, H.H. (1991). "Reliability in rock engineering", *Proc., Geotech. Engrg. Congress, Geotech. Spec. Publ. #27*, ASCE, 1, 608–633.

Einstein, H.H. (1996). "Risk and risk analysis in rock engineering", *Tunnelling and Underground Space Technology*, 11(2), 141–155.

Einstein, H.H., and Baecher, G.B. (1983). "Probabilistic and statistical methods in engineering geology – Specific methods and examples part I: Exploration", *Rock Mechanics and Rock Engineering*, 16(1), 39–72.

Einstein, H.H., Veneziano, D., Baecher, G.B., and O'Reilly, K.J. (1983). "The effect of discontinuity persistence on rock slope stability", *International Journal of Rock Mechanics and Mining Sciences*, 20(5), 227–236.

ENV 1997-1. (1994). *Eurocode 7: Geotechnical Design – Part 1: General Rules*, CEN, European Committee for Standardization, Brussels.

Fenton, G.A., Griffiths, D.V., and Zhang, X. (2008). "Load and resistance factor design of shallow foundations against bearing failure", *Canadian Geotechnical Journal*, 45(11), 1556–1571.

Fiessler, B., Neumann, H.J., and Rackwitz, R. (1979). "Quadratic limit states in structural reliability", *Journal of the Engineering Mechanics Division*, ASCE, 105(4), 661–676.

Foye, K.C., Salgado, R., and Scott, B. (2006). "Resistance factors for use in shallow foundation LRFD", *Journal of Geotechnical and Geoenvironmental Engineering*, 1208–1218. doi:10.1061/(ASCE)1090-0241(2006)132:9(1208)

Gilbert, R.B., Habibi, M., and Nadim, F. (2016a). "Accounting for unknown unknowns in managing multi-hazard risks" (Book Chapter), *Multi-Hazard Approaches to Civil Infrastructure Engineering*, Springer International Publishing, pp. 383–412.

Gilbert, R.B., Wright, S.G., and Allen, J.M. (2016b). "Lessons learned about the stability of waste containment systems", *Geotechnical Special Publication* (GSP 274), pp. 45–59.

Griffiths, D.V., and Fenton, G.A. (2004). "Probabilistic slope stability analysis by finite elements", *Journal of Geotechnical and Geoenvironmental Engineering*, 130(5), 507–518.

Griffiths, D.V., Huang, J., and Fenton, G.A. (2009). "Influence of spatial variability on slope reliability using 2-D random fields", *Journal of Geotechnical and Geoenvironmental Engineering*, 135(10), 1367–1378.

Harrison, J.P. (2018). "Eurocode 7: Genesis, development and implications for rock engineering", *ISRM International Symposium – 10th Asian Rock Mechanics Symposium*, 29 October–3 November, International Society for Rock Mech. & Rock Engrg, Singapore. ISRM-ARMS10-2018-253.

Harrison, J.P., and Hudson, J.A. (2010). "Incorporating parameter variability in rock mechanics analyses: Fuzzy mathematics applied to underground rock spalling", *Rock Mechanics and Rock Engineering*, 43(2), 219–224.

Hohenbichler, M., and Rackwitz, R. (1988). "Improvement of second-order estimates by importance sampling", *Journal of Engineering Mechanics*, ASCE, 114(12), 2195–2199.

Hong, H.P. (1999). "Simple approximations for improving second-order reliability estimates", *Journal of Engineering Mechanics*, ASCE, 125(5), 592–595.

Jaksa, M.B., Brooker, P.I., and Kaggwa, W.S. (1997). "Inaccuracies associated with estimating random measurement errors", *Journal of Geotechnical and Geoenvironmental Engineering*, 123(5), 393–401.

Jaksa, M.B., Kaggwa, W.S., Fenton, G.A., and Poulos, H.G. (2003). "A framework for quantifying the reliability of geotechnical investigations", *9th International Conference on Applications of Statistics and Probability in Civil Engineering*, San Francisco, California, 6–9 July 2003.

Ji, J., and Low, B.K. (2012). "Stratified response surfaces for system probabilistic evaluation of slopes", *Journal of Geotechnical and Geoenvironmental Engineering*, 138(11), 1398–1406.

Kahiel, A., Najjar, S., and Sadek, S. (2017). "Reliability-based design of spread footings on clays reinforced with aggregate piers", *Georisk: Assessment and Management of Risk for Engineered Systems and Geohazards*, 11(1), 75–85.

Kitch, W.A., Wright, S.G., and Gilbert, R.B. (1995). "Probabilistic analysis of reinforced soil slopes", *Proc., 10th Conf. on Engrg. Mech.*, ASCE, 1, 325–328.

Koyluoglu, H.U., and Nielsen, S.R.K. (1994). "New approximations for SORM integrals", *Structural Safety*, Elsevier, 13(4), 235–246.

Kulatilake, P.H.S.W., Chen, J., Teng, J., Shufang, X., and Pan, G. (1996). "Discontinuity geometry characterization in a tunnel close to the proposed permanent shiplock area of the Three Gorges Dam site in China", *International Journal of Rock Mechanics and Mining Sciences & Geomechanics Abstracts*, Elsevier, 33(3), 255–277.

Lacasse, S., and Nadim, F. (1996). "Model uncertainty in pile axial capacity calculations", *Proc. 28th Offshore Technology Conference*, Houston, Texas, pp. 369–380.

Lacasse, S., Liu, Z., Kim, J., Choi, J.C., and Nadim, F. (2017). "Reliability of slopes in sensitive clays", *Advances in Natural and Technological Hazards Research*, 46, pp. 511–537.

Lesny, K. (2019). "Probability-based derivation of resistance factors for bearing capacity prediction of shallow foundations under combined loading", *Georisk: Assessment and Management of Risk for Engineered Systems and Geohazards*, 13(4), 284–290.

Li, D. (2017). "Incorporating spatial variability into geotechnical reliability based design" (Chapter 7), *TC205/TC304 Joint Report on Statistical/Reliability Methods for Eurocodes*.

Li, D., Chen, Y., Lu, W., and Zhou, C. (2011). "Stochastic response surface method for reliability analysis of rock slopes involving correlated non-normal variables", *Computers and Geotechnics*, 38(2011), 58–68.

Li, H-Z., and Low, B. K. (2010). "Reliability analysis of circular tunnel under hydrostatic stress field." *Computers and Geotechnics*, Elsevier, 37(1–2), 50–58.

Li, D.Q., Zheng, D., Cao, Z.J., Tang, X.S., and Phoon, K.-K. (2016). "Response surface methods for slope reliability analysis: Review and comparison", *Engineering Geology*, 203, 3–14.

Li, K.S., and Lumb, P. (1987). "Probabilistic design of slopes", *Canadian Geotechnical Journal*, 24(4), 520–535.

Lin, P., and Bathurst, R.J. (2018). "Influence of cross correlation between nominal load and resistance on reliability-based design for simple linear soil-structure limit states", *Canadian Geotechnical Journal*, 55, 279–295.

Liu, W.F., and Leung, Y.F. (2018). "Characterizing three-dimensional anisotropic spatial correlation of soil properties through in situ test results", *Geotechnique*, 68(9), 805–819.

Liu, H.X., and Low, B.K. (2017). "System reliability analysis of tunnels reinforced by rockbolts", *Tunnelling and Underground Space Technology*, 65, 155–166.

Liu, H., and Low, B.K. (2018). "Reliability-based design of tunnelling problems and insights for Eurocode 7", *Computers and Geotechnics*, Elsevier, 97(42), 51.

Low, B.K. (2008). "Efficient probabilistic algorithm illustrated for a rock slope", *Rock Mechanics and Rock Engineering*, Springer-Verlag, 41(5), 715–734.

Low, B.K. (2014). "FORM, SORM, and spatial modeling in geotechnical engineering", *Structural Safety*, Elsevier, 49, 56–64.

Low, B.K., and Teh, C.I. (1999). "Probabilistic analysis of pile deflection under lateral loads", *International Conference on Applications of Statistics and Probability (ICASP8)*, A.A. Balkema Publishers, Rotterdam, The Netherlands, Vol. 1, pp. 407–414.

Lü, Q., Chan, C.L., and Low, B.K. (2012). "Probabilistic evaluation of ground-support interaction for deep rock excavation using artificial neural network and uniform design", *Tunnelling and Underground Space Technology*, 32, 1–18.

Lü, Q., Sun, H.-Y., and Low, B.K. (2011). "Reliability analysis of ground-support interaction in circular tunnels using the response surface method", *International Journal of Rock Mechanics and Mining Sciences*, 48(8), 1329–1343.

Luckman, P.G., Der-Kiureghian, A., and Sitar, N. (1987). "Use of stochastic stability analysis for Bayesian back calculation of pore pressures acting in a cut slope at failure", *Proc., ICASP5, the Fifth International Conference on Applications of Statistics and Probability in Soil and Structural Engineering*, Vancouver, BC, Canada, Vol. 2, pp. 922–929.

Massih, D.Y.A., Soubra, A.H., and Low, B.K. (2008). "Reliability-based analysis and design of strip footings against bearing capacity failure." *Journal of Geotechnical and Geoenvironmental Engineering*, ASCE, 134(7), 917–928.

Matlock, H. (1970). "Correlations for design of laterally loaded piles in soft clay", *Proc., Offshore Technology Conference*, Houston, Texas, Paper OTC 1204.

Mitchell, J.K., Seed, R.B., and Seed, H.B. (1990). "Kettleman hills waste landfill slope failure. I: Liner-system properties", *Journal of Geotechnical Engineering*, 116(4), 647–668.

Mostyn, G.R., and Li, K.S. (1993). "Probabilistic slope analysis – State of play", *Proc., Conf. on Probabilistic Methods in Geotechnical Engineering*, Balkema, pp. 89–109.

Nadim, F., Einstein, H.H., and Roberds, W. (2005). "Probabilistic stability analysis for individual slopes in soil and rock", *International Conference on Landslide Risk Management*, Vancouver, BC, Canada, 31 May–2 June 2005, pp. 63–98.

Najjar, S., Shammas, E., and Saad, M. (2017). "A reliability-based approach to the serviceability limit state design of spread footings on granular soil", *Geotechnical Special Publication* (GSP 286), pp. 185–202.

National Research Council. (1995). *Probabilistic Methods in Geotechnical Engineering*, National Academy Press, Washington, DC.

Oka, Y., and Wu, T.H. (1990). "System reliability of slope stability", *Journal of Geotechnical Engineering*, ASCE, 116(8), 1185–1189.

Orr, T.L.L. (2019). "Honing safety and reliability aspects for the second generation of Eurocode 7", *Georisk: Assessment and Management of Risk for Engineered Systems and Geohazards*, 13(3), 205–213.

Orr, T.L.L., and Breysse, D. (2008). "Eurocode 7 and reliability-based design" (Chapter 8), *Reliability-Based Design in Geotechnical Engineering*, edited by K.-K. Phoon, Taylor & Francis Group, Milton Park, UK, pp. 298–343.

Paikowsky, S., Canniff, M., Lesny, K., Kisse, A., Amatya, S., and Muganga, R. (2010). "LRFD design and construction of shallow foundations for highway bridge structures", *NCHRP Report 651*. Transportation Research Board, Washington, DC.

Phoon, K.-K., and Kulhawy, F.H. (2008). "Serviceability limit state reliability-nased design" (Chapter 9), *Reliability-Based Design in Geotechnical Engineering: Computations & Applications*, edited by K.-K. Phoon, Taylor & Francis, London, pp. 344–384.

Phoon, K.-K., Prakoso, W.A., Wang, Y., and Ching, J.Y. (2016). "Uncertainty representation of geotechnical design parameters" (Chapter 3), *Reliability of Geotechnical Structures in ISO2394*, edited by K.-K. Phoon and J.V. Retief, CRC Press/Balkema, London, pp. 49–87.

Randolph, M.F., and Wroth, C.P. (1979). "An analysis of the vertical deformation of pile groups", *Géotechnique*, 29, 423–439.

Rosenblatt, M. (1952). "Remarks on a multivariate transformation", *Annals of Mathematical Statistics*, 23(3), 470–472.

Simpson, B., and Driscoll, R. (1998). *Eurocode 7: A Commentary*, ARUP/BRE, Construction Research Communications Ltd, London.

Simpson, B., and Hocombe, T. (2010). "Implications of modern design codes for earth retaining structures", *Geotechnical Special Publication 384* (GSP 208), pp. 786–803.

Song, L., Li, H-Z, Chan, C.L., and Low, B.K. (2016). "Reliability analysis of underground excavation in elastic-strain-softening rock mass". *Tunnelling and Underground Space Technology*, 60, 66–79.

Tang, W.H. (1993). "Recent developments in geotechnical reliability", *Probabilistic Methods in Geotechnical Engineering*, edited by K.S. Li and S.-C.R. Lo, CRC Press/Balkema, Rotterdam, pp. 3–27.

TC205/TC304 Working Group. (2017). "Discussion of statistical/reliability methods for eurocodes", https://www.icsmge2017.org/download/19th%20ICSMGE_Workshop_TC205&304.pdf

Tvedt, L. (1988). "Second-order reliability by an exact integral", *2nd IFIP Working Conference on Reliability and Optimization on Structural Systems*, Springer-Verlag, Berlin, Germany, pp. 377–384.

Tvedt, L. (1990). "Distribution of quadratic forms in normal space: Application to structural reliability", *Journal of Engineering Mechanics*, ASCE, 116(6), 1183–1197.

Vanmarcke, E.H. (1977). "Reliability of earth slopes", *Journal of Geotechnical Engineering*, ASCE, 103(11), 1247–1266.

Veneziano, D., Agarwal, A., and Karaca, E. (2009). "Decision making with epistemic uncertainty under safety constraints: An application to seismic design", *Probabilistic Engineering Mechanics*, 24(3), 426–437.

Wang, Y., Cao, Z., and Li, D. (2016). "Bayesian perspective on geotechnical variability and site characterization", *Engineering Geology*, 203, 117–125.

Wathugala, D.N., Kulatilake, P.H.S.W., Wathugala, G.W., and Stephansson, O. (1990). " A general procedure to correct sampling bias on joint orientation using a vector approach", *Computers and Geotechnics*, Elsevier, 10, 1–31.

Whitman, R.V. (1984). "Evaluating calculated risk in geotechnical engineering", *Journal of Geotechnical Engineering*, ASCE, 110(2), 145–188.

Whitman, R.V. (1996). "Organizing and evaluating uncertainnty in geotechnical engineering", *Proc., Uncertainty in Geologic Environment: From Theory to Practice*, ASCE Geotechnical Special Publication #58, 1, 1–28.

Wu, Y.T., and Wirsching, P.H. (1987). "New algorithm for structural reliability estimation", *Journal of Engineering Mechanics*, ASCE, 113(9), 1319–1335.

Zhang, J. (2017). "Bayesian method: A natural tool for processing geotechnical information", (Chapter 6), *TC205/TC304 Joint Report on Statistical/Reliability Methods for Eurocodes*.

Zhang, L.L., Zhang, J, Zhang, L.M., Tang, Wilson H. (2011). "Stability analysis of rainfall-induced slope failure: a review", *Proceedings of the Institution of Civil Engineers - Geotechnical Engineering*, 164(5), 299–316, Thomas Telford Ltd.

Zhao, Y.-G., and Ono, T. (1999). "New approximations for SORM: Part 1", *Journal of Engineering*, 25(1), 79–85.

Zhao, Y.-G., and Ono, T. (2001). "Moment methods for structural reliability", *Structural Safety*, 23(1), 47–75.

Index

@RISK software, 17, 29, 41, 48, 50, 52, 65, 75, 102, 111, 122, 205, 252, 296

A

Ang, Alfredo H-S., 30, 35, 50, 53, 69, 71, 90, 128, 258
anisotropic shear strength, 233, 255
applications of RBD-via-FORM in civil and environmental engineering
  arriving on time, 50–53
  beam on Winkler medium, 76–79
  column with axial loading and biaxial bending, FORM and SORM, 73–75
  design of travel time, 53–57
  earthquake-induced shear stress, FORM and SORM, 35–38
  harbour breakwater, 66–69
  increasing spillway capacity, cost comparisons, 71–73
  irrigation water, 57–61
  reinforced concrete beam, 76
  spillway capacity, 69–71
  spring suspending a load, 108–110
  storm sewer system, 63–66
  strut with complex supports and implicit performance function, 79–82
  thermal pollution of river, 61–63
  traffic capacity, 48–50
  traffic congestion, 45–48
applications of RBD-via-FORM in soil engineering
  earth retaining structures

anchored sheet pile wall, 214–222
  search for critical quadrilateral wedge, 222–229
  semi-gravity wall, 194–214
  vertical drainage layer behind retaining wall, 229–232
pile foundations
  laterally loaded piles in soil with nonlinear Matlock p-y curves
    fully embedded pile,189–191
    long cantilever length, 177–188
  pile in sand, 166–172
  pile settlement, 172–177
  pile in stiff clay below a jetty, 157–166
soil improvement
  Chek Lap Kok test fill, 93–108
soil slope stability
  excavated slope in London Clay at Bradwell, 243–249
  Norwegian slope with spatial autocorrelation, 253–256
  a slope in layered soil, 233–243
  slope system reliability analysis, 256–258
  sloping core dam, 258–263
  Spencer method for non-circular slip surface, 264–274
  underwater slope in San Francisco Bay mud, 249–253
spread foundation
  bearing capacity of spread foundation: vertical load only, 116–119

consolidation settlement of soft clay, 30–35

footing width design by EC7 and RBD-via-FORM, 128–134

foundation of retaining wall involving load-resistance duality, 135–145

foundation settlement by Burland and Burbidge method, 145–155

RBD of foundation settlement, 90–93

Rotational stability of a lightweight structure and correlated sensitivities, 24–29

applications of RBD-via-FORM in rock engineering

plane sliding in rock slopes

a failed slope in a limestone quarry, 303–307

Sau Mau Ping slope with combined slope reprofiling, reinforcement and drainage, 303–304

Sau Mau Ping slope, Hong Kong, FORM and SORM, 297–303

slope with active and passive blocks, 288–297

slope with a discontinuity plane and context-dependent sensitivities, 277–288

rock slopes with 3D tetrahedral wedges

from dip directions and dip angles to intrinsic angles β_1, β_2, δ_1 and δ_2, 312, 322, 332

a more critical sliding mode behind the mean-value sliding mode, 328–332

tetrahedral rock wedge with a sliding mode along both planes, 310, 314, 324–328

tetrahedral wedge with sliding mode on plane 1 only, 322

verification of closed-form solutions with vectorial procedure, 323

water pressures in terms of G_{w1} and G_{w2}, or in terms of u_1 and u_2, 314–315, 317–318

underground excavations in rock

FORM and SORM of tunnel roof wedge, 346–349

ground-support interaction of a shaft in sandstone, 335–342

roof wedge in tunnel, a tale of two factors of safety, 342–346

tunnel reinforced with rock bolts, FORM and SORM, 349–355

Asadollahi, P., 343–344

autocorrelation, 164–166, 168, 184, 247, 253–255, 271

B

Baecher, G.B., 9, 256, 344

Bathurst, R.J., 136

Bayesian, 9, 53–54

Benjamin, J.R., 9

beta general distribution, see probability distributions

Bobet, A., 349–353

Bowles, J.E., 118, 136

Brady, B.H.G., 342–343

Bray, J., 312, 320, 322–324, 342

breakwater, 66

breasting dolphin, 179, 185

Brown, E.T, 342–343

Burbidge, M.C., 145–146, 148–150, 152

Burland, J.B., 145–146, 148–150, 152

C

CDF (cumulative distribution function), 9, 45–46, 103–108, 195, 358, 360

Chan, C.L. 38–39

characteristic value, 83, 87, 129–131, 164, 169, 171, 174, 208–209, 218–219, 339–340, 353

Chek Lap Kok, Hong Kong, 93

Chen, Zu-Yu, 269

Ching, J.Y., 256

Cholesky decomposition, 23, 359

Christian, J.T., 9, 256

Clayton, C.R.I., 230

coefficient of variation (CV, or c.o.v.), 40, 45, 61, 68, 73, 75,

109–110, 131, 160, 177, 184, 246
consolidation settlement of soft clay, 30–35, 95–98
Cornell, C.A., 9, 204
correlated sensitivities, 24–29
correlation coefficient (ρ), 15, 24, 28, 37, 59, 125, 287, 300, 351
correlation matrix $\mathbf{R}$, 13, 18, 21, 37, 40, 55, 58, 215, 248, 357
cost-benefit analysis, increasing spillway capacity, 71
covariance matrix $\mathbf{C}$, 13, 120, 199, 357
Craig, R.F., 128, 146, 166, 172
cumulative distribution function, see CDF

D

dam, 69–73, 258–264
Davis, E.H., 172, 191
Der Kiureghian, 38, 73, 75
design point, 15–16, 20, 25, 84, 92, 117, 137, 164, 188, 196, 241, 254, 301, 354
dip angle, 283, 285, 311, 318, 332
dip direction, 89, 309, 312, 318, 320, 332, 334
direction cosine α, 34–35, 63, 328
discretized random field, 165, 170, 184, 246, 248, 255, 273
Ditlevsen, O., 40, 204, 257, 357
Duncan, J.M., 233, 243–244, 246, 249

E

EC7 (Eurocode 7), 88, 129–130, 147, 166–167, 171–174, 206–211, 215, 218–220, 297, 302, 339
Economist, the, 9
Einstein, H.H., 309, 344–345, 349–353
equivalent normal mean of a nonnormal distribution, 18, 21, 25, 65, 126, 132, 137, 241, 272, 358
equivalent normal standard deviation of a nonnormal distribution, 18, 22, 65, 137, 241, 272, 358
Eurocode 7, see EC7
Evans, M., 306

excel solutions for hands-on experience, www.routledge.com/ 9780367631390
expanding dispersion ellipsoid perspective, 14–17
expanding equivalent dispersion ellipsoid perspective, 18, 20, 23–24, 97, 240, 284, 302
exponential distribution, see probability distributions
extreme value distribution type 1 (Gumbel), see probability distributions

F

failure probability estimated from β of FORM, 17
favourable-unfavourable duality, see load-resistance duality
first-order reliability method, see FORM
fissured clay, 246–247
flood, 69
FORM (first-order reliability method), 17–21, 357–362
FOSM (first-order second moment), see MVFOSM
foundations
   pile foundation, 157–192
   spread foundation, 115–156

G

gamma distribution, see probability distributions
Gaussian distribution, see probability distributions
geology, geological, 164, 238, 246
Goodman, R.E, 278, 288–289, 294
Gram-Schmidt orthogonalization, 38
ground improvement/stabilization, 93–99, 303
ground-support interaction, 335–342
Gumbel distribution (extreme value distribution type 1), see probability distributions

H

Haldar, A., 76–77, 141
harbour breakwater, 66
Harr, M.E., 45–46

Harrison, J.P, 342
Hasofer, A.M., 10, 12, 14–15, 46, 116, 199
Hasofer-Lind index, 12, 15, 18, 33, 35
Hoek, E., 298, 312, 320, 322–324, 335–337
Hong Kong, 93, 297
Hudson, J.A., 342
human error, 86

I

importance sampling, 41, 256, 341, 352, 354–355
irrigation water, 58–61

J

Jaksa, M.B., 246
Javankhoshdel, S., 136
jetty, 157

K

Kerisel-Absi passive earth pressure coefficient, 214, 216
Kim, D., 84
Knappett, J., 128, 146, 166, 172
Kottegoda, N.T., 57, 59, 61, 66, 69
Kounias-Ditlevsen bimodal bounds, 204, 257
Kraft, L.M., 256

L

Lambe, T.W., 198, 232, 258
LaRochelle, P., 243–244
La Spezia Harbour, Italy, 66
land reclamation, 94, 96
laterally loaded pile, 177, 189, 191
limit state surface, see LSS
Lind, N.C., 10, 12, 14–15, 46, 116, 199
load and resistance factor design, see LRFD
load factor, see LRFD
load-resistance duality, or stabilizing-destabilizing duality, 135–139, 143, 193, 214, 218–219, 284
lognormal distribution, see probability distributions

London Clay, 243, 246–247
Low and Tang 2004 spreadsheet FORM method, 18–19, 132, 271, 358
Low and Tang 2007 spreadsheet FORM method, 18–22, 25, 44, 80, 135, 170, 215, 248, 357–362
lower hemisphere stereographic projection, 281, 311, 319
LRFD, 84, 141–145, 173, 185–188, 209, 211, 213, 217, 220, 285, 297, 299, 302
LSS (limit state surface, defined by $g(x) = 0$), 15–16, 18, 23–26, 30, 38–39, 60, 65

M

Madsen, H.O., 333
Mahadevan, S., 76–77, 141
marine clay, 94–99, 249–253, 255
Matlock p-y curves, 179–180
mean-value point, 16, 20, 29, 48, 56, 89, 99, 144, 156, 203, 212, 220, 284, 288, 330, 345, 358
Melchers, R.E., 23, 258
method of slices for slope stability analysis
    Bishop simplified method, 268, 270
    ordinary method of slices, 233, 243, 257
    Spencer method, noncircular slip surface, 250, 255, 264–269
model factor, model correction factor, 63, 69, 91, 135–136, 227, 231, 263
Monte Carlo simulation, 17, 29, 48, 52, 65, 75, 102, 111, 122, 203, 252, 296
Morgenstern, N.R., 269
most probable point (of failure), see MPP of failure
MPP (most probable point) of failure, 15–16, 22–23, 25
multivariate normal distribution, 15
MVFOSM (mean-value first-order second moment method), 59, 66, 87, 140

N

negative reliability index, 29, 155, 220
non-circular slip surface of soil slopes,
        250, 255, 264–269
non-Gaussian, *see* probability
        distributions
nonnormal, *see* probability distributions
nonnormal distribution, *see* probability
        distributions
normal distribution, *see* probability
        distributions
Norway, 253

P

partial factors, 88, 90–93, 147, 166–
        167, 171–174, 206–211,
        215, 218–220, 283, 286,
        296, 329
PDF (probability density function), 15,
        46, 74, 105, 120, 177, 306,
        357–358
performance function g(x), 13, 21–22,
        73, 76, 79, 136, 238, 282,
        300, 325, 351
PERT distribution, *see* probability
        distributions
Phoon, K.K., 136
pile foundation
    laterally loaded piles, 177, 189, 191
    vertical piles, 157, 166, 172
pollution, 61
positive-definite correlation matrix, 40
Poulos, H.G., 172, 191
prefabricated vertical drains, 94–98
probability density function, *see* PDF
probability distributions
    beta general distribution, 22, 37,
        78–80, 177, 306, 361
    exponential distribution, 61–62, 361
    extreme value distribution type 1,
        see Gumbel distribution
    gamma distribution, 61–62, 80, 361
    Gaussian distribution (normal
        distribution), 11, 15, 33,
        47, 49, 55, 58, 77,
        176, 361
    Gumbel distribution (extreme value
        distribution type 1), 63, 67,
        70, 72, 74, 360–361

lognormal distribution, 25, 37, 47,
        51–55, 77, 80, 152–155,
        176, 215, 271, 360–361
    normal (Gaussian) distribution, *see*
        Gaussian distribution
    PERT distribution, 80, 177, 307,
        361
    triangular distribution, 22, 44, 80,
        307, 361
    uniform distribution, 23, 359, 361
    Weibull distribution, 22, 44, 74–75,
        361
probability of failure, 17, 86, 140, 154,
        166, 205, 252, 257
p-y curves
    constant with depth, 191
    nonlinear Matlock, 180

R

Rackwitz-Fiessler equivalent normal
        transformation, 18–20, 22
random field, *see* discretized random
        field
Randolph, M.F. 172, 175
RBD, 24, 48, 53, 71, 87, 90, 116, 133,
        135, 141, 160, 168, 214,
        222, 240, 281, 324
RBD-via-FORM, *see* RBD and FORM
reliability index (Hasofer-Lind for
        correlated normals, FORM
        for correlated nonnormals)
    applications in CEE, 24, 30, 35, 45,
        48, 50, 53, 57, 61, 63, 66,
        69, 71, 73, 76, 79, 108
    applications in rock engineering,
        283, 287, 295, 299, 305,
        307, 325, 330, 340, 347,
        352
    applications in soil engineering, 117,
        135, 152, 162, 170, 176,
        194, 215, 227, 231, 248,
        263
    defined, 10, 14, 17
reliability-based design, *see* RBD
resistance bias, 135–136
resistance factor, *see* LRFD
response surface method (RSM), 41–42,
        274
retaining walls
    anchored sheet pile walls, 214

quadrilateral soil wedge, 222
semi-gravity walls, 194–214
steady-state seepage towards vertical
    drainage layer, 229, 232
risk, 35, 58, 60, 238
Rocscience.com, 343–344
Rosso, R., 57, 59, 61, 66, 69

S

Salgado, R., 84
San Francisco Bay mud, 249–253
Sau Mau Ping slope, Hong Kong,
    297–304
second-order reliability method,
    (SORM), 38–41, 73–75,
    299, 301, 347–348, 352,
    354–355
sensitivity analysis of @RISK software,
    29
sensitivity indicators n, 21–22, 25, 28,
    65, 68, 121, 123, 132, 139,
    171, 186–187, 196, 283,
    287
serviceability limit state (SLS), 31–33,
    99–108, 145, 149–154,
    172, 175–178, 186–187
settlement
    pile foundation, 172
    soft clay consolidation settlement,
        30–35, 90–99
    spread foundation, 145
shaft excavated in sandstone, 335
Shinozuka, M., 15, 206
sign of reliability index, 29, 155, 220
Skempton, A.W., 243–244
SLS, *see* serviceability limit state
slope stability
    embankment on soft ground,
        264–268
    excavated slope in fissured London
        Clay, 243–249
    Norwegian slope with spatial
        autocorrelation, 253–256
    rock slopes, plane sliding, 277–307
    rock slope with tetrahedral wedge,
        309–334
    sloping core dam, 258–264, 269
    underwater slope in San Francisco
        Bay mud, 249–253
sloping core dam, 258–264
Sofianos, A.I., 342–343

Solver (a built-in Microsoft Excel tool),
    11, 14, 33, 44, 101, 239,
    267
SORM (second-order reliability
    method), 38–41, 73–75,
    299, 301, 347–348, 352,
    354–355
spatial autocorrelation, 164–166, 168,
    184, 247, 253–255, 271
Spencer method, reformulated,
    noncircular slip surface,
    264–269
spillway, 69–73
spread foundation, 115–156
stabilizing-destabilizing duality, *see*
    load-resistance duality
standard normal cumulative
    distribution, 20–21, 67,
    360
standard penetration tests, 145–148,
    150
stereographic projection, 281, 309, 311
storm sewer system, 63–66
strike direction, 278, 311, 321, 324
system reliability, 204–205, 256–257,
    330

T

Tang, W.H., 30, 35, 50, 53, 69, 71, 90,
    128, 258
Terzaghi, K., 146–148, 154, 222,
    225–226, 233, 238, 246
Tichy, M., 332
Tomlinson, M.J., 135, 145–146,
    157–158, 179
Tonon, F., 343–344
traffic capacity, congestion, 45, 48
triangular distribution, *see* probability
    distributions
tunnel, 342–344, 346–351
Tung, Y.K., 63, 66

U

ULS, *see* ultimate limit state
ultimate limit state, 40, 75, 117, 128,
    135, 139, 157, 187, 194,
    240, 245, 250, 259, 295,
    299, 336, 343
underground excavation in rocks,
    335–355

uniform distribution, *see* probability
        distributions
unknown unknowns, 86
u-space classical FORM procedure, 23,
        38–39, 121, 126–127, 301,
        358

V

Vanmarcke, E.H., 165
variance reduction factor, 165–166, 169
VBA (Visual Basic for Applications), 35,
        180, 216, 267, 271, 361
Veneziano, D., 357

Visual Basic for Applications, *see* VBA

W

Weibull distribution, *see* probability
        distributions
Whitman, R.V., 198, 232, 258
Wikipedia, 9
Winkler foundation, 76, 78
Woodward, J., 179
Wright, S.G., 249
Wroth, C.P., 172, 175
Wu, T.H., 141, 256
Wyllie, D.C. 303, 319–320, 324

Printed in the United States
by Baker & Taylor Publisher Services

Printed in the United States
by Baker & Taylor Publisher Services